Management of High-Technology Research and Development

Management of High-Technology Research and Development

John H. Dumbleton

Executive Director R & D Worldwide, Howmedica, Rutherford, N.J., U.S.A.

ELSEVIER

Amsterdam – Oxford – New York – Tokyo 1986

ELSEVIER SCIENCE PUBLISHERS B.V.
Sara Burgerhartstraat 25
P.O. Box 211, 1000 AE Amsterdam, The Netherlands

Distributors for the United States and Canada:

ELSEVIER SCIENCE PUBLISHING COMPANY INC.
52, Vanderbilt Avenue
New York, NY 10017, U.S.A.

Library of Congress Cataloging-in-Publication Data

Dumbleton, John H.
Management of high technology research and development.

Bibliography: p.
Includes index.
1. Research, Industrial--Management. I. Title.
T175.5.D86 1986 658.5'7 85-25437
ISBN 0-444-42572-1

ISBN 0-444-42572-1

Printed in The Netherlands

This book is dedicated to

Eilis, Grainne, Siobhan and Maire.

Preface

There have been many books written on the subject of management but few concerning the management of research and development (R & D). Writings in the management area have been by individuals with a background in specific management disciplines such as marketing, finance or general management. Those writings on the management of R & D have come from individuals with a background in R & D who have moved into R & D management. It is true to say that few of the top positions in corporations are filled by individuals from the technical areas. Hence those writings on the management of R & D have come from the perspective of middle management rather than from top management. One consequence of this is that the treatment of R & D has been based on local rather than global considerations as far as the company is concerned. A further weakness in the treatments of R & D has been the lack of attention to interpersonal considerations (individuals, groups and interactions between groups). In general, individuals engaged in R & D are not strong in the area of human relationships and have learned about management considerations on the job. In view of the above situation, the role of R & D in the firm is generally not appreciated either by top management or by those in the R & D organization. At the very least, this will lead to the waste of needed resources and in the extreme can threaten the very existence of the firm.

The aim of the present book is to remedy this situation by discussing R & D in terms of its position in the strategy-structure relationship of the firm. The intent is to present an integrated approach that will benefit those in R & D, especially managers, and also those involved in the management of the firm. It is not intended that the treatment be complete, but five areas are covered: the nature of R & D, strategy and structure relationships, creativity, the R & D process and the R & D interface. To a large extent these sections are self-contained and may be read independently. However, all these subjects have a direct bearing on the effectiveness of industrial R & D.

Part 1 deals with the nature of R & D. After a short historical introduction, definitions of R & D are given, along with statements on related and non-related R & D activities. Models of increasing complexity are considered for the R & D process; each type of model illustrates a different perspective of the R & D process. Finally, a model is presented illustrating the positioning of R & D in the firm, along with its relationship to the internal company environment and the external environment. This model highlights the importance of the R & D interface. Part 1 ends with a discussion of ways to evaluate the performance of R & D. This subject is especially germane as management is constantly frustrated in its efforts to quantify the return on investment in R & D.

Part 2 covers the subject of strategy and structure - how the structure of an organization reflects company strategies. The management literature is reviewed as to the implications for the structure of R & D and the choice of R & D strategies. Also discussed is the concept of the product life cycle and the application of this concept to consider company stages of growth and decay. The subject of "make or buy" is considered, i.e. whether new products should be developed internally in the R & D organization or by external acquisition of products by licensing or by purchase of companies. This section, therefore, places R & D in the organizational hierarchy and illustrates that R & D is not an appendage that periodically delivers new products but an integrated part of the new product and business generation mechanism of the firm.

Although the positioning of R & D is important, the requirement that products emanating from R & D be novel and useful mandates at least an element of creativity in the R & D organization. Part 3 covers the area of creativity. After a discussion of the current and past investigations of the creativity area, the creativity process is considered. The relationship of intelligence to creativity is discussed. Personality aspects of the creative person are emphasized. The subject of creativity, intelligence, personality and general ability tests is covered. Methods of enhancing creativity usage are discussed in the R & D context with the illustration of an actual intervention.

At the center of new product development is the R & D project, which formalizes and directs activity towards the development of a new product or process or the development of an area of technology. Part 4 covers this subject. Methods of project generation are discussed, including technological forecasting and systematic methods of delineating project opportunities. After the development of a list of projects, it is necessary to select the project portfolio. Methods of project selection are covered in detail, including financial evaluation of project alternatives. Once the project selection has been made, it remains to ensure that progress is to plan; accordingly, project planning and control is discussed.

Part 5 discusses personal and interpersonal factors starting with the results of studies carried out in R & D organizations. The areas of motivation, leadership and group dynamics are discussed. The need for good relationships across the R & D interface is emphasized. Statistics are presented indicating that new product success depends on close relationships with marketing and production units; the need for good relationships with corporate management is indicated. R & D studies are discussed concerning the interface with the external company environment. This leads to the subject of stars and gatekeepers. Management literature often deals with ideal cases. Rationality is assumed in organizations. The reality is very different, with individuals having private agendas. This whole subject is covered under the heading of political activity.

As mentioned above, there have been many books written in the management area. It is true to say that no one book provides all the answers, and this is true of the present work. However, the material presented herein should be useful and should stimulate insights and connections based on the reader's experiences in the R & D and management area.

ACKNOWLEDGEMENTS

I would like to express my sincere thanks to Mrs. Carolyn Inglesby, who carefully typed and formatted the manuscript and tables. Also my thanks are due to Mr. Ted Helder, who skillfully prepared the figures.

We are grateful to the following for permission to reproduce copyright material: AMACOM, a division of American Management Associations, New York, for figures on pp. 27, 28 and 39 of 'How to Select Successful R & D Projects' by D. Bruce Merrifield, Management Review (Dec. 1978), c 1978 AMACOM, all rights reserved and figure on p. 72 of 'What is the Right Organizational Structure?' by Robert Duncan, Organizational Dynamics (Winter 1979), c 1979 AMACOM, all rights reserved; Basil Blackwell Publisher Ltd. for Tables 1, 2, 3, 6 and 7 on pp. 159, 160, 162 and 163 of R & D Management, Vol. 4, No. 3 (1974) by W. H. Gruber, O. H. Poensgen and F. Prakke, figures on pp. 23-24 of R & D Management, Vol. 1, No. 1 (1970) by D. G. S. Davis; tables on pp. 62, 63 and 65 of R & D Management, Vol. 7, No. 2 (1977); Butterworth Scientific Ltd. and the author, Tudor Rickards, for Table 3 on p. 270 of Design Studies, Vol. 1, No. 5 (1980); John Calder (Publishers) Ltd. for table on p. 12 and figure on p. 16 of 'Innovation' by W. Kingston; Cornell Univ. Grad. School of Bus. & Pub. Admin. for figures on pp. 588-9 of Admin. Sci. Quart. Vol. 22, No. 4 (Dec. 1977) by M. L. Tushman; Elsevier Science Publishers for figure 9.2 on p. 88 and figure in Chapter 9 of 'Innovation in Big Business' by L. W. Steele; Prof. C. Freeman for figures in 'Economics of Innovation', Penguin Books Ltd. (London, 1974), J. R. Galbraith for figure on p. 17 of Organizational Dynamics (Winter 1982); Gower Publishing Company Limited for Illustrations 1.1, 1.5 and 1.11 in 'Problem Solving Through Creative Analysis' by Tudor Rickards (1974); Harper & Row, Publishers, Inc. for Figures 1.1, 3-2, 4.8 and 5.7 on pp. 14, 48, 82, 107 of 'Organizations and Their Members: A Contingency Approach' by Jay W. Lorsch and John J. Morse, Copyright c 1974 by Jay W. Lorsch and John J. Morse; Harvard Business Review for Fig. 1 on p. 64 of 'How to Evaluate Research Output' by James B. Quinn, Harvard Business Review, (Mar./Apr. 1960) Copyright c 1960 by the President and Fellows of Harvard College, all rights reserved; Charles W. Hofer for Fig. 5.1 and 5.2 on pp. 104 and 108 of 'Strategy Formulation: Analytical Concepts' (1978); IEEE for Tables 5, 6, 7 and Fig. 1 of 'The R & D Marketing Interface' by W. E. Souder and A. Chakrabarti, IEEE Transactions in Engineering Management Vol. 25, Part 4, 1978 (1978 IEEE); Richard D. Irwin, Inc. for Fig. 18-4 of 'Management Accounting' 5th ed., by R. N. Anthony and J. S. Reece (1975); John Wiley & Sons for Table 1 of 'Scientific Creativity: Its Recognition and Development', C. W. Taylor and F. Baron, editors, Figs. 10.7, 10.8 and 9.4 of 'Managing Research and Development' by J. E. Gibson (1981), Tables 13.2 and 13.4 from 'Technological Innovation: Government/Industry Cooperation', Arthur Gerstenfeld, ed.; Longman Group Ltd. for Figs. 1.4, 2.3, 8.9, 5.6, 5.7 from 'Managing Technological Innovation' 2nd ed., by Brian Twiss; McGraw-Hill Book Co. for Figs. on pp. 109 and

132 of 'Corporate Strategy' by H. I. Ansoff (1965) and Fig. 2-1 from 'Organizational Strategy, Structure and Process' by R. E. Miles and C. C. Snow (1978); North Holland Publishing Company for Fig. 9.2 and Figure in Chapter 9 of 'Innovation in Big Business' by L. W. Steele (1976); Oelgeschlager, Gunn & Hain, Inc. for 11 figures from 'Overseas R and D. Activities of Transnational Companies' by J. N. Behrman and W. A. Fischer (1980); Penguin Books Ltd. for Fig. 5 and Table 3-1 from 'Understanding Organizations' by C. P. Handy (1976), Fig. 3 and Exhibit 2 from 'Technological Forecasting' by G. Wills et al (1972) and Fig. 8 from 'Creativity in Industry' by P. R. Whitfield; Pergamon Press for table on p. 28 of 'Climate for Creativity', C. W. Taylor, ed. (1972); Regents of the University of California for Tables 1 and 3 in 'The Performance of Innovation: Managerial Roles' in California Management Review Vol. XX, No. 3 (Spring 1978) by A. L. Frohmann; Research Management for Tables 1-7 from 'Promoting an Effective R & D/Marketing Interface' in Research Management Vol. XXIII, No. 4 (1980) by W. E. Souder; West Publishing Company for Table 5.7 and Figure 2.2 from 'Strategy Formulation: Analytical Concepts by C. F. Hofer and D. Schendel (1978), Fig. 8.3 and Illus. 8.1 from 'Strategy Implementation: The Role of Structure and process' by J. R. Galbraith and D. A. Nathanson (1978), Table 2.2 from 'Organizational Goal Structures' by M. D. Richards (1978); Vanderbilt Univ. Press for illustration on pp. 190-192 from 'Patterns of Organizational Adaptation: A Political Perspective' by E. Harvey and R. Mills, in 'Power in Organizations', N. A. Zald, ed. (1970); George S. Welsh for Table A-1 on pp. 200-202 of 'Creativity and Intelligence: A Personality Approach' by George S. Welsh.

CONTENTS

PART 1

RESEARCH AND DEVELOPMENT

1.1 INTRODUCTION

Research is concerned with the search for new knowledge, with ideas and with invention. Research is carried out in many different locations such as universities, government laboratories, independent laboratories and industry. Research is often associated with developmental work and this association is especially relevant for industry which relies on technological innovation for survival. This is especially true for industries with high rates of technological change. There is a distinction between invention and innovation. An invention is an idea or drawing, or model, for a new or improved device, product, process or system. In the economic sense, an innovation is only accomplished with the first commercial transaction. Thus, not only invention but also innovation is required to generate profits. A similar though not entirely analogous distinction may be drawn between research and development. Research is that part of the process dealing with invention whilst development is concerned with taking the idea and turning it into a commercial product. Although not exclusively confined to industry, Research and Development (R & D) may be viewed as the main method by which a company promotes growth through technological innovation from within. It is the various aspects of industrial R & D and the means for improving the effectiveness of this function that are to be discussed in the following.

1.2 EVOLUTION OF R & D

Research and Development did not always exist in the structured, highly organized form found today in the modern firm. For example, the growth of R & D for several British industries may be described (Burns and Stalker, 1966a). The early discoveries at the time of the Industrial Revolution were made by individuals, often working alone. In the latter half of the eighteenth century "coteries and clubs" were developed which served to aid communication between individual workers. The diffusion of knowledge became more formalized with the appearance of technical journals. In 1800 there were less than ten scientific journals in the British Isles, but by 1900 there were over one hundred and thirty. By the end of the century, science was the province of groups of specialists working in universities or other institutions. Science was organized into the specialities. Scientists were salaried and professional men by 1900.

Changes had also occurred in industry. Development by improvement and new application became too difficult for workers trained only in traditional methods. Advances in Germany made British goods obsolete or inferior as was shown at the 1867 Paris Exhibition. To combat technological pressures, Britain introduced technical training specifically for the industrial setting. This system of education was distinct and separate from the system evolving around the universities to the disadvantage of both academia and industry. Many innovations were developed abroad in France, Germany or America as these countries had seen the benefits of technological innovation. Industrial research in Britain did not grow until the danger of overseas threats was fully recognized. The growth in research in British industry may be ascertained from the increase in expenditure from 2 million pounds in 1938 to 183 million pounds in 1955 (Burns and Stalker, 1966b).

Despite the cited innovations from the United States, formalized R & D is of rather recent origin. In general, R & D has only existed as an entity for the past 30 years. There are some exceptions, however. DuPont, founded in 1802, became involved in formalized R & D 100 years later and by 1920 had 3000 people involved in that activity. Dow Chemical was involved in R & D by 1909. Bell Telephone Laboratories was incorporated for R & D in 1925.

The role of R & D in the growth of certain industries has been described by Freeman (Freeman, 1974a). The chemical process industry was stimulated by technological innovation which facilitated the growth of professional in-house industrial R & D. The great success of the German chemical industry in the latter half of the nineteenth century was due to employment of technical staff and to the close relationships between industry and the universities. In 1880 the German industry accounted for one-third of the world dye-stuffs production, but by 1900 it accounted for four-fifths of the production. The role of R & D in the growth of the oil, plastics and electronics industries has also been demonstrated (Freeman, 1974a).

With the advent of faster rates of technological change, it has become more important than ever for a company to carry out R & D so that a continual improvement in products can occur. As has been mentioned above, not only must R & D be carried out as a separate function in industry, but adequate integration between industry and educational institutes must occur.

1.3 R & D EXPENDITURES

Although the discussion here will focus on industrial R & D, it is helpful to stand back and examine the relationship between R & D and GNP (Gross National Product) and to examine expenditures on R & D. Royce has indicated how R & D fits into the components of GNP (Royce, 1968a). As shown in Figure 1, R & D is committed to the development of new concepts, processes and techniques, but is integrated into the whole process since all

FIGURE 1

THE PLACE OF R&D AS A COMPONENT OF GNP

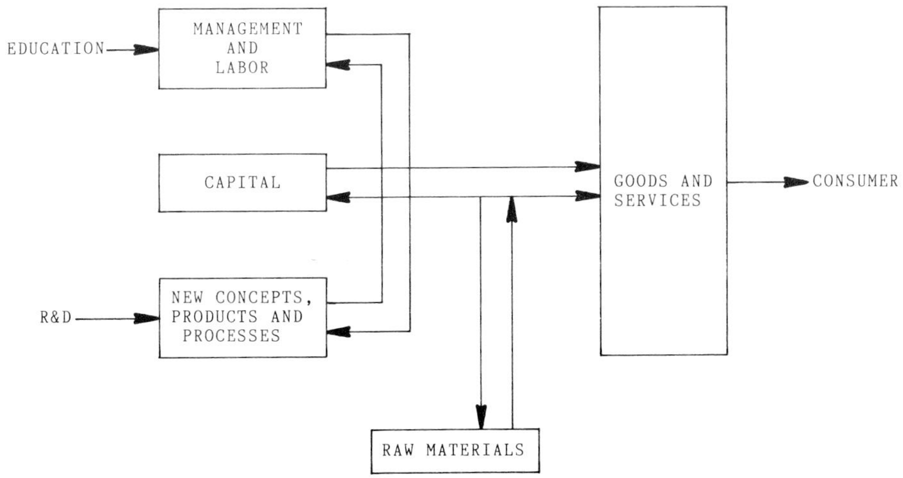

the components are linked together. Not shown is the link between education and R & D and the need for capital; capital expenditures on education can lead directly to new discoveries, but in any event are needed for the training of persons for R & D activity. Royce also cites studies that indicate a high correlation between the level of R & D expenditures and the level of exports from the United States. A recent report by Arthur D. Little Inc. highlights the importance of innovation. Three industries were cited: television, jet aircraft and digital computers. Non-existent in 1945, these industries contributed more than $13 billion to the GNP of the United States and were responsible for the employment of 90,000 people just 20 years later (Gerstenfeld, 1979a). Similar comments will no doubt be made in the future concerning silicon technology, biotechnology and other areas of current rapid change. Also for the United States, about 85 percent of the nation's productivity increase and about 45 percent of the total national growth from 1929 to 1969 were due to technological innovation contributions. Expenditures on R & D are considerable. Table 1 gives government expenditures on science and technology for 1975 (Gerstenfeld, 1979b). A substantial part of the budget was spent on research by European countries, Japan and the United States. The percent of sales spent on R & D by industry is also substantial as shown in Table 2. It is interesting to note that the low expenditures in the iron and steel industry may be reflected in the wholesale disappearance of large sections of this industry in Europe and the USA some ten years later.

TABLE 1

Rate of government expenditure on science and technology for 1975

Country	Budget for science and technology ($ millions)	Percentage of Budget
USA	19,317	6.0
Germany	4,574	3.9
France	3,710	6.0
U.K.	2,655	4.2
Japan	2,286	3.2

TABLE 2

Percentage of sales expenditures on research and development for industries in different countries.

Industry	Japan (1974)	U.S.A. (1973)	Germany (1973)	France (1969)
Average	1.5	3.0	2.6	3.0
Food	0.5	0.4	0.2	0.4
Chemical	2.3	3.5	-	3.4
Petroleum	0.2	0.7	-	0.9
Iron and steel	1.0	0.5	0.6	0.4
Machinery	1.9	3.8	3.1	2.2
Electrical/electronics	3.7	7.1	5.3	3.8
Transportation	2.1	3.5	3.1	3.2
Aerospace	0	13.5	34.4	27.6

The foregoing has indicated the pervasiveness of R & D throughout society and the connection between successful R & D and economic health. The R & D process is not, however, understood. In the United States and other countries cost and time over-runs on weapons systems, on aircraft development such as the Concorde, and on nuclear reactors has been severely criticized. Concern has been expressed in the United States that innovation has diminished in recent years and that the technological base for innovation has eroded. There is also concern that industrial R & D in the United States is concentrated in a few industries and in a relatively small number of companies within each industry. Some 31 companies accounted for more than 60 percent of all R & D expenditures. A survey in the United States has indicated that as of 1977 industrial funding of research was barely keeping up with inflation.

More disturbing, much research is not directed at new product innovation, but at satisfying government regulations and controls on existing products. In the medical area, the introduction of a new drug costs $50 million and takes five years; the rate of new drug introductions has slowed dramatically. Similarly, manufacturers must now contend with increased expenditures and time scales to bring new implants to the market. More and more research is being directed for short-term payoffs (Gerstenfeld, 1979a). An interesting study has divided R & D expenditures into basic research (long-term) and short-term research

(Mansfield, 1980). One hundred nineteen firms engaged in different areas of technology were surveyed retrospectively and were also asked to forecast future changes in the pattern of R & D expenditures. The results showed that the proportion of R & D expenditures devoted to basic research declined between 1967 and 1977 in almost every industry. According to the company forecasts, there was no evidence of further drops during the period 1977-1980, but there was little evidence of substantial increases in expenditure. In 80 percent of the industries, there was a decline between 1967 and 1977 in the proportion of R & D expenditures for high-risk projects (especially ones with less than a 50 percent probability of success), but this was expected to recover in the period 1977-1980. Between 1967 and 1977 there was a decline in expenditures aimed at entirely new products, and this situation was not expected to change in the period 1977-1980. The study reinforces the conclusions given above.

Further information on the state of science in the United States and elsewhere is to be found in the Science Indicator series published by the National Science Board (Gibson, 1981). Some 2100 journals were surveyed containing 279,000 articles. The United States was responsible for 40 percent in physics and 41 percent in mathematics for 1978. However, the trends are not encouraging. The United States is behind the Soviet Union in the number of scientists in R & D on a per capita basis and in national expenditures for R & D as a fraction of GNP (2.2 percent versus 3.4 percent). The trend in the United States is downward (at least up to 1978) and tends towards the levels spent by the United Kingdom, France, Germany and Japan. In terms of the GDP (Gross Domestic Product), expenditures by the above-mentioned countries are approaching that of the United States. The decline in GNP percentage spent on R & D since 1968 is almost entirely due to declining federal investment; industry has maintained a constant investment in R & D as a proportion of GNP. However, this may not be true in the recent long-drawn-out recessionary period.

1.4 SOME DIFFICULTIES WITH R & D

One reason why expenditures on R & D have been under pressure has been the difficulty encountered by companies with the R & D process. Many problems in carrying out R & D have been mentioned in the literature, and publication still continues indicating that these problems are by no means solved. Due to the high cost of R & D, there is a keen interest in ensuring that work is carried out on the right projects and that time and cost controls are properly implemented. The concern expressed on a national scale regarding innovativeness has also been voiced at the level of the individual company. There is anxiety over the difficulty of transferring projects from one stage to the next along the chain from research to development to production and to the marketplace. R & D personnel have been accused of elitist attitudes, of being inward-looking and of being unresponsive to the needs of marketing. In all of the above, there seems to be a hierarchy of problems. Problems regarding the mission of R & D and specifically the projects tackled lie at the level of the fit between R & D and the company in terms of corporate, business and functional objectives. The problem of innovativeness deals with the processes within R & D and with the interaction of R & D with the external environment of the company. The other problems deal with the interface between R & D and the marketing, manufacturing and other groups. Although it is perhaps an oversimplification to consider these problems as falling into three non-interacting groups, the approach is attractive in that the major elements may be considered first with the secondary interactions considered at a later stage if necessary.

1.5 THE CLASSIFICATION OF R & D

The classification of research and development has always posed something of a problem and there is no universally accepted classification scheme. There are many dimensions to the R & D process and various classification schemes may be suggested depending on the focus adopted, and on the emphasis placed on any particular dimension. For example, the classification could be drawn up around the type of organization involved in R & D. This would give a classification into academic research, government research, independent laboratories and industrial research activities with the associated development work where appropriate.

An alternative classification could be drawn up for industrial research in which a market strategy was employed as the basis for the classification. Thus, research could be regarded as offensive or defensive depending on the intent of the work with regard to the actions and products of competitors. The time orientation of research could be used to define the class of research work and the division would be made between long-term and short-term returns which would give a goal-oriented division. The intent of the research could be employed to divide research work according to the reason for undertaking the studies. The divisions here would be between directed research with the intent of giving a new product and undirected research which would simply lead to an increase in the store of knowledge in a particular area. All of these classifications have merit in specific instances; some classifications combine the elements from the different examples given. The classification recommended here is from the Organization for Economic Cooperation and Development (OECD) report on "The Measurement of Scientific and Technical Activities" (Freeman, 1974b).

In the OECD classification, three categories of R & D are distinguished: basic research, applied research and experimental development.

(i) Basic research is original investigation undertaken in order to gain new scientific knowledge and understanding. It is not primarily directed towards any specific practical aim or application. In pure basic research, it is generally the scientific interest of the investigator which determines the subject studied. In oriented basic research the investigator is directed towards a specific field by the employing organization.

(ii) Applied research is original investigation undertaken in order to gain new scientific or technical knowledge. It is directed primarily towards a specific practical aim or objective.

(iii) Experimental development is the use of scientific knowledge in order to produce new or substantially improved materials, devices, products, processes, systems or services.

From the above classification, it would appear that in industry most activities fall under applied research or experimental development.

It is remarked in the OECD report that R & D activities do not necessarily fall neatly into the three successive and distinct categories and that, on occasion, arbitrary distinctions may have to be made in what is more or less a continuous process.

There are many activities related to R & D but which nevertheless are not R & D activities. The inclusion or exclusion of an activity must be based on the presence or absence of an appreciable element of novelty. Activities to be excluded from R & D are scientific education, the activities of scientific and technical information personnel, general purpose data collection, testing and standardization, feasibility studies for engineering projects and patent and license work.

There is usually great difficulty in defining the boundary between experimental development and production or technical services. Table 3 gives the recommended treatment of borderline cases.

TABLE 3

The separation of experimental development from production or technical services

Item	Treatment	Remarks
Prototypes	include in R & D	as long as the primary objective is to make further improvements
Pilot plant	include in R & D	so long as the primary purpose is R & D. If it is subsequently used as a production unit or is sold, deduct the sales price from the capital account of the original year of investment.
Design and drawing	divide	include design required during R & D. exclude design for production process
Trial production and tooling up	exclude	except 'feed-back' R & D
After-sales service and 'trouble-shooting'	exclude	except 'feed-back' R & D
Patent and license work	exclude	all administrative and legal work connected with patents and licenses
Routine tests	exclude	even if undertaken by R & D staff

Figure 2 gives a diagramatic presentation with examples of the conventions regarding the boundaries between R & D, scientific activities related to R & D and non-R & D activities. It is interesting to note that innovation is regarded as a non-R & D activity in the OECD classification.

FIGURE 2

DIAGRAMATIC PRESENTATION OF THE BORDERLINES BETWEEN R&D, R&D RELATED ACTIVITIES AND NON-R&D ACTIVITIES

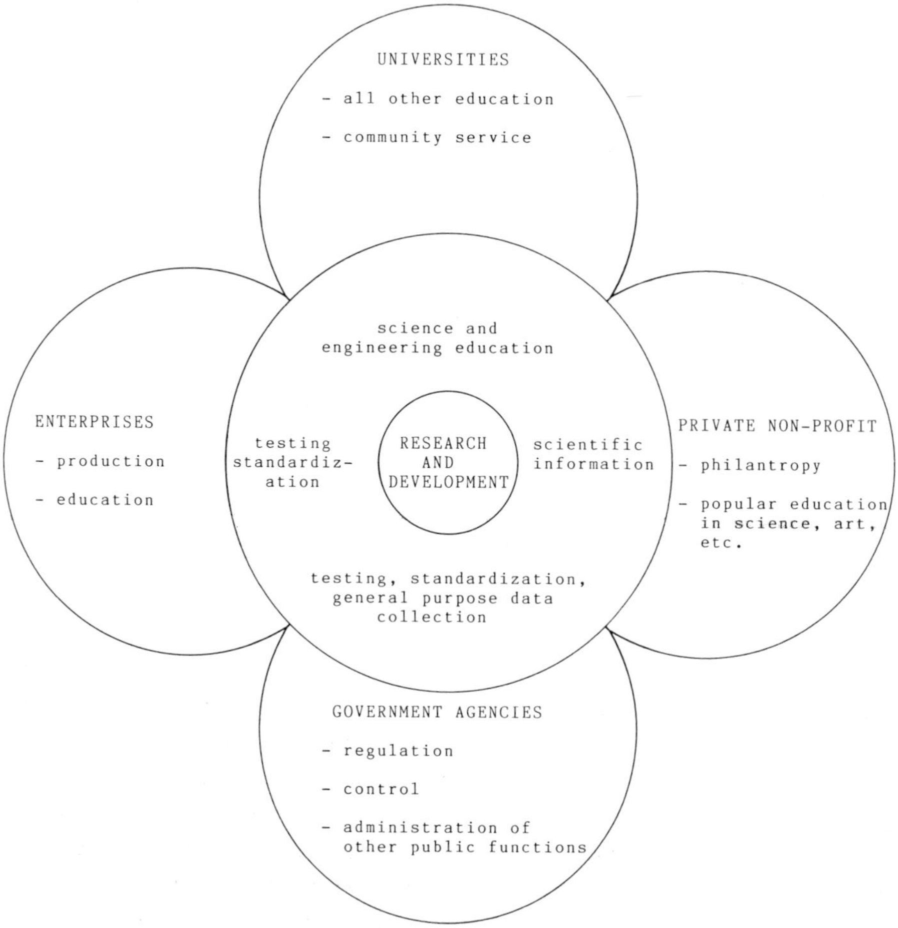

The activity of innovation has been addressed by Kingston who has placed innovation within a spectrum of activities as shown in Figure 3 (Kingston, 1977a). The approach is one of personalizing the innovation process in terms of the characteristics of the inventor, innovator and so on. Table 4 gives the characteristics of the personalities ranging from the Artist to the Trader. The Artist is concerned with seeing but only a little with doing, whereas the Trader is concerned with doing, but only a little with seeing. The other characteristics follow in a similar fashion. The classifications of Dreamer and Mandarin are to be regarded as the limiting cases of the spectrum.

FIGURE 3

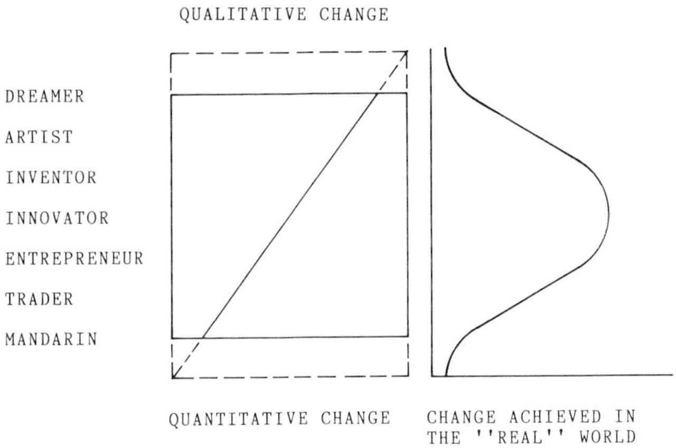

THE SPECTRUM OF ACTIVITIES LEADING FROM ''PURE THOUGHT'' TO ''PURE ACTION''

It is interesting to see how R & D fits into this spectrum of activities. It would appear that R & D fits into the upper part of the spectrum of Figure 3 with basic research at the level of the Artist (perhaps the level of the Dreamer would be more appropriate for an Einstein) and applied research at the Inventor level. The correspondence is clearly imperfect, but the exercise does show the much closer relationship of innovation to invention, of innovation to research, than does the boundary diagram of Figure 2. The desirable action of R & D as a driving force for innovation is clearly brought out in this analogy. The analogy also brings out the importance of the interface with R & D. Whether or not innovation is regarded as an R & D activity is of little importance as long as there is a close coupling between R & D and the innovation activity.

TABLE 4

The characteristics of personalities engaged in activities from artist to trader

Characteristic	
A	B
Seeing	- Doing
Personal	- Social
Individual	- Crowd
Imagination	- Verifiable facts
Ideas	- Concrete realities
New things	- Familiar things
World as it might be	- World as it is
Untried methods	- Established techniques
Long time-scale	- Short time-scale
Element of chance	- Predictable
Spontaneity	- Learned behavior
Inarticulate need	- Formulated want
Emotion	- Intellect
Revolution	- Evolution
Concern with future	- Concern with past or present

The OECD classification of R & D into basic research, applied research and experimental development gives a framework for the basic description of R & D activities. Since the intent is to describe only industrial R & D, it is not necessary to qualify the description as to place of activity. However, as required, qualification of the basic description to indicate market purpose and other points of focus will be made. The question of innovation and the link with R & D can be taken up later. For the moment, invention can be taken as the conception of an idea and innovation as the process by which an invention is translated into the economy. At the very least, the output from the R & D process greatly influences the innovation process.

1.6 THE FUNCTION OF R & D IN THE FIRM

The classical view of R & D in the firm is as an internal mechanism for growth. By introducing new products, processes or services R & D can ensure that the competitiveness of existing products is maintained through property improvements and cost reduction and that new products are introduced on a phased basis to satisfy existing market needs and to anticipate or create new market needs. The classical view is thus one of conversion in which ideas are generated within R & D or are taken from outside and turned into new products, in the broadest sense, ready for the next stage in the journey to the marketplace. This concept fits with the spectrum of activities given by Kingston and also with the OECD division of R & D into linking activities.

However, there must be a large contingency element in the actual function of R & D adopted in any given company. This would be expected on no other grounds than that the structure/strategy relationships of firms are different and would demand different functional requirements of R & D to obtain the best fit. Thus, the demands on R & D will depend upon the structure of the company and how the R & D activity is located in the hierarchy of reporting relationships. The size of the company and the degree of diversification will also play a role in determining the function of R & D. Large size and high diversity of products call for considerable specialization of any particular R & D unit. At the very least, the functions will broadly include basic research, applied research and experimental development with appropriate weighting. In some cases, a particular unit may be limited to a segment such as basic research alone or to a selected range of products. The sophistication of the products will also affect the function of R & D at least as to the level of science and technology employed. In an unsophisticated market, what passes for R & D may be dismissed in a more sophisticated market as no more than related R & D activity or even non-R & D activity. The important point to remember is that it is immaterial, as far as the company is concerned, whether the activity is true R & D or not. For what is regarded as R & D will be treated as R & D and similar problems will be found in interunit relationships as in more sophisticated technology companies. It is only from the outside that the difference in technical level and function between true

R & D and perceived R & D will be identified, but even here it may be necessary to go outside the particular industrial sector to obtain absolute comparisons. Obviously, if a company that perceives that R & D is being undertaken is carrying out this activity to the industry norm, then the perception will be reinforced.

R & D may be regarded as one of the change mechanisms in the company and may be the major mechanism for technological change. This view of R & D removes the untidyness over whether or not a company is truly carrying out R & D. Only one difficulty remains. A company carrying out or perceiving that it carries out R & D to the industry norm may lose all of its market to an innovation from outside of the industry. It is arguable whether true R & D would have saved such a company and even in rather sophisticated markets R & D has not always served to protect the company from such a situation.

The view of R & D as a change mechanism, and the classical view would have it that it is only an internal change mechanism, emphasizes the contingency nature of the R & D process. In many cases the R & D group carries out tasks which are only related R & D or even non-R & D. This is done in addition to the true R & D activities. As pointed out, in some companies the R & D group is engaged almost exclusively on non-R & D activities. Having said this, it is true to say that the R & D group is capable of carrying out tasks other than true R & D, and that there are advantages for the company that employs R & D in other capacities additional to those classically regarded as the R & D preserve.

There are many examples of assistance which the R & D group can give. The group is a reservoir of technical knowledge which can be used in the preparation of environmental forecasts to indicate the state-of-the-art in a particular area, the future trends of technology and likely competitive threats based on technological innovation. Where the intent is to promote growth from without through acquisition, licensing or the adoption of products developed external to the company, R & D can be employed to indicate the advantages and disadvantages of the offering in the technical area. Freeman argues that one of the requirements of a threshold level of R & D is to ensure that licensing and know-how agreements are efficiently implemented (Freeman, 1974c).

In addition to the involvement of R & D in the generation of growth by external means and in charting the course of likely technological developments, there are other areas in which R & D expertise may be employed. Specific assistance may be given as technical service to sales, marketing, customers and in other areas to quality assurance in the form of standards and test development and in failure analysis and in process improvements and troubleshooting. R & D may also become involved in the innovation process directly further towards the marketplace than usual and might even carry out manufacturing in specific instances. The involvement of R & D in any or all of these areas will depend on the demands of the company and on the orientation of the R & D group.

The primary aim of R & D must be to promote internal growth through basic research, applied research and experimental development. But there is a considerable contingency element to the actual makeup of the mission of R & D which will determine the activities of the R & D group. The R & D group may become involved in related or non-R & D activities simply because there is no other mechanism available in the company to carry out that specific activity. In other words, R & D moves to fill an activity gap. The actual function of R & D is, therefore, specific to the particular company under consideration. Undoubtedly, comparisons may be made between companies in similar circumstances (matched for industrial sector, for example) to give a picture of the appropriateness of the R & D activity, but the view obtained probably could not be translated to a company in a different sector. The pragmatic approach that R & D is a change mechanism in the technical area to promote company viability and growth indicates a philosophy without rigidly confining R & D to a predetermined list of activities.

1.7 MODELS OF THE R & D PROCESS

It is useful to examine models of the R & D process since this examination can indicate those factors of greatest importance in success of the process. In addition, interactions between factors may be more easily visualized. It is interesting to speculate as to whether there is one model of the R & D process. At the present stage of knowledge, it would appear that there is no one general model which is applicable to each R & D situation. In order to cope with each specific case, the general model would need to be highly complex and this complexity itself would act as a barrier to the adaptation of the model to specific cases. Instead of attempting a general formulation of the R & D process which is then applied to a particular case, it appears to be preferable to begin with simple, specific models which illustrate different facets of the process. As knowledge of the R & D process increases, abstraction from the particular to the general may be undertaken. Thus, the approach to be adopted here will be to examine specifics of the R & D process by way of simple models. The formulation of a general model will not be attempted.

Figure 4a gives a simple linear model of the R & D process. An input to R & D results, after processing, in an output to the design function then to manufacturing and ultimately results in a new product or other innovative result on the output side. This model regards R & D as a "black box" in which undefined processes occur which result in an output. Although the model is simple, it does illustrate that R & D is coupled to the rest of the firm. In this case, the coupling on the input side is undefined, but on the output side there is coupling to the design group. The model is limited because it does not consider the input and output in more detail, it does not consider feedback or multi-unit coupling; environmental coupling is not considered and the R & D unit is not examined in more detail. These factors will be considered below.

A natural way to look at the R & D unit is to sub-divide the nature of the process according to the classification given earlier, that is into basic research, applied research and experimental development. A further sub-classification of basic research into pure research and oriented basic research may be made. Figure 4b gives the sequential flow with R & D. Again there is an input and an output to the process which starts with the creation of basic knowledge which may be applied to the development of a product or applied to the attainment of some

FIGURE 4

(a) A SIMPLE LINEAR MODEL OF THE R&D PROCESS

(B) THE SUB-DIVISION OF R&D TASKS

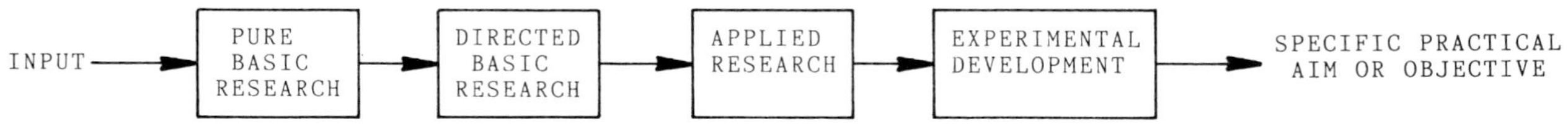

MODELS OF R&D PROCESS

other specific aim or objective. Although these sequential steps may be required, the model does not represent the situation within R & D. In industry, the attention is usually directed towards applied research and experimental development. The basic research component often lies outside the company and the flow from basic research to applied research represents a flow of information across the boundary of the firm from the external environment into R & D. Thus, in the firm, the R & D process is not self-contained, but represents only a part of the total R & D process. The time scale must also be considered. It is quite probable that the basic research, which led to applied research and ultimately to a product, was carried out some time earlier, perhaps many years earlier. The basic research results exist as a fund of knowledge which can be utilized any time following discovery provided the conditions are right. Basic research knowledge may also be used many times over in widely diverse applied research projects. Therefore, the model might better be represented with the basic research at the center with arms radiating out into different applied research areas.

Table 5 illustrates the inputs and outputs from the R & D system. Also given are feedback outputs (Freeman, 1974d). The inputs are of two types, information and resource. Resource inputs include manpower, equipment, supplies, buildings and so on. Although these inputs are important and represent money spent on R & D, for which a return on investment is desired, it is more interesting to consider the information inputs. Leaving aside the resource question, R & D can be regarded as a flow of information. The R & D process generates and transforms information in response to the input stimulus. The output is information. Even a specific output such as a new product can be regarded in terms of the information needed to design and fabricate that product. That the information is more powerful than the product itself is well illustrated by the process of the licensing of a product into a firm. The provision of the product itself without information on design and fabrication would make production extremely difficult. It would in fact be necessary to carry out experiments to rediscover knowledge embodied in the product itself. This process would only be a shortcut if there were a sufficient body of knowledge in the firm so that the presentation of the product supplied the last

TABLE 5

Inputs and outputs of the R & D system

	Illustrative inputs		Illustrative outputs	
Stage	Feedback inputs	Other inputs	Feedback outputs	Other outputs
Basic research	orders from entrepreneurs basic research inventive work development work 'bugs'	scientists laboratories nonscientific labour materials, power, fuel	new scientific problems laboratory results	hypotheses and theories research papers-formulas
Inventive work and applied research	orders from entrepreneurs inventive work development work 'bugs'	output of basic research scientists engineers laboratories nonscientific labour materials, power, fuel	new scientific problems laboratory results unexplainable successes and failures	patents nonpatentable inventions (memoranda, working models, sketches)
Experimental development work	orders from entrepreneurs development work 'bugs'	inventive output engineers draughtsmen other labour	new scientific problems need for inventions unexplainable successes and failures	blueprints specifications samples pilot plants prototypes patents manuals
New-type plant construction	orders from entrepreneurs 'bugs'	development output resources of an ordinary construction firm	'bugs'	new-type factory

link in the discovery process. It may be remarked that sometimes the realization that a product can be produced is sufficient to allow many companies in different geographical areas to solve the problems necessary for production of that product almost simultaneously.

The inputs given at the different stages of the R & D system are not specific to the industrial system, but here the inputs from the market are of prime importance to the R & D process. This is illustrated in Figure 5. The R & D process is represented as the technological concept which results from research, with an input of scientific knowledge, in response to a customer need. As has been mentioned, in most cases the R & D process utilizes knowledge gained outside the firm. Only the largest companies can engage in substantial basic research effort and even here the time delay between basic knowledge generation and the use of that knowledge means that the current applied research is using basic knowledge generated much earlier. Since in most cases, basic knowledge is freely shared, it is likely that this basic knowledge exists outside the company even if the knowledge was generated inside the firm at an earlier time. The lag between the generation of basic knowledge and its employment in the production of new products via applied research and development means that decreased spending on basic research may not produce a noticeable impact in the marketplace until quite some time later.

The general treatment of inputs to the R & D process may now be given. The inputs are as follows:

(i) output from a previous stage in the R & D system which may or may not lie within the firm.

(ii) requirement from the marketplace for fulfillment of a need.

(iii) requirement from a company unit other than marketing for a need to be fulfilled, e.g. development of an improved process for manufacturing.

(iv) requirement for an action in support of corporate objectives.

(v) feedback from company units.

FIGURE 5

MARKET NEED AS AN INPUT TO THE R&D PROCESS

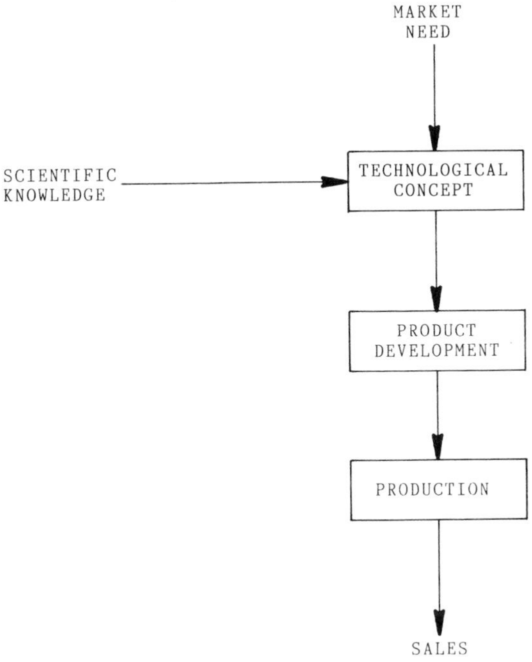

The flow of knowledge into R & D in connection with a specific project will co-exist with a general non-specific knowledge flow from the outside. Requirements for fulfillment of a need should mainly come from the marketing group, but other units of the company may request action from R & D and not all of the externally requested projects are at the request of marketing. Corporate objectives must be considered as an input to the R & D system if only as a constraint to ensure that the actions of the R & D unit are not contrary to these objectives. On a more positive note, the corporate objectives may require R & D to adopt a specific line of research irrespective of the inputs from the other units of the company. Such factors are best considered under the general heading of strategy and so the influence of corporate objectives will not be considered further at this time.

As indicated in Table 5, there are many outputs from R & D, and these outputs depend on the stage of the research. These outputs may be classified as follows:

(i) output to the next stage of the R & D process. The output may not be directed as for basic research and may simply add to the store of knowledge.

(ii) research papers, patents or other formal indications of research output.

(iii) feedback to earlier parts of the R & D system. This is treated as an internal input, and so is not strictly an output from R & D unless the research functions are formally separated in the organization.

(iv) output to company units in the form of designs, products, processes and so on.

(v) outputs in response to corporate requirements.

There is indeed a relationship between input and output but it will be seen that this relationship is complex as there are many inputs to the R & D system and many ways in which the output manifests itself. It may be noted that the outputs may be external to the company as in the presentation and publication of a scientific paper. The relationship between R & D and the company units is complicated by the presence of feedback. In practice the initiation of a project does not end the contact between R & D and the initiating company unit as there should be continuing dialogue during research and development. The movement of a new product out of R & D does not eliminate the need for communication since contact must be maintained with the company units further downstream towards the marketplace as well as with the initiating unit. This network of communications must be maintained throughout the life of the project with emphasis placed on different parts of the network as necessary. Failure to maintain this network can lead to failure of the project as a whole. Hence a failure to maintain contact with marketing can result in a product no longer needed in the marketplace, while failure to keep good contact with design and manufacturing can give rise to cost and time over-runs which can again result in a product that does not fit a customer need.

Although the route to innovation has been taken as in response to an input stimulus, innovation can occur as a result of an internal R & D stimulus. A new product or process can be produced without the need for an input from outside except general information flow. A distinction must be made between demand-induced innovation and supply-pushed innovation. It is demand-induced innovation which has been discussed up to now. This type of innovation is carried out in response to an external stimulus. Since there is a recognized need for the innovation, there are forces acting to pull the project along. The risk of failure is frequently lower than for supply-pushed innovation and the disruption of the organization is manageable. According to Steele, most of the innovations in the United States fall into the demand-pull category (Steèle, 1975a). The complete reliance on demand-pull projects may not be wholly satisfactory for the R & D organization since R & D becomes a tool of the organization without the outlet of creative project commitment.

Supply-pushed innovation is essentially opposite to demand-pull innovation. The forces which give rise to supply-pushed innovation arise within R & D as a development of the areas of expertise which grow over time. The tendency in this type of innovation is to aim for a major achievement or to produce a discontinuity in technology. In these terms the distinction between demand-pull and supply-pushed innovation is similar to the distinction between Schumpeters A-phase and B-phase innovations (Kingston, 1977b). Accomplishments of this magnitude are likely to be far riskier than for demand-pull innovation.

For R & D and for the company as a whole, it is necessary to have both demand-pull and supply-push projects. There is a balance which must be struck between the two types of project. The company which concentrates entirely on demand-pull projects certainly does well in the short and medium term, but has the risk of being overtaken by a substantial technology change and may also miss promising new areas of technology. The company which concentrates on supply-push innovations runs the risk of failing to satisfy the marketplace and may never be able to justify the investment in R & D with a substantial new product innovation. In the former case, there will be intra-R & D conflict due to the loss of autonomy while in the latter case, there will be hostility towards R & D from the rest of the company since R & D will have an "ivory tower" image. Steel suggests that a 70/30 ratio is the right balance with 70 percent of the projects of the demand-pull type and 30 percent of the supply-push type. The ratio will depend on the industrial sector.

The position of R & D with regard to the complex input and output conditions is shown in Figure 6 which is based on the domain idea of Thompson (Thompson, 1967). This type of model shows the input, output and feedback relationships, but does not give any indication of the processes going on within the R & D domain. There is no indication of the position of R & D in the company structure. However, the model does focus on three areas. These are the input to R & D, the output and the R & D process itself within the domain. It appears to be critical to provide communication paths so that input from the units of the company is used to define projects. An adequate interaction with the external environment must be maintained to allow the state-of-the-art in science and technology to be appreciated. Corporate objectives must be borne in mind when deciding on projects. Adequate communications must be maintained on the output side towards units of the company and also to the external environment. As far as the domain of the company is concerned, R & D must sit near the domain boundary for only in this position can R & D maintain the required internal and external contacts. R & D is a boundary-spanning unit. Three factors are of crucial importance, the boundary of R & D with the company units, the boundary with the external environment and the internal R & D processes which include supply-push innovation.

It is interesting here to examine the idea of an R & D domain in more detail. In the context of the firm Thompson proposed that a technological core could be identified. This core would represent the protected part of the company, for example manufacturing, in which the emphasis was on efficiency. The core activities were to be insulated against fluctuations which would be felt if the activity were close to or on the domain boundary.

FIGURE 6

A DOMAIN VIEW OF R&D

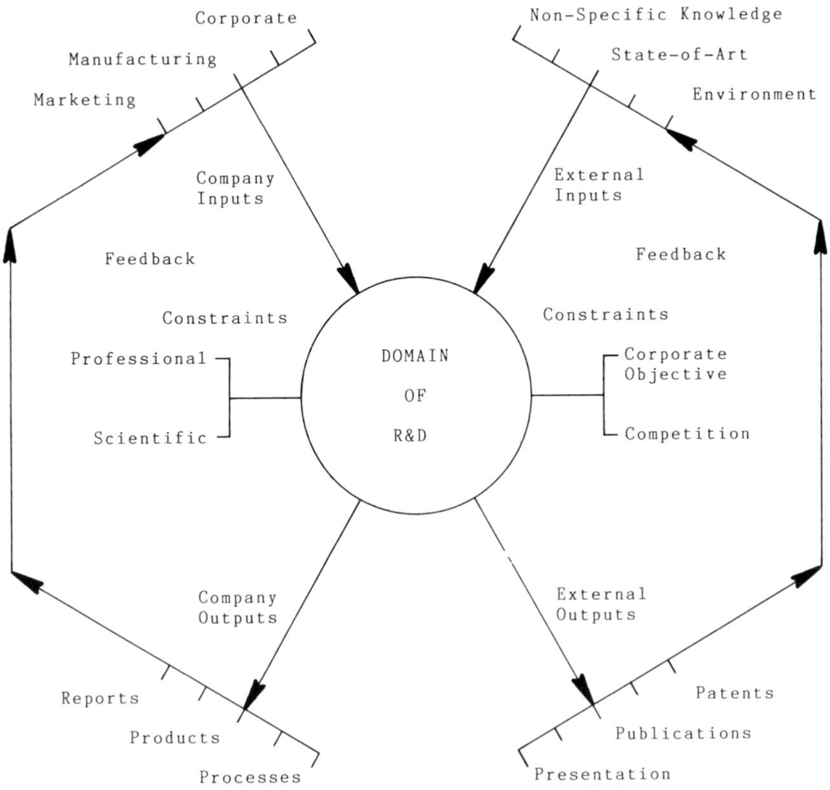

It is tempting to suggest that R & D should have a core in which the day-to-day operations of R & D progress without interference from outside. This core would be protected from the fluctuations which would result if the system were entirely open and which would reduce the efficiency of the R & D process. On the other hand, it could be argued that effectiveness is of prime importance in the R & D system and that the nature of the interactions with the company and the environment militate against the provision of a core in the R & D domain. The extreme example is where the entire R & D domain is a core activity as sometimes happens with a central R & D laboratory divorced from the line divisions of the company. In this case research of high efficiency and quality may be done, but which is totally irrelevant to the needs of the company. However, if the R & D system is too open, reaction will be the major response and coherent programs will either not be set up or will not be given sufficient time to come to fruition. It appears that a core R & D activity within the domain of R & D can be set up and prove effective if individuals are available to link the core activity with the activities in the company units and in the external environment. This is the gatekeeper approach (Tushman, 1977).

The models considered have focused little on the internal processes of R & D. Apart from the techniques of project selection and program control, for which there is a reasonable literature available, team building, motivation and leadership must be mentioned. There is rather less work reported in these latter areas. The subject of creativity should also be discussed. The nature of R & D demands that creative individuals be involved and there is a considerable literature extending back some 30 years on the nature of creativity, its measurement and stimulation. This subject will be considered later.

1.8 SUCCESS FACTORS IN R & D

Apart from the approach of model building to identify the important components necessary for success of R & D in an organization, the examination of actual case histories of successful and less successful companies can be done to identify the important factors. In this approach, a model for the R & D system is not required. Three studies have been published concerning industry in Britain and these have been summarized by Twiss (Twiss, 1980a).

The first study investigated 84 innovations granted Queen's Awards for technological innovation in 1966 and 1967. Several important factors contributing to success were identified:

(i) the presence of an outstanding person in a position of authority to back the project,

(ii) the presence of some other type of outstanding individual,

(iii) clear identification of a need,

(iv) the realization of the potential usefulness of the discovery,

(v) good cooperation,

(vi) availability of resources,
(vii) help from government sources.

Six factors which caused delay to innovation were:
(i) some other technology not sufficiently developed,
(ii) no market or need,
(iii) potential not recognized by management,
(iv) resistance to new ideas,
(v) shortage of resources,
(vi) poor cooperation or communication.

In project SAPPHO at the University of Sussex, 29 pairs of similar projects were examined; in each pair one project was successful while the other was less so. Five statements summarized the difference between success and failure:

(i) successful innovators had a much better understanding of user needs,

(ii) successful innovators paid much more attention to marketing.

(iii) successful innovators performed development work more efficiently but not necessarily more quickly,

(iv) successful innovators made more effective use of outside technology and outside advice, even though more of the work was done in-house,

(v) the responsible individuals in the successful attempts were usually more senior and had greater authority than their counterparts who failed.

The third study was undertaken by the Centre for Industrial Innovation and examined 53 projects which were abandoned during development. Environmental factors causing failure were: unattractively small market, uncertainty with monopsomistic buyers, unattractive level of competition, uncertainty of suppliers and obsolescence. Organizational factors involved in failure were: lack of marketing capacity or expertise, lack of production capacity or expertise, faulty communications with associated firms, R & D cost escalation and shortage of R & D resources.

FIGURE 7

TECHNOLOGICAL INNOVATION AS A RESULT OF COMPLEX INTERACTIONS

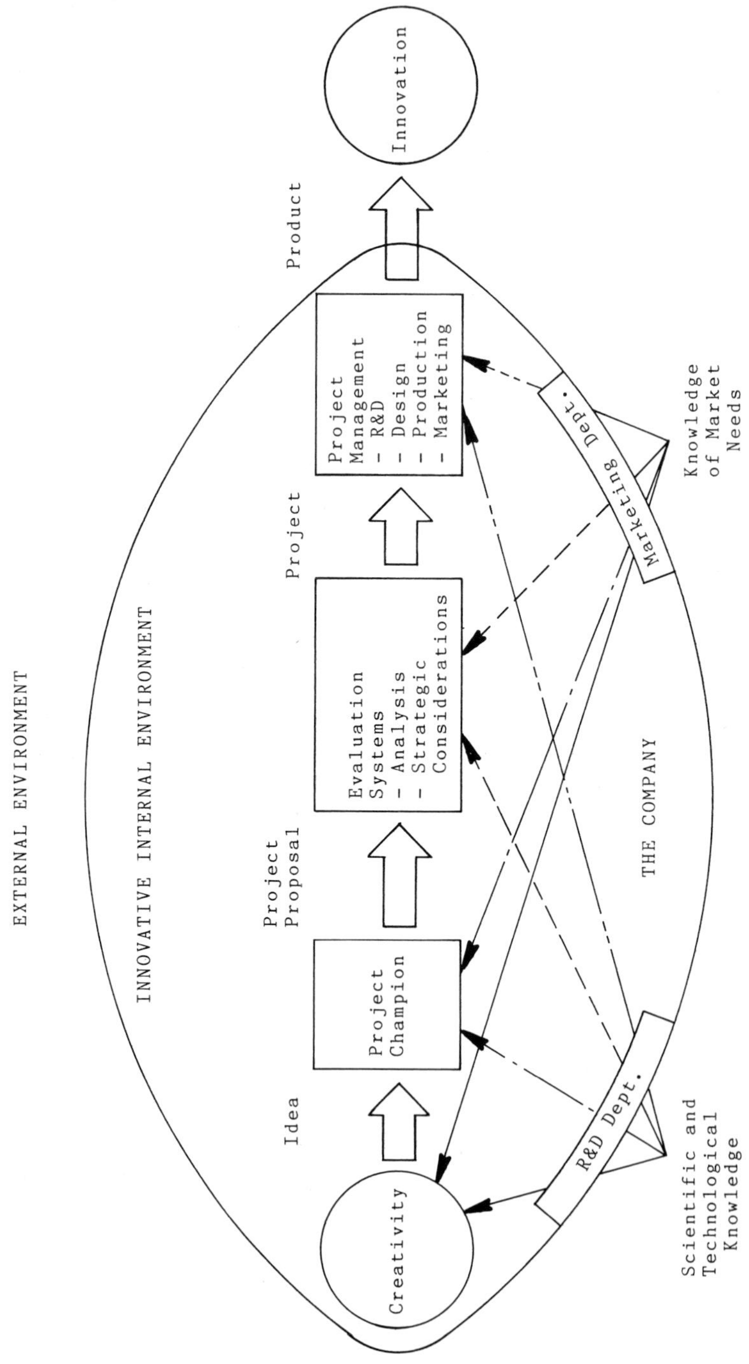

*Reprinted by permission from "Managing Technological Innovation", 2nd ed., by Brian Twiss, Longman Group Limited, publisher.

Although these factors refer to the innovation process as a whole, and not specifically to R & D, there are lessons to be learned. The results are summarized in a model of the innovation process (Twiss, 1980b). Again this model is not a complete description, but Figure 7 does illustrate the need for creativity. Both R & D and marketing are shown at the boundary of the organization domain, but this does not mean that the orientations of R & D and marketing are similar for R & D is in contact with the external scientific environment while marketing is in contact with the marketplace. According to Twiss, the most critical factors for successful innovation are as follows: (Twiss, 1980c)

(i) a market orientation,
(ii) relevance to the organization's corporate objectives,
(iii) an effective project selection and evaluation system,
(iv) effective project management and control,
(v) a source of creative ideas,
(vi) an organization receptive to innovation,
(vii) commitment by one or a few individuals.

These factors are broadly as discussed earlier with the different models of the R & D process and reflect the importance of the input to R & D, the activities within the R & D domain and the importance of the transfer of results from R & D to the company.

1.9 THE MEASUREMENT OF R & D ACTIVITY

As has been mentioned above, a substantial fraction of the GNP is spent on R & D. The activities of R & D have come under an increasing scrutiny because of the cost of these activities, and inevitably, the cost of R & D has been compared to the benefits which accrue from undertaking this activity. Attention has been directed at the efficiency (optimum resource usage) with which an R & D goal is achieved, and in this regard cost over-runs have occasioned visible public reaction. Effectiveness (optimum achievement) has also been addressed, and here the attention is on the nature of the goals themselves and whether the appropriate objectives are being concentrated on for social and economic progress. What is required is a way of measuring the success of R & D.

The measurement of the success of R & D activity is by no means confined to the national level, for interest exists in the efficiency and effectiveness of R & D in industry, academia and elsewhere. In industry, measurement may be confined to an industrial segment, to a company or to a unit of a company. Thus, the evaluation may proceed at different levels from the macroscopic to the microscopic.

At the outset, it must be stated that there is no one satisfactory way of evaluating the success of R & D activity. The models of the R & D process developed earlier do, however, suggest various classifications of methods for carrying out this evaluation. The models indicate that measurement may be made by utilizing the amount of resources, an input method, the control of the R & D process itself or the measurement of results on the output side. If reliable and meaningful methods are available for measuring input and output variables, then relationships between input and output may be established.

With a company, the focus must be on both efficiency and on effectiveness. Effectiveness relates to the worth to the company of the R & D goals pursued while efficiency relates to the utilization of resources in achieving these goals.

The discussion here will be done mainly at the level of R & D in the individual company, but before doing this it is well to consider the reasons for evaluating R & D activity. At the outset, the focus must be on the best return on the R & D funds expended, but a close examination of the R & D activity can lead to a re-examination of the goals and strategies of the company itself, the behavior of the competition and even an evaluation of other areas of the company. It may turn out that potential new products do issue from R & D, but that these new products are not developed further and introduced to the marketplace. Evaluation of R & D activity is of benefit to the research personnel themselves for it gives an opportunity to demonstrate the contribution of R & D over and above the development of new products directly. A demonstration of the utility of R & D has great benefit in showing other units of the company that research is a worthwhile activity.

The use of input measures of R & D activity is employed at the macroscopic level to indicate the level of funding relative to funds expended elsewhere. For example, at the national level the amount spent on basic research may be compared to the amount spent on applied research. The spending on research in the health area may be compared to spending on research for military purposes. R & D spending may be expressed as a fraction of GNP. The assumption which is implicit in employing spending as a measure of R & D activity is that the higher the spending, the greater the return in new knowledge, products and other outputs. Even if this assumption were true, it would still be necessary to take account of the fact that the return for unit input will be different in different research areas, and that for some research there is a threshold below which new product development is not possible (Freeman, 1974e). It is also necessary to note that different input measures than money might also be employed, such as type and level of manpower, equipment and so on. It is, however, valuable to reduce inputs to monetary values, and it is interesting to compare R & D spending levels of companies in the same business area. The Strategic Planning Institute has given guidelines for the amount of R & D spending taking into account market share, new product introductions and market growth rate (SPI, 1978). Input measures can provide a determination of the level of R & D activity, but do not provide a determination of the success of the activity. Input measures are often used where there is no other method available for the determination of R & D activity and its success.

Measurement of activity can be undertaken within the R & D unit itself. Techniques are available for program planning and control (Twiss, 1980d). These techniques may be employed to compare the status of projects to the original plan and also to compare spending to planned spending. A measure of the time and cost performance is thus available. In general terms such a measure indicates the efficiency of the technical core of R & D. Such a procedure is very necessary, especially in view of the tendency for projects to escalate in both time and cost. However, it is also necessary to examine the relevance of the project portfolio at frequent intervals to determine if projects are still matched to corporate objectives and market needs. This is an effectiveness measurement. A project may still be dropped, however, even if it meets the efficiency and effectiveness criteria, if forecasting indicates that there will be an unacceptable increase in cost in bringing the product to the marketplace. The evaluation process must, therefore, include project forecasts for cost, time and other factors.

Many publications have focused on output measurements of R & D activity. Thus, Quinn has discussed in detail the ways in which research output may be evaluated and has given three key factors (Quinn, 1960). These factors are as follows:

(i) The economic value of the technology produced as opposed to the cost of the research which produced it.

(ii) The amount of technological output per unit of scientific effort expended (productivity).

(iii) The degree to which the program's technology supports company goals.

The first two factors strictly refer to output-input relationships but do depend on the measurement of the amount and economic value of R & D outputs. The third factor refers to the effectiveness of the program. The measurement of the economic value depends on the valuation of the technology. Quinn maintains that most applied and development and some fundamental research can be quantitatively evaluated in monetary terms and recommends the use of Present Value or Return on Investment (ROI) approaches. Qualitative methods are to be used where value and return calculations are not possible. The amount of the output should also be measured to determine whether the results could have been achieved at lower cost. Quinn also suggests the use of an Exploitation Ratio which indicates how well the company has exploited any particular technology.

A different approach to the evaluation of the output from R & D has been taken by Hodge (Hodge, 1963). Here, it was pointed out that in general the valuation of research was difficult because of variation regarding what was included in the research budget, the fact that basic research was occasionally a by-product of a product-oriented effort and the long and yet-to-be-measured time lag between input and research output. As research output measurement, Hodge recommends the use of scientific publications and patents issued. The use of such measures has been reviewed by Freeman (Freeman, 1974f) who examined the use of various measures of R & D and concluded that scientific publications, "discoveries" and peer evaluation were the three main methods used to rate basic research; patents issued had been used as the measure for applied research. Although patents and scientific papers have been widely used, there are obvious difficulties. The procedures do not measure the quality of publication, but only the quantity. There may be restrictions on publication in many areas such as in industry. Similarly, some companies do not patent discoveries because it is felt that there is greater secrecy without patent protection. Differences in patent law between countries determine the ease of obtaining a patent and the cost of maintaining the patent. A recent development has been to examine the frequency with which a given published work is cited in the literature; this may be a measure of the quality of the work.

Variations on the above measures have been suggested. The treatment of R & D as a profit center rather than as a cost center has been outlined (Galloway, 1971). This approach starts with a review of the latest company product list. A list is then made of those products where R & D made a critical contribution. A total is made of the profits from products on this list and this total is compared with the previous year's results. This approach depends upon R & D being in existence for a sufficient time for comparative data to be generated.

The use of an "Opportunity Criterion" has been suggested by Gee (Gee, 1972). This approach depends upon the calculation of the value-in-use for a new product in a specified end use. The extent to which this value-in-use is greater than the projected product price is a measure of the driving force for adoption of the product in the marketplace. The disadvantage is that pricing policy is involved and that R & D output is measured by a single criterion.

Whelan has recommended the use of Project Profile Reports to measure R & D effectiveness (Whelan, 1976). This approach looks at different criteria including the type and nature of the project, the probability of success, the cost and the estimated income. The calculation of the ratio of defensive to offensive

TABLE 6

Frequencies and percentages of R & D effectiveness measures

Effectiveness measure	I	G	A	S	M	Total Frequency	Total Percentage
Number refereed articles	2	2	15	0	3	22	12
Number citations to articles		1	9	9	1	20	11
Total research performance	10	5	3			18	11
Peer evaluation index		1	11	5		17	9
Dept. quality/prestige		1	14	1		16	9
Productivity	2	1	7	1		11	6
Awards/prizes		1	6	1		8	4
Number of books		1	3		1	5	3
Number of research reports	1	1	1		2	5	3
Number of patents/inventions	3		2			5	3
Evaluated contributions	1		2		2	5	3
Number of invited papers		1	4			5	3
Number of referred papers			3		1	4	2
Number of doctoral dissertations supervised			4			4	2
Grant efficiency index			4			4	2
Satisfaction	2				2	4	2
Integration	2	2				4	2
Goal achievement	2		1		1	4	2
Self-evaluation index	1		2			3	2
Evaluated usefulness	2				1	3	2
Adaptability/flexibility	2	1				3	2
Number of commercial publications			2			2	1
Number of copyrights			2			2	1
Research evaluators			2			2	1
Visibility awareness			2			2	1
Financial performance	2					2	1
Visitors to work with researcher			1			1	1
Recognized benefits			1			1	1
Base salary			1			1	1
Job position			1			1	1

Table 6 (continued)

Effectiveness measure	I	G	A	S	M	Total Frequency	Total Percentage
Planning	1					1	1
Reliability	1					1	1
Name order on articles			1			1	1
Journal quality index			1			1	1
TOTAL:	34	18	105	17	14	188	
PERCENTAGE:	18	10	56	9	7		100

NOTE: I = Industrial, G = Governmental, A = Academic, S = Science, M = Mixed research is suggested to give an indication of the overall posture of R & D.

A multi-faceted approach to R & D evaluation has also been suggested by Royce (Royce, 1968b). The format includes the reporting of R & D expenditures, income from new products and processes, employee information and details on publication and patents. The difficulty with this approach is similar to that of Whelan in that the data may not be available to fill in the evaluation forms. The effort required to obtain the information is considerable.

A review of research on research effectiveness has recently been published (Birnbaum, 1980). This review examines the techniques for evaluating R & D in industry, government, academia and science in general since 1956. Table 6 gives the summary of the frequencies of the different effectiveness measures. Birnbaum makes the point that most evaluations have focused on the short-term; few studies have directly linked the relationship between research outputs and the long-term consequences of the research. Evaluations have generally focused at the technological core. Birmbaum states that further work is needed on the relationship of short-term to longer-term outcomes and that there is a need for the study of the technical core activities with those of the higher organizational level including the structure design of the organization and strategic management. Sharper criteria are also needed for measuring effectiveness.

The studies undertaken on the evaluation of R & D activity show a great diversity of opinion and methods of evaluation. Despite the lack of agreement, there are conclusions which can be drawn. At the outset there appears to be a dichotomous attitude towards the evaluation of R & D. This was highlighted by Quinn who stated that one attitude views technical programs as only capable of evaluation by scientists. A further consequence of this attitude is that there are no numerical measures of research effectiveness and that broad composite judgments are the only way to evaluate research performance (Quinn, 1960). Gee listed arguments why R & D must be evaluated and also listed reasons why such an evaluation was extremely difficult (Gee,

1972). The important point made was that R & D is evaluated and that this evaluation is done all the time in organizations. The evaluation may not be formal and may not be the result of conscious deliberation, but it is done. In a thorough review of the literature, Freeman concluded that the evaluation of R & D was difficult, but rejected the position that output measurement was intrinsically unattainable. The position seems to be that an evaluation process is needed and is possible despite the immense difficulties (Freeman, 1974f).

The difficulties stem from the complex nature of R & D and from the many interactions between R & D and the company and the outside environment. Figure 8 shows one way of viewing the R & D process (Quinn, 1960). This is a simplified view because the process is viewed as linear with a flow from research to technology, exploitation and the end result. Even here there are difficulties in quantifying some of the outputs and in relating R & D outputs to the end goals. Other units of the company are responsible for exploitation of the R & D results, and the time lag between R & D success and commercialization will determine the value placed on the output. Such a scheme does not take into account the related activities in which R & D often engages (Figure 2). Other factors which must be considered are the use of R & D as a resource and other paths for the R & D output such as licensing, joint ventures and so on.

Of the different measures employed to evaluate R & D, there can be no unanimous choice. Undoubtedly measures of economic worth should be used whenever possible, but reliance on a single index for the evaluation of R & D is unrealistic. The impression gained from the literature is that many of the schemes only exist on paper and have never been tried in practice, or if tried, have been quietly dropped. Mechlin and Berg have recently examined the evaluation of R & D and their conclusion best summarizes the approach which should be taken (Mechlin and Berg, 1980). "There can be no shortcut, no easy formula, for assessing the value of corporate research. The complex expectations of its sponsors affect the measures those sponsors use to determine whether their expectations have been met. And the choice of methods affect, in turn, what kind of expectations they can have in the first place. Every accounting procedure, no matter how

FIGURE 8

A VIEW OF THE DIVERSE OUTPUTS FROM R&D

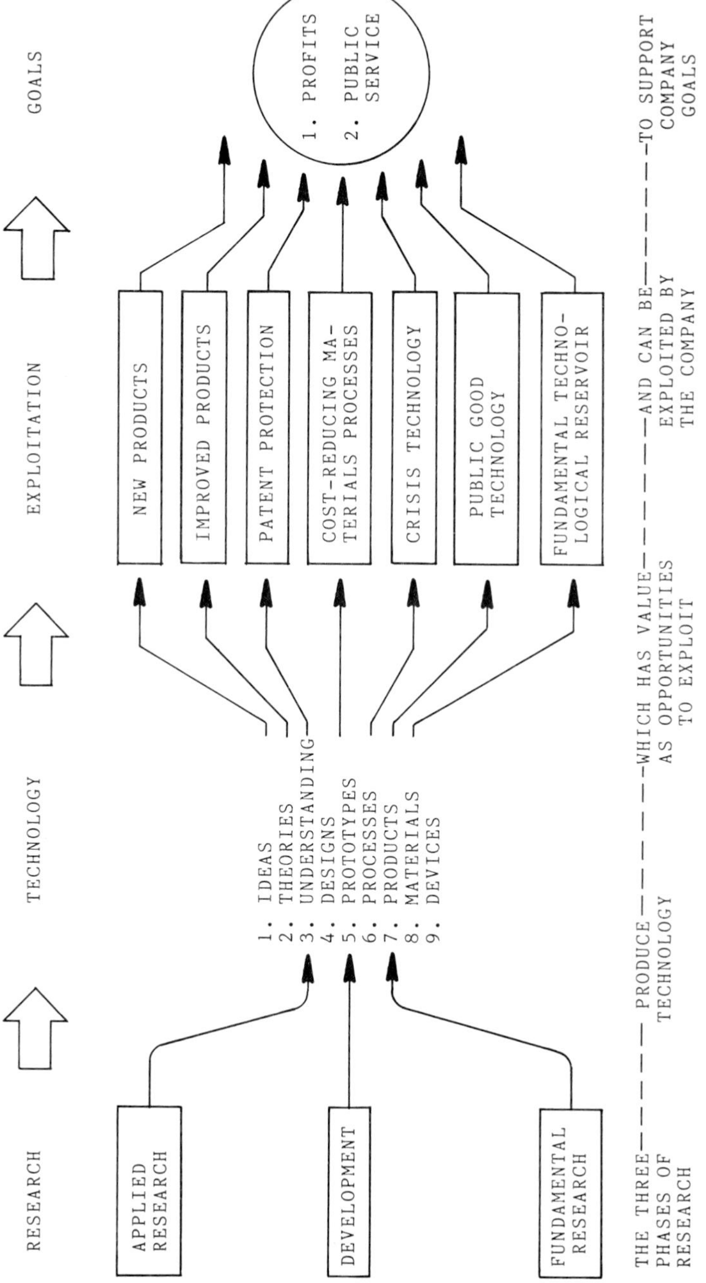

*Reprinted by permission of the Harvard Business Review. An exhibit from "How to Evaluate Research Output" by James B. Quinn (March/April 1960).

carefully employed, becomes a hidden determinant - and not just a measurement - of corporate policy.

"What each company needs, therefore, is a reasonable, deliberately thought-through set of expectations for its research activity. What we must not have is a return to the euphoric illusions of the post-World War II era when there seemed to be no limit to the social and financial benefits of industrial research. Companies must use a technique like ROI only to the extent that it conveys significant information and never allow themselves to become the captives of their own accounting procedures. Finally, they must supplement ROI calculations with a host of more subjective judgments. With the evaluation of research - as with so much else managers do - there is just no substitute for disciplined professional judgment."

In fact, there appears to be a need for an adversary method of evaluating R & D in which progress is evaluated by corporate, R & D and other company units such as marketing and manufacturing. Only in this way can the expectations of the different actors in the R & D process be matched with the results.

1.10 ALTERNATIVES TO R & D

R & D should not be regarded as a separate, autonomous activity, but as one of the integrated activities required for the survival and growth of a successful business. It must be realized that business strategies do not automatically revolve around R & D, but that investment in R & D is merely one of the ways in which a company can grow and prosper. In specific instances there may be very attractive alternatives to R & D. These alternatives have been discussed by Steele (Steele, 1975b).

Four fundamental options or combinations of these options may be considered for business strategy. The first, and most limited, is the case where little opportunity is perceived for active growth. The strategy here is to optimize the generation of profits from a stable business. This strategy can be challenging as the harvesting of the present business may extend over many years. Thus, the time scale is important as is the competitive position of the business. At minimum, the choice of this strategy indicates that no effort will be made to growing the business. Investment in additional resources is only made to give the best possible return and to maintain competitive advantages in cost, product leadership and customer loyalty. R & D, if required at all, is directed to incremental cost-effective changes to the product line or production process. The emphasis is on maintaining the status quo. Funds are expended on R & D, with the emphasis on development, only to achieve short-term well-defined objectives. Clearly, in such a situation, the employment of R & D as an institutional change mechanism is not envisaged.

A second strategy is to grow the existing business in its current form. In this strategy increased profitability and/or increased market penetration is sought by selling more of the same product, either to the same customers or to new customers, but in the same markets and using the same channels of distribution. The emphasis will be on reducing costs, improving product performance or on achieving product differentiation through special features. These changes need not be accomplished through R & D. Cost reduction would appear to be an area where technological innovation offers great benefit, but there are alternatives to R & D programs which often appear to be lengthy, costly and of high risk. For example, pricing strategy may increase volumes sufficiently to greatly reduce the unit cost by spreading high fixed costs. Volumes may be increase through a good advertising campaign. Increase in the efficiency of the distribution system may also lead to increased volume and reduced costs. Manufacturing and distribution costs may be reduced by reducing the width of the product range. Improvement in warehouse location can result in reduced transportation costs. Inventory control to avoid excessive levels of inventory can greatly reduce cost. Even in the production area, costs may be reduced without recourse to R & D simply by purchase of new equipment or better supervision, employee training and improved motivation.

In the case of product differentiation, there may be R & D input, but again there may not. Color, configuration and style are important in consumer goods. Service and delivery schedules can play a great role in product differentiation. Advertising may be used to differentiate the product from the products of competitors. All of these options may succeed at lower cost and have lower risk of failure than investment in R & D. Even in those cases where improved technical performance is required, it is not always necessary to achieve this goal through the internal generation of technological innovation. Needed technology may be purchased and the internal focus is then on application of this technology. This approach is of the "make or buy" type and can well involve lower risk than internal R & D. However, R & D should be involved in the evaluation of technology and may be involved in the application.

The third option is to extend the present business into new areas either by selling new products to old markets or old products to new markets. Here it might seem that there is a guaranteed role for R & D, but again analysis shows that there are other options. Products may be acquired externally by purchasing the manufacturer, or by buying the product itself. The option of selling into new markets may involve no more than a geographic expansion of the sales force or the acquisition of distributors. Decisions may be made to go direct rather than through distributors and so gain direct access to the marketplace. New classes of customers may be identified and new types of salesmen hired to deal with these new customers.

Even when the decision to enter a new market involves the development of special products, it frequently involves no more than the application of existing technology and may involve the custom building of equipment or the addition of special features. Often such moves would not involve R & D.

The business may be extended by vertical integration either backwards or forwards. Technology will probably be involved but such technology will almost certainly exist already. Vertical integration is usually achieved by acquisition of companies or by the acquisition of knowledge by licensing agreements. This is preferable in most cases to the development of technology internally to allow vertical integration. The purchase of a company at the same time removes a competitor while the purchase of technology involves less risk, but is not risk-free, than the internal generation of the technology. Although R & D would not be overwhelmingly involved in the vertical integration there would be some involvement on the evaluation side and in the feasibility studies with bought-in technology.

The fourth strategic option for growth is to create entirely new business in which the technology, markets and customers are new. One attractive way to do this is through acquisition rather than through the development of proprietary technology (Salter and Weinhold, 1979). The growth of conglomerates has demonstrated that financial manipulation may provide a faster path to growth than by the development or technology. Licensing for acquiring technology and products also represents an alternative to in-house development.

In some situations the company may choose a low-technology approach in a new business by emphasizing other characteristics of the product, such as appearance and convenience, over performance. New business may be generated with new products based on existing technology simply by perceiving a gap and exploiting this market need.

The discussion on the four options has deliberately de-emphasized the role of R & D to highlight the other options open to a business. These other options may well be chosen because of the risk and uncertainty inherent in R & D. In some discussions the terms "risk" and "uncertainty" are used synonymously, but it is more useful to regard these as being qualitatively different. Risk is quantifiable in that the probability of an event occurring can be estimated. Business situations deal with risk, whether or not the probability of success is explicitly evaluated. Uncertainty represents a non-quantifiable situation in which the outcome cannot be probablistically forecast. R & D projects may fall into either of these two categories, but the uncertainty is greater the closer the project is to the limits of knowledge. Basic research programs often are associated with a high degree of uncertainty. On the other hand, development projects are more closely associated with risk. The reason for the business executive to prefer other options to R & D is explained by the fact that non-R & D actions usually lie in the area of risk. Where R & D is to be utilized, the emphasis again is on the risk end of the spectrum and so the emphasis is on development rather than research. The preference for the other options may be responsible for the following statistics for the U.S.A. in the decade ending 1978 (Mechlin and Berg, 1980):

(i) R & D as a fraction of GNP had decreased 20%.

(ii) Basic research as a fraction of GNP down 24%.

(iii) A reduction of 13% in the proportion of the labor force involved in R & D.

(iv) Industrial investment in basic research as a function of net sales declined by 32%.

There is a general realization that this trend must be halted if the United States is to retain technological leadership. As mentioned earlier, technological innovation was responsible for 45% of the nation's economic growth from 1929 to 1969 and so must be encouraged. It must be realized that there is a connection between R & D and business growth. Thus, the president of Arthur D. Little has stated "...of 100 leading American companies that were leading companies in 1909, only 36 remained in the top 100 forty years later ... these 36 remained in the top 100 through aggressive research and continued development and introduction of new products", (Hodge, 1963).

That there are alternatives to in-house R & D cannot be disputed and this fact must be realized by R & D personnel. However, R & D does have a place in the growth and survival of the firm and should be used in the appropriate circumstances. In specific cases it may well be more desirable to buy technology outside but this must not become a general attitude for industry in a particular country or even in a particular industrial sector. The fact that the technology exists means that someone developed it somewhere and the developer usually has a competitive edge. If industry does not carry out a certain level of internal R & D, the licensing strategy becomes a dependency strategy which in the long run will only be to the detriment of industry.

1.11 SUMMARY

The role of R & D has grown to assume great importance in modern society. This is shown by the high spending on R & D in the developed countries. R & D is conducted in many sectors including government, academia and industry. Despite the high expenditure on R & D and the general desire for invention and for innovation, the R & D process is imperfectly understood.

In part, the difficulty with understanding R & D is due to the overlapping of other tasks with R & D tasks, related R & D and non-R & D tasks (Freeman, 1974c). Within the R & D system itself there are contradictory and confusing divisions of work, but again a rational proposal for the classification of R & D tasks has been made (Freeman, 1974b). There is, as yet, no general acceptance of these different classification criteria.

The function of R & D can be defined in terms of the classification of what is an R & D task but in practice it appears that a contingency approach should be taken. R & D is an agent of change; in an unsophisticated company R & D may therefore be involved in relatively mundane tasks which would not be defined as R & D in a sophisticated company.

There is no general model of the R & D process, but different models may be used to illustrate different facets of the process. The emphasis in modeling is for a linear flow between different stages of R & D and then a linear progression through company units. However, feedback and complex multiple and iterative interactions with the units of the company and from outside must be considered. The model is a dynamic one with the time dimension assuming some importance. The importance of marketing input and the satisfaction of corporate objectives must be stressed. Creativity is important as is the interface between R & D and the company units for optimal transfer of results.

The diverse nature of R & D tasks and the different outputs from R & D is very evident in the case of the evaluation of R & D. Measures which have been employed include those concerned with the input, such as amount expended on R & D, those concerned with the output, such as patents, publications and new products, and those concerned with the output in relation to a given input, such as ROI. Efficiency measures within the R & D system may be used. The method preferred for R & D evaluation depends on the location of the work. Thus, different methods apply in academia in contrast to industry. In any area a balanced application of criteria should be used. Reliance on a single measure of performance is not recommended.

Although R & D has been considered as the main mechanism of change, it has to be accepted that change is not required in circumstances such as harvesting the present business. Even when change is desired, there are many alternatives to R & D. Emphasis may be placed on cost reduction, advertising, new designs using existing knowledge, licensing or new acquisitions. At the business level the expenditure on R & D must be justified against that on other alternatives which may give a better return, at least in the short term. However, the benefits of in-house R & D in allowing more effective evaluation of outside technical opportunities must be stressed.

REFERENCES

Birnbaum, P.H.	1980	"Research on Research Effectiveness: Where Do We Stand in 1980?", R and D Management, 10th Annual Conference on Industrial R & D Strategies, Manchester Business School, England, 30 June - 2 July.
Burns, T. and Stalker, G.M.	1966a	"The Management of Innovation", Tavistock Publications (2nd edition). Chapter 2.
Burns, T. and Stalker, G.M.	1966b	Ibid, p. 32.
Freeman, C.	1974a	"The Economics of Industrial Innovation", Penguin, Harmondsworth, Middlesex, England. Chapters 2, 3, 4.
Freeman, C.	1974b	Ibid, 313-315.
Freeman, C.	1974c	Ibid, p. 152.
Freeman, C.	1974d	Ibid, p. 23.
Freeman, C.	1974e	Ibid, 152-157.
Freeman, C.	1974f	Ibid, 322-387.
Galloway, E.C.	1971	"Evaluating R and D Performance - Keep It Simple", Res. Mgmt. XIV, #2, 50-58.

Gee, R.E.	1972	"The Opportunity Criterion - A New Approach to the Evaluation of R and D", Res. Mgmt., XV, #3, 64-71.
Gerstenfeld, A.(ed)	1979a	"Technological Innovation: Government/Industry Cooperation", Wiley Interscience, New York, Chapter 7.
Gerstenfeld, A.(ed)	1979b	Ibid, Chapter 13.
Gibson, J.E.	1981	"Managing Research and Development", John Wiley, New York, Chapter 6.
Hodge, M.H.(Jr.)	1963	"Rate Your Company's Research Productivity", HBR Reprints, Nov.-Dec. 1963-68, 39-52.
Kingston, W.	1977a	"Innovation", John Calder, London, p. 16.
Kingston, W.	1977b	Ibid, p. 30.
Mansfield, E.	1980	"Basic Research and Productivity Increase", American Economic Review, 70, Pt. 5, 863-873.
Mechlin, G.F. and Berg, D.	1980	"Evaluating Research - ROI is Not Enough", HBR, 58, 93-99.
Quinn, J.B.	1960	"R and D Management", HBR Reprints, March-April 1963-68, 63-74.
Royce, W.S.	1968a	"R and D and Economic Growth", Conference on Industrial Research, Institute for Industrial Research and Standards, Dublin, Ireland.
Royce, W.S.	1968b	"On Evaluating the Contribution of R and D to Corporate Performance", Ibid.
Salter, M.S. and Weinhold, W.A.	1979	"Diversification Through Acquisition", The Free Press, New York.
SPI	1978	"How Much to Spend on R and D", Pimsletter, Cambridge, Mass.
Steele, L.W.	1975a	"Innovation in Big Business", Elsevier, New York, p. 10-18.
Steele, L.W.	1975b	Ibid, 54-62.

Thompson, J.D.	1967	"Organizations in Action", McGraw-Hill, New York, Pt. 1.
Tushman, M.L.	1977	"Special Boundary Roles in the Innovation Process", Admin. Sci. Quart., 22, 587-605.
Twiss, B.C.	1980a	"Managing Technological Innovation", Longman, London, 16-17.
Twiss, B.C.	1980b	Ibid, p. 18.
Twiss, B.C.	1980c	Ibid, p. 6.
Twiss, B.C.	1980d	Ibid, Chapter 6.
Whelan, J.M.	1976	"Project Profile Reports Measure R and D Effectiveness", Res. Mgmt., XIX, #5, 14-16.

PART 2

STRATEGY AND STRUCTURE

2.1 INTRODUCTION

The growth and survival of an organization depends on certain key strategies. The earliest work in this area identified the strategies of volume, geographic dispersion, vertical integration and product diversification as key strategies (Chandler, 1962). The volume strategy relates to an increase in the quantity of goods produced. As volume increases so it is necessary to alter the organization to cope. Geographic dispersion indicates that the goods are sold in a wider area than previously. For example, a local company can produce for the immediate surrounding area, for the state and then for the whole country. A further stage would be overseas expansion. Vertical integration refers to changes in the scope of the business; backward integration is concerned with expansion in the direction of supply (the input side) whilst forward integration implies expansion towards the market (the output side). Very often companies come into being with a single product but over time product diversification is mandated both to broaden the range of products and to introduce improved products.

Chandler showed that each strategy gave a different type of difficulty which was addressable by a different form of organization structure. This initial study has led to much research on the role of strategy and structure on the growth of the firm. The concept of "fit" has been introduced to describe how well the structure of the company matches the adopted strategy. The implication is that companies with a good fit, in other words with a consistent strategy and structure, prosper compared to those companies with a mis-match or non-optimal fit. However, an adopted strategy does not exist in isolation but is influenced by the environment in which the organization exists. Environmental factors such as rate of change in technology, competitive pressures, economic forces and many others greatly influence the success of a chosen strategy and therefore must modify or entirely determine the choice of strategy. Naturally a good fit is a necessary but not sufficient condition for success since the adopted strategy may be inappropriate. In order to explore the strategy-structure relation it is necessary to examine the concepts of strategy, structure and environment in more detail. It should be noted that the discussion is ideal in the sense that it is assumed a company will choose the best strategy given a choice. In the real world this assumption may not be valid. For example a strategy may be chosen because of the organizational structure which is the reverse of the logical choice.

2.2 STRATEGY AND STRUCTURE

According to Hofer and Schendel strategy is the set of basic characteristics of the match an organization achieves with its environment (Hofer and Schendel, 1978a). Strategy is a means for coping with both external and internal changes. Strategy is the path charted for the organization and is linked to the organizational goals and objectives which are to be achieved. Hofer and Schendel go on to discuss the different definitions of strategy which have been given in the literature (Hofer and Schendel, 1978b). It is pointed out that some authors do not differentiate between strategy as a concept and the formulation process itself. In addition there is major disagreement over whether strategy is a broad or a narrow concept. The broad concept of strategy includes not only the ends, the goals and objectives, but also the means used to achieve these ends. The narrow view of strategy is that it is a description of the means employed to achieve goals and objectives set in a separate process. Hofer and Schendel choose the narrow concept and consider goal setting and strategy formulation as two distinct but interrelated processes. A similar view has been taken by Richards who has considered organizational goal structures in detail (Richards, 1978a). According to Richards, "A strategy is a set of actions, policies governing actions and plans of activity to achieve given results; a goal is a result aspired to". This narrow definition of strategy is recommended here.

A hierarchy of strategies exists in an organization (Hofer and Schendel, 1976b). Three major levels of organizational strategy may be distinguished. These are corporate strategy, business strategy and functional area strategy. The corporate-level strategy is concerned primarily with the question of what business should we be in? Scope and resource deployments among businesses are the primary components of corporate strategy. At the business level, strategy is directed to the subject of how to compete in the particular industry or product/market segment. Scope is less important at the business level and choices are more concerned with segmentation and with the stage of product/market evolution. At the functional level the principal focus of strategy is on the maximization of resource productivity. Scope is even less important than at the business level. Synergy is of great importance at the functional level but is of lesser importance at the business level and is not important at the corporate level.

Since there exists a hierarchy of strategies there must also exist a hierarchy of goals (Richards, 1978b). If goals are the ends and strategies the means, there exists a means-ends chain. The first step is to set the goals for the highest level and this then defines the strategies to be employed; an iterative process is used between the goals and strategies until a consistency is reached. Once this task is accomplished for this level, the goals for the next level down are set. The goal and strategy hierarchies are shown as related to the managerial and organizational hierarchies in Figure 9.

Organizations are purposeful social units which consist of people who carry out differentiated tasks which are coordinated to contribute to the goals of the organization (Dessler, 1976a). Structure has been defined as "those aspects of behavior and organizations subject to existing programs and controls" (Lawrence and Lorsch, 1967). Structure in an organization thus refers to information flow and to the hierarchy of decision making. Examples of different structure are given elsewhere (Dessler, 1977b). Factors to be taken into account in the design of an organizational structure include centralization or decentralization, line and staff function, organization by product or by geographical area, and many others. There are many different arrangements of company units which can be adopted. Contingency theory would state that there is no one best way of organization but that the structure should reflect the strategy.

FIGURE 9

THE RELATIONSHIP OF GOAL, STRATEGY, MANAGERIAL AND ORGANIZATIONAL HIERARCHIES

ORGANIZATIONAL HIERARCHY	MANAGERIAL HIERARCHY	GOAL HIERARCHY	STRATEGIC HIERARCHY
Board of Directors Corporate Officers	Founder, Chairman, President	Economic and social values and balance among them	Guiding Philosophy
Corporate	President	Overall goals corporate objectives	Master Strategy
Business Unit and Corporate	President; Division Manager	Business goals	Business Strategies
Functions and Business unit	Function Manager	Functional goals	Functional Strategies

*Reprinted by permission from "Organizational Goal Structures" by M.D. Richards.

Although there are many choices of structure depending on the strategy to be adopted there are main classes of structures which can be distinguished. Thus a company may be functionally organized (organized around functions such as marketing, manufacturing or R & D) or organized by product. In a matrix organization, dual reporting relationships by function and by product or business are used as the basis for the organizational design. It may be mentioned that mixed modes of structure may be adopted. For example, geographic decentralization of the company by location may occur while centralized decision-making is retained. In considering the structure of an organization it suffices to assume that, using the descriptive variables available, an unambiguous picture of the inter-dependent arrangement of the units of the company can be given. This is certainly true in theory but in practice modification may be necessary to allow for political and career factors in the organization. Such factors would also influence the implicit assumption that organizations act in a rational manner.

2.3 THE ENVIRONMENT

A considerable amount of study has been undertaken to define the environment of the firm. In early work four classes of environment were defined by examination of actual organizations using case studies (Emery and Trist, 1963). These environmental types are: placid, randomized; placid, clustered; disturbed, reactive; turbulent field. Uncertainty increases in moving from the placid to the turbulent environment. The latter environment involves continuous change. Burns and Stalker examined companies in the United Kingdom and focussed on the predictability of the environmental demands (Burns and Stalker, 1966a). Five different types of environment were distinguished which ranged from stable to least predictable. Duncan has summarized the studies on the environment up to 1973 and, from a study of three manufacturing and three R & D organizations, found that there are two dimensions to the environment, simple-complex and static-dynamic (Duncan, 1972). Prior to that report an uncertainty scale was constructed to measure environmental uncertainty (Lawrence and Lorsch, 1969). This scale, which was presented in the form of a questionnaire, has had a great influence on the direction of research on environmental uncertainty. However, attempts to repeat environmental assessments using the scale have been disappointing (Tosi et al, 1973). The approaches of Lawrence and Lorsch and also of Duncan have been further examined for conceptual and methodological adequacy (Downey et al, 1975). It was found that the two scales had only a small overlap. Comparison of the scales with criterion measures produced disappointing results. Replication of the Duncan complexity-dynamism study produced contradictory results. At the present time the concepts of environmental complexity and dynamism would appear to be valid dimensions for the description of the environment; the relationship of these dimensions to the degree of environmental uncertainty is problematic. A satisfactory scale for evaluating the environment has yet to be demonstrated.

The above studies on the environment have examined measures which are related to the tasks of the organization. This so-called task environment is the one which immediately influences the organization. However there is a broader environment which will include socioeconomic, political and technological factors which may only influence the organization in the long run. The boundary between the organization and the environment is not sharp. Thus the organization spills over into the environment and the environment intrudes into the organization (Galbraith, 1977). All these factors complicate the definition of organization and environment.

The model which results from the above discussions is of the organization as an entity which has a diffuse boundary beyond which lies the immediate task environment and then the wider environment. Internally the organization is structured so that there is an appropriate division of task activity, flow of information and decision making. A means-end chain results in the setting of goals and of strategies to achieve these goals via an iterative process in a hierarchy of goals and objectives. Having developed these concepts the strategy-structure relationships resulting from the study of case histories can now be examined.

2.4 ORGANIZATIONAL GROWTH

Growth and survival of an organization has been examined in

the light of the earliest studies which indicated a connection between strategy and structure. A large number of inquiries have been made in the area of model building to illustrate how growth of an organization occurs. This whole literature has been well reviewed (Galbraith and Nathanson, 1978a).

Metamorphosis models are typified by the approach of Greiner (Greiner, 1972). Here it is assumed that growth is not smooth but involves discontinuities when the degree of change is too large for the organization to handle within the existing structure. Different factors such as age, size or complexity have been regarded as the driving force for metamorphic change. The difficulty with such models is that it is an oversimplification to regard the change in form to be a direct response to change in one, two or even three factors. However, of the factors cited, complexity has been the basis for many of the models.

The complexity models also grew out of the work of Chandler, whose work was extended by Scott (Scott, 1971). Scott envisages three stages to growth. Stage I represents a simple functional organization involved with a single product line. Stage II represents growth of the Stage I organization by vertical integration while still retaining a single product line. Stage III involves the diversification into multiple product lines with the introduction of a multi-divisional structure in which each division is a functional organization with a single product line.

The Scott model has been analyzed by Salter (Salter, 1970) who suggested that Stage III should be divided into two stages, a Stage 3 which represents a geographic form and a Stage 4 which represents a product form. Galbraith and Nathanson, however, feel that the geographic product centre does not represent a distinct stage.

Another extension of the Scott model has been proposed by Stopford to account for the international expansion of American firms (Stopford, 1968; Stopford and Wells, 1972). It was found that initially the firms formed an international division and attached it to the domestic product division. Then the international division was disbanded and either worldwide product divisions or area divisions were adopted. Thus Stage III was split into Stage III with an international division and Stage III other. However, the introduction of global operations gives rise to significant complexity which cannot simply be handled by a division of Stage III as proposed by Stopford.

The question of international operations is topical, has represented a significant expansion and increase in complexity and has involved many modifications in the models of organizational growth. Smith and Charmoz reported on the problems that arise in the growth of multi-national corporations (Smith and Charmoz, 1975). From their discussion a five-phase model for overseas expansion was postulated. Other empirical studies have focussed on the growth of American and European multi-nationals (Stopford and Wells, 1972; Franko, 1974). These studies indicated the preferential growth paths for Continental and American enterprises; these growth paths were different.

Based on their review of the literature Galbraith and Nathanson have given a revised model for growth and development which takes into account global expansion. It is assumed that, starting with a simple form, any source of diversity may be added to move to a new form. In other words, there are alternative paths to development and different paths may be taken which will all end at a global multi-national structure. There is no one best way (contingency approach). Figure 10 shows this model (Galbraith and Nathanson, 1978a). Indicated on the model is the dominant growth path taken by American companies as determined by the above empirical studies. This model reflects a contingency theory of growth; although the majority of firms pursued a set path, under other circumstances a different path could be taken. Thus, the step of vertical integration cited by Chandler has turned out to be the preferred step but is not necessary since a company could proceed directly to a multidivisonal form by internal growth and a related diversification of the product line. Alternatively the global multi-national state could be achieved by unrelated diversification leading to a holding company, then to a global holding company and finally to the global multi- national structure through internal consolidation.

FIGURE 10

A MODEL OF GROWTH AND DEVELOPMENT TO THE GLOBAL MULTI-NATIONAL FORM

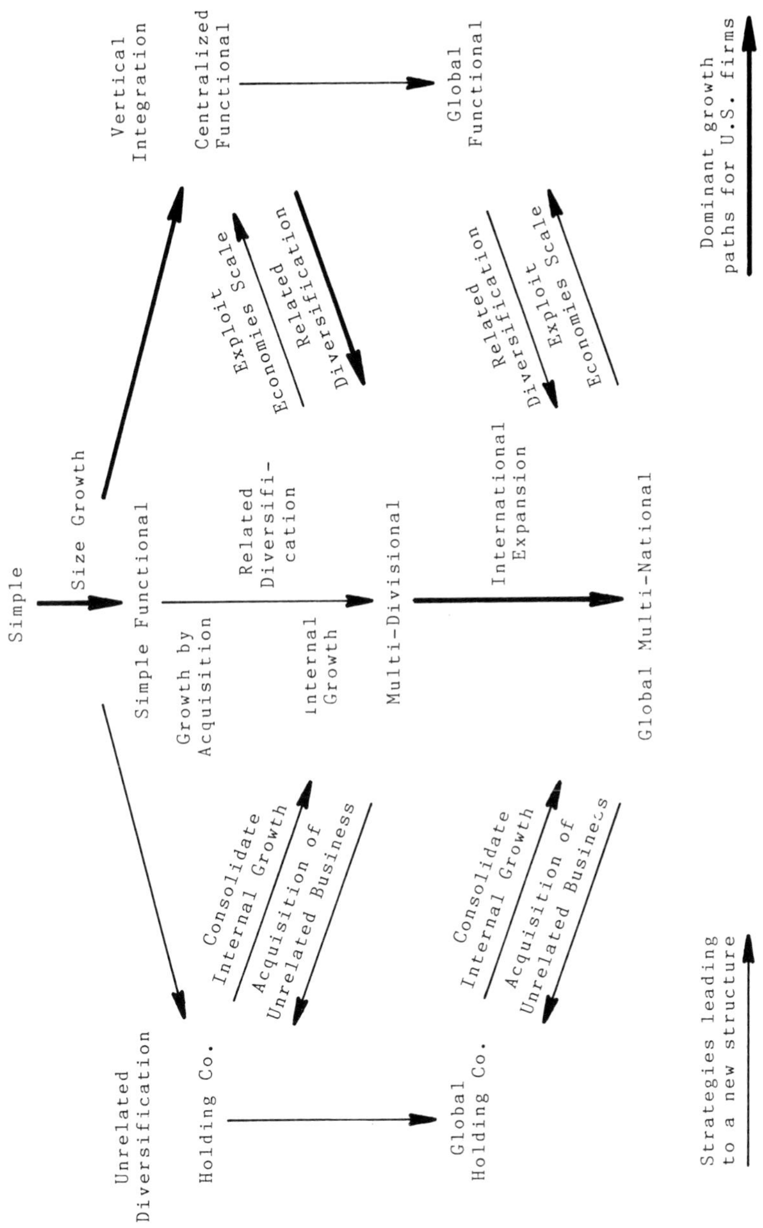

*Reprinted by permission from "Strategy Implementation: The Role of Structure and Process" by J.R. Galbraith and D.A. Nathanson, Copyright © 1978 by West Publishing Company. All rights reserved.

The other point to make is that a complete progression to the final global multi-national structure is not mandated. Depending on the circumstances a company can find a niche and stay there, it can progress or it can reverse its steps. Also, while the overall structure of an organization is reflected by its position in the model, within the organization there will be a hierarchy of structures. Thus in a global structure based on areas, within an area the structure might be based on products and within a product sub-structure the organization could be based on market sector. A functional organization could be found within the market sector. All that this means is that each level is structured according to the source of diversity which is faced (contingency approach).

The model of Galbraith and Nathanson consists of five types of organization as shown in Table 7. These types are the simple, functional, holding, multi-divisional and global forms. Each type of organization has specific characteristics regarding strategy, structure and internal processes. The extension over the model of Scott is in the provision of specific expansion paths to reach the different stages represented by different organizational types. The extension over the mainstream of strategy-structure research is in the emphasis on the need for congruity in the internal processes of the organization as an additional element in the strategy-structure formulation. Galbraith and Nathanson also speculate on the existence of stages beyond the global multi-national form. Further increases in volume, diversification or changes in technology could lead to a new form. These authors believe that the grid or matrix structure could be a new type of structure for the organization (Galbraith and Nathanson, 1978b).

In contrast to the above approach which is concerned with the stages of organizational growth and with the characteristics of each stage, a less specific but potentially more powerful formulation was adopted by Thompson (Thompson, 1967). The Thompson picture of an organization is as a domain which exists in an environment which is described in terms of two dimensions, stability and homogeneity. Organizations are assumed to act in a rational manner and the goal of the organization is to reduce uncertainty. Depending on the nature of the uncertainty different strategies may be undertaken such as contracting, co-opting or coalescing. The strategies adopted may lead to growth. For example, vertical integration to obtain control of a raw material source would reduce uncertainty and would at the same time lead to growth of the organization. The approach of Thompson emphasises the power aspects of organizational strategies; an increase in control over an aspect of the environment gives a reduction in uncertainty in exchange interactions with that environmental segment. Another driving force for growth is the degree to which capacity is utilized. There will be a move to enlarge the domain of the organization to satisfy excess capacity.

The subject of organizational structure is dealt with in terms of internal and external dependencies. Organizations are assumed to group positions in order to minimize cost. Functions involving high efficiency needs are placed within a technological core for protection from uncertainty. Boundary spanning units are used to interact with the environment. The number and distribution of these units depends upon the complexity and stability of the environment. For complex environments the boundary spanning units may be segmented so that each unit faces a homogeneous environment. Units will require more autonomy in the face of dynamic environmental conditions whereas the behavior can be governed by specific rules in a stable environment. The conditions under which there is a dynamic, complex environment call for decentralization and local autonomy of action.

The predictions of Thompson on organization growth can be shown to fit into the growth model of Scott (Galbraith and Nathanson, 1978c). This was done on the basis of groupings by interdependence to create a hierarchy. However, it would appear preferable to accept the formulation of Thompson as a general model which, because of its generality, can be applied to many different situations. The power and dependency aspects of growth are complementary forces to those considered so far.

The approach of Thompson has been extended by Miles and Snow to specify four types of organizations based on adaptive

TABLE 7

Model illustrating the five organizational types

Type / Characteristic	Single	Functional	Holding	Multi-divisional	Global-M
Strategy	Single product	Single product and vertical integration	Growth by acquisition unrelated diversity	Related diversity of product lines-internal growth acquisition	Multi products in multiple countries.
Organization structure	Single functional	Central functional	Decentralized profit centres around product divisions. Small headquarters	Decentralized product or area division profit centers	Decentralized profit centres around worldwide product or area divisions
Research and development	Not institutionalized. Random search	Increasingly institutionalized around product and process improvements	Institutionalized search for new products and improvements - Decentralized to divisions	Institutionalized search for new products and improvements - Centralized guidance	Institutionalized search for new products which is centralized and decentralized in centres of expertise
Performance measurement	By personal contact subjective	Increasingly impersonal based on cost, productivity but still subjective	Impersonal based on return on investment and profitability	Impersonal based on return on investment profitability with some subjective contribution	Impersonal with multiple goals like ROI profit tailored to product and country

*Reprinted by permission from "Strategy Implementation: The Role of Structure and Process" by J.R. Galbraith and D.A. Nathanson,

Table 7 (Continued)

Rewards	Unsystematic paternalistic based on loyalty	Increasingly related to performance around productivity and volume	Formula based bonus on ROI or profitability - Equity rewards	Bonus based on profit performance but more subjective than holding. Cash rewards	Bonus based on multiple planned goals. More discretion. Cash rewards
Careers	Single function specialist	Functional specialists with some generalist interfunctional moves	Cross function but intra-divisional	Cross function inter-divisional and corporate-divisional moves	Cross functional intersubsidiary. Subsidiary/corporate moves
Leader style and control	Personal control of strategic and operating decisions by top management	Top control of Strategic decisions. Some delegation of operations through plans, procedures	Almost complete delegation of operations and strategy within existing businesses. Indirect control through results and selection of management and capital funding.	Delegation of operations with indirect control through results. Some decentralization of strategy within existing businesses	Delegation of operations with indirect control through results according to plan. Some delegation of strategy within countries and existing businesses. Some political delegation
Strategic choices	Need of owner v's needs of firm	Degree of integration. Market share. Breadth of product line	Degree of diversity. Types of business. Acquisition targets. Entry and exit from businesses	Allocation of resources by business. Exit and entry from businesses. Rate of growth	Allocation of resources across businesses and countries. Degree of ownership and type of country involvement

strategy (Miles and Snow, 1978a). The adaptive cycle is shown in Figure 11. This is a contingency model in that the strategy may be chosen from a range of options. Following the choice of strategy, the structure and process parts of the organization must be rationalized to preserve the structure-strategy-process fit. The four types of organizations are Defender, Reactor, Analyzer and Prospector.

Defenders are organizations which have narrow product-market domains. These organizations have a narrow focus and do not tend to search for new opportunities outside their domains. Primary attention is directed towards improving the efficiency of the existing operations.

FIGURE 11

THE ADAPTIVE CYCLE

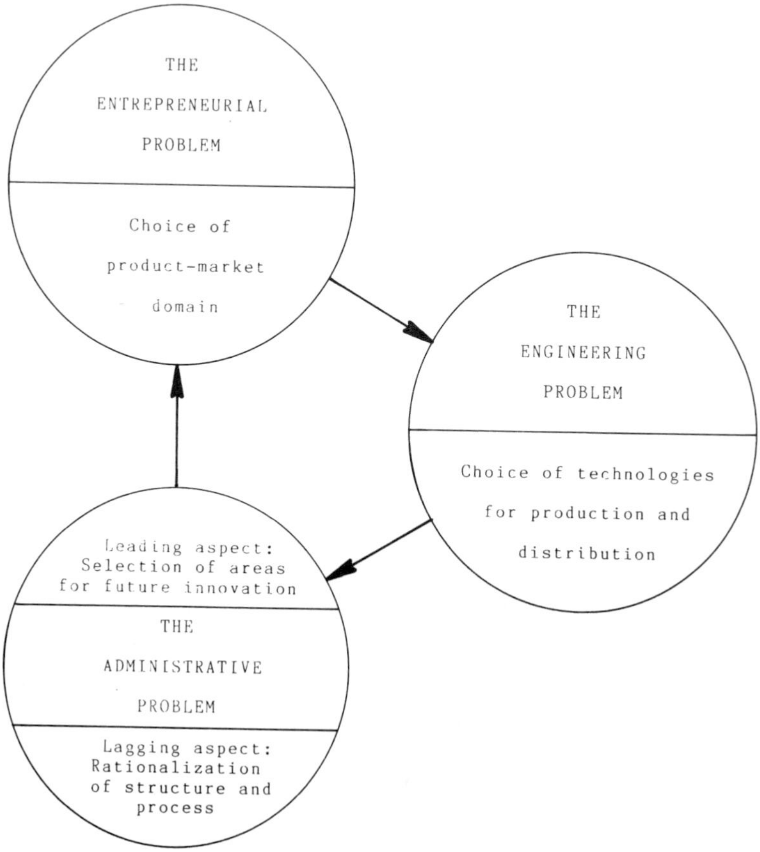

Prospectors are organizations which almost continually search for market opportunities and respond to emerging environmental trends. These organizations are often the creators of change and uncertainty themselves. There is a concentration on product and marketing innovation and so these organizations are not completely efficient.

Analyzers are organizations which operate in two types of market domain, one relatively stable and the other rapidly changing. In turbulent areas the progress of competitors is closely watched and the most promising ideas are rapidly adopted.

Reactors are organizations which cannot respond effectively to the change and uncertainty in the environment. There is no consistent strategy-structure relationship and adjustments are not made until absolutely forced by environmental pressures.

For each of three of the four types of organization, Miles and Snow give entrepreneurial, engineering and administrative characteristics. Characteristics cannot be given for the Reactor organization since there is no consistent set of relationships.

Miles and Snow believe that mixed strategies are developing in some industries and that the matrix organization represents such a mixed strategy having characteristics of the Defender organization in functional form and the Prospector organization in project form (Miles and Snow, 1978b).

Recently the work has been extended to consider the distinctive competence of organizations which were classed by the management as one of the four types using a questionnaire (Snow and Hrebiniak, 1980). This study claimed that Defenders, Prospectors and Analyzers all have distinctive competence in the general and financial management area. Beyond this Defenders and Prospectors have other but different areas of competence. Analyzers' special capabilities are less apparent while Reactors have no consistent pattern of distinctive competence. From the Miles and Snow formulation the pattern and behavior of an organization can be predicted once the organization has been placed in one of the four categories of Defender, Prospector, Analyzer and Reactor.

The approaches so far considered have discussed some of the pressures for growth and models for growth and development. The underlying assumption summarized in the statements of Thompson, "under norms of rationality", is that rational logical decisions are the basis for organizational growth and development. There are, however, other factors that influence decision-making in the company. There are a plurality of social systems within the organization (Burns and Stalker, 1966b). Earlier research had indicated two polar types of organization, the mechanistic organization governed by rules and procedures, and the organic organization which could adapt to a highly fluid environment. Organizations which were mechanistic but which should adopt the organic form did not do so even when it was recognized that change should be made. Evidently other forces than logic are present in the organization.

The change from a mechanistic to an organic organization should in theory be undertaken because a dysfunction exists. Burns and Stalker make the point that this is a management dysfunction but that working communities are much more than management systems and that looked at from another viewpoint dysfunctional types of management system may be seen as highly effective if other ends of the social organization are served irrespective of the ends of the organization. The organizational hierarchy of rank and power, the career structure and the political ends of individuals all play a role in modifying and sometimes completely altering the rational decisions which should be made on behalf of the organization.

There is a tripartite set of commitments in an organization, the first is to the firm itself, the second to political groups and the third to the career prospects of the individual himself. Neither political nor career preoccupations operate overtly, since this is against the rules, but these aspirations give rise to intricate manoeuvres expressed in terms of discussions and decisions concerning the organization and policies of the firm. It may ultimately turn out that the organization merely exists to serve the political and career system and that actions of growth and development only take place to serve the ends of some clique rather than as a natural progression of the organization in terms of some basic organizational dynamics.

Weick would argue that the concepts of organizational domain and environment are misplaced in that it is extremely difficult in practice to separate the organization from its environment (Weick, 1969). Weick states that organizations determine the environment to a much greater extent than is generally realized but that often an organization does not know what it is doing until it is done. This view suggests that organizations must continually experiment with the environment and that there is no reality except that which exists as the perception of the environment by the organization.

Studies of strategy and structure have evolved to include intra-organizational approaches. One theme has been the development of models of organizational growth. These models have been extended to include development of global multi-national forms. The different stages of growth have been described in terms of strategy, likely structure and process within the organization. Table 7 provides one example but there are others (Salter, 1970; Scott, 1971). If an organization can be located as to stage of development, then predictions can be made about the likely development of the company. This approach has implications, for example, in terms of predicted strategies for R & D.

A second theme concerning strategy and structure has been that of Thompson. Again, employing the propositions of Thompson the behavior of an organization in specific circumstances might be predicted. The formulation of the strategy-structure-process approach by Miles and Snow might also be useful since identification of a company as a Defender, Prospector, Analyzer or Reactor allows predictions to be made.

Intrusion of the power and career structures into an organization and its influence on organizational behavior may modify the conclusion reached in any particular use of the strategy-structure models and in the use of the Thompson approach. It is difficult to see how, in general, this can be allowed for but in specific instances it may be possible to obtain information on the situation in a particular firm. If strategy-structure concepts are to be employed for a particular company it is important that the power and career structures be understood.

The viewpoint of Weick seems to be the antithesis of the view that predictions can be made regarding the growth and development of the firm. A modification of the view that organizations can only know what has been done after the fact would recognize that the knowledge of the environment is imperfect. Rather than sitting passively an organization can gain more knowledge about the environment by actively experimenting with the environment. At least locally such experimentation may well change the environment itself and in this sense the statement that organizations determine their own environment is partly correct. Although not incorporated directly into the use of the strategy-structure approach as a predictive tool of organization behavior some of the concepts of Weick are useful to bear in mind.

2.5 THE INFLUENCE OF TECHNOLOGY

The studies on strategy and structure have not explicitly focussed on the role of technology and on R & D. However, there are implications for R & D which will be discussed later. At this time the role of technology in corporate strategy will be addressed.

Ansoff has considered the situation of technology in the diversification of a company's products (Ansoff, 1965). The first step is to examine the product/mission matrix of the organization as shown in Table 8. Technology may appear in any of the four boxes but is explicitly required in new product development. Diversification involves new products in new markets and the options are described in the diversification matrix shown in Table 9. The role of technology is shown in the diversification matrix. Thus the approach of Ansoff indicates that strategy, especially in diversification, involves technological considerations but the mechanisms for generating new products and new technology are not given and the mechanisms may not involve R & D. As discussed in Part 1 there are many alternatives to the internal generation of knowledge. It is arguable that such decisions are not the role of the corporate strategist but should be more correctly made at the business level (Hofer and Schendel, 1978c).

TABLE 8

A product/mission matrix

Mission	Product: Present	Product: New
Present	Market penetration	Product development
New	Market development	Diversification

TABLE 9

A diversification matrix

	Customers \ Products	New products: Related technology	New products: Unrelated technology
	Firm its own customer	Vertical integration	
New missions	Same type	Horizontal diversification	
New missions	Similar type	Marketing & technology related concentric diversification	Marketing related concentric diversification
New missions	New type	Technology related concentric diversification	Conglomerate diversification

At the business level, Ansoff and Stewart have discussed strategy and technology (Ansoff and Stewart, 1967). In technically intensive businesses the marketing strategy involves a technological component. Four strategies were identified. In the "first to market", strong R & D, technical leadership and risk taking are required. The "follow the leader" strategy is based on strong development resources and an ability to react quickly as the market starts its growth phase. "Application engineering" is based on product modifications to fit the needs of particular customers in a mature market. "Me-too" strategy is based on superior manufacturing efficiency and cost control.

Ansoff and Stewart also pointed out that technological change can exert a major influence on the nature of effective competitive strategies in particular industries. The two aspects of technological change that are important are the overall rate of change and the variations that occur at different stages of the product market evolution (Hofer and Schendel, 1978d). Hofer and Schendel have related the rate of technological change in a field to the type of variation that could occur in the cases of product design, process design and breakthrough. This is shown in Table 10. For example, in industries with high rates of technological change the major challenge will involve the types of design change and the time needed to mass produce a design once the design has been frozen. Major breakthroughs in product form will be the principal type of technological threat to firms in industries with low overall rates of technological change.

TABLE 10

Hypothesized variations in the major types of technological challenges particular businesses will face

		Type of technological change		
		Product design	Process design	Breakthrough
Overall rate of technological change	High	Major	Intermediate	Moderate
	Medium	Moderate	Major	Intermediate
	Low	None	Moderate	Major

*Reprinted by permission from "Strategy Formulation: Analytical Concepts" by C.W. Hofer and D. Schendel.

The major challenge facing firms in industries with intermediate rates of technological change is the problem of changing from a product to a process focus in the engineering and R & D activities. Single product companies face special difficulties because of resource limitations which do not allow adequate numbers of product and process engineers. In the early stages as the industry develops such a company usually hires reasonable numbers of design engineers but as the industry matures the firm is unable to meet its commitments in the process engineering area.

The appropriate response to a technological challenge depends on the rate of technological change in the industry and on the technological strategy of the business. A breakthrough in an industry of low technological change is a great threat to the existence of the firm and there are only three options; namely, to acquire a firm with the new technology, to become a distributor for the new products or to liquidate the business in an orderly manner before being forced to do so by the market. As Hoffer and Schendel state, however, ".... we still know very little about technological innovation and its effects on competitive business strategy, except that it can have a profound impact on a business's chance for long-run survival". (Hofer and Schendel, 1978d).

Another aspect of the technological nature of a business concerns the balance between research and development. Companies may be classed as R-intensive or D-intensive depending on the relative weight given to research and to development (Ansoff and Stewart, 1967). R-intensive organizations display six characteristics:

(i) work is done with indefinite design specifications,
(ii) objectives are broadcast among technical people rather than information following specific channels,
(iii) work assignments are non-directive,
(iv) continuing project evaluation and selection is maintained,
(v) the perception of significant results is stressed,
(vi) innovation is stressed over efficiency.

On the other hand D-intensive organizations are typified by four characteristics:

(i) there are well-defined design objectives,
(ii) there is highly directive supervision,
(iii) tasks are sequentially arranged,
(iv) there is vulnerability to change.

There will be cases of intermediate mixes of research and development. However, it is clear that the correct choice for the particular business is important since the way in which an R-intensive organization is managed is quite different from the management of a D-intensive organization.

In summary Ansoff and Stewart see the technological profile of a company in terms of five factors. There is the balance between research and development, the product life cycle, the coupling between R & D and other company units, the proximity to the state of the art and the R & D investment ratio.

The approach of Ansoff has been extended recently by Abell (Abell, 1980). Whereas, Ansoff defines the business in terms of a product/market mission (Table 8), Abell adds an extra dimension to define the business along three coordinate axes labelled customer groups, customer functions and alternative technologies. Thus the present business can be defined in three-dimensional space. This analysis will indicate obvious gaps that can be filled. For example plotting the present position could indicate that with the existing technology and functional use another group of customers could be served. The possibilities for diversification are indicated quite graphically. Often in diversification attempts companies move far away from the known product/market relationships of the existing business. The definition of the present business along the three dimensions will give a three-dimensional picture which indicates the distance from the existing business and hence gives a measure of the risk and of the opportunities. The existence of a customer/function/technology domain can be used to analyze distinctive competence which is another indicator of where the business should go next. The actions of competitors may also be plotted and this may suggest where competitive pressures will next be found.

Although there is no definitive work on the role of technology in the strategy of the firm there are some significant pointers to the way in which strategy and technology interact and the growth and survival of the firm can depend upon the technological choices made. Hofer and Schendel, in characterizing strategies as corporate, business or functional would place technological policies as part of corporate strategy and would include technology as a possible distinctive competence at the corporate level. R & D policies are mentioned at the business level only. This classification indicates that technology covers a much broader area than R & D; R & D is merely one of the components, albeit an important component, of the technological profile of the firm.

2.6 LIFE CYCLE CONCEPTS

Implicit in much of the foregoing has been the idea that a product has a finite lifetime. The concept of the product life cycle was introduced some 30 years ago but it is only rather recently that the concept has been broadened to include the idea that a firm which stays in the same business also has a finite lifespan. Strategies at the corporate, business and functional levels are enriched by consideration of the lifetime concept.

The review of Rink and Swan indicates that although there has been much research in the area of product life cycles much remains to be done (Rink and Swan, 1979). However, a useful form of the product life cycle curve has been given (Hofer and Schendel 1978e). Figure 12 gives the fundamental stages in the product life cycle while Table 11 illustrates some of the characteristics of each stage. Some seven stages of product/market evolution are identified. These are market development, growth, shake-out, maturity, saturation, decline and petrification. The basic nature of competition changes during development, shake-out and decline stages of product/market evolution and major changes in competitive position are accomplished most easily during these stages. It may be noted that a product may stabilize in the marketplace so that decline and petrification do not occur.

The study of the product life cycle indicates some of the differences in focus required at the different stages of product/market evolution. During the early stages the emphasis is on innovation, then engineering, production and marketing, finance and distribution. The early emphasis is on effectiveness but this shifts to an emphasis on efficiency as the market matures.

FIGURE 12

THE FUNDAMENTAL STAGES OF PRODUCT/MARKET EVOLUTION

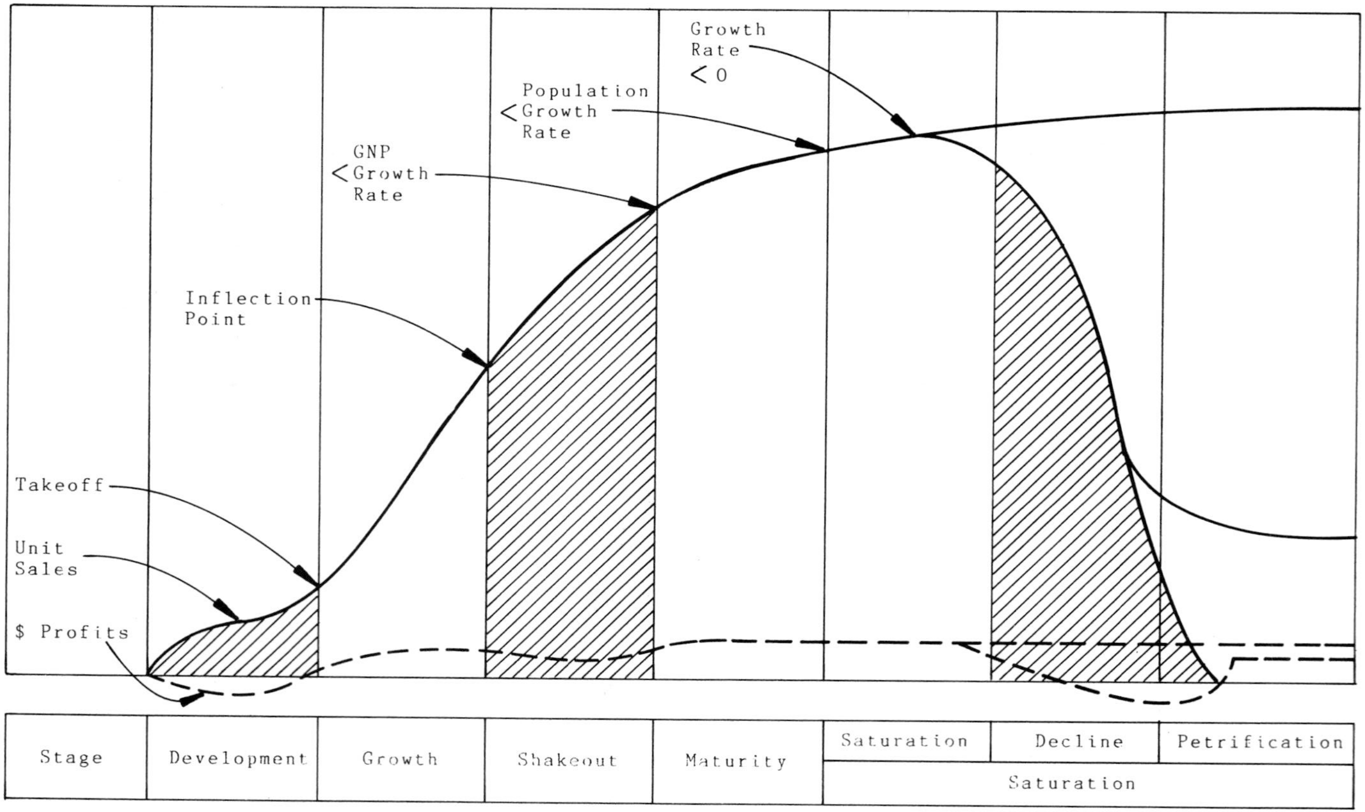

TABLE 11
The fundamental stages of product/market evolution

Stage	Development	Growth	Shakeout	Maturity	Saturation		
					Saturation	Decline	Petrification
Market Growth Rate	Slight	Very Large	Large	GNP Growth	Population Growth	Negative	Slight to none
Change in Growth Rate	Little	Increases Rapidly	Decreases Rapidly	Decreases Slowly	Little	Decreases Rapidly, Then Slow. May increase then slow.	Little
# of Segments	Very Few	Some	Some	Some to Many		Few	Few
Techno-logical Change in Product Design	Very Great	Great	Moderate	Slight	Slight	Slight	Slight
Techno-logical Change in Process Design	Slight	Slight/ Moderate	Very Great	Great Moderate	Slight	Slight	Slight
Major Func-tional Concern	R and D	Engineer-ing	Production	Marketing-Distribution-Finance		Finance	Marketing and Finance

Attempts have been made to link the product life cycle concept to areas of action for the firm. Life cycles have been quantitatively studied to determine the link between innovation and the life cycle stage. The length of time spent at different stages has been correlated with the "degree of product newness". An innovative new product gives an extended early period with a late peak in the volume of units sold (de Kluyver, 1977). From the degree of newness a forecast can be made over the shape of the product life cycle curve. Hayes and Wheelwright have focussed on the link between the life cycle and manufacturing processes (Hayes and Wheelwright, 1979a). This approach emphasizes manufacturing rather than marketing concepts and seeks to fit the production process to the stage in the life cycle. Further development of this approach has been made (Hayes and Wheelwright, 1979b). The use of an "inverted product life cycle", has also been advocated (Weber, 1976). This approach looks at the gap between the firm's sales, competitor's sales and the industry market potential sales. Apart from a usage gap, product line and distribution gaps are employed to break down the areas in which improvement is possible.

Although there are examples of the use of the product life cycle concept in the manufacturing, innovation and other areas it is in the strategic planning area that the concept has had considerable utility. Figure 13 illustrates recommended investment strategies as a function of the relative competitive position of the firm and the stage of the market evolution. Strictly speaking the strategies apply to a single product company but the approach can be generalized for a multi-product concern by considering each segment of the business. The strategies may or may not include R & D. Another approach to the same subject has been adopted by considering similar criteria with some additions and the question posed is how to decide on growth, holding or harvesting (SPI, -). The product life cycle is also one of the concepts employed in deciding on the ease of competitive entry into an industry (SPI, 1979).

At the corporate strategy level the product life cycle concept is employed to determine the balanced portfolio of businesses. The simplest approach is the BCG matrix shown in

FIGURE 13

INVESTMENT STRATEGIES AT THE BUSINESS LEVEL

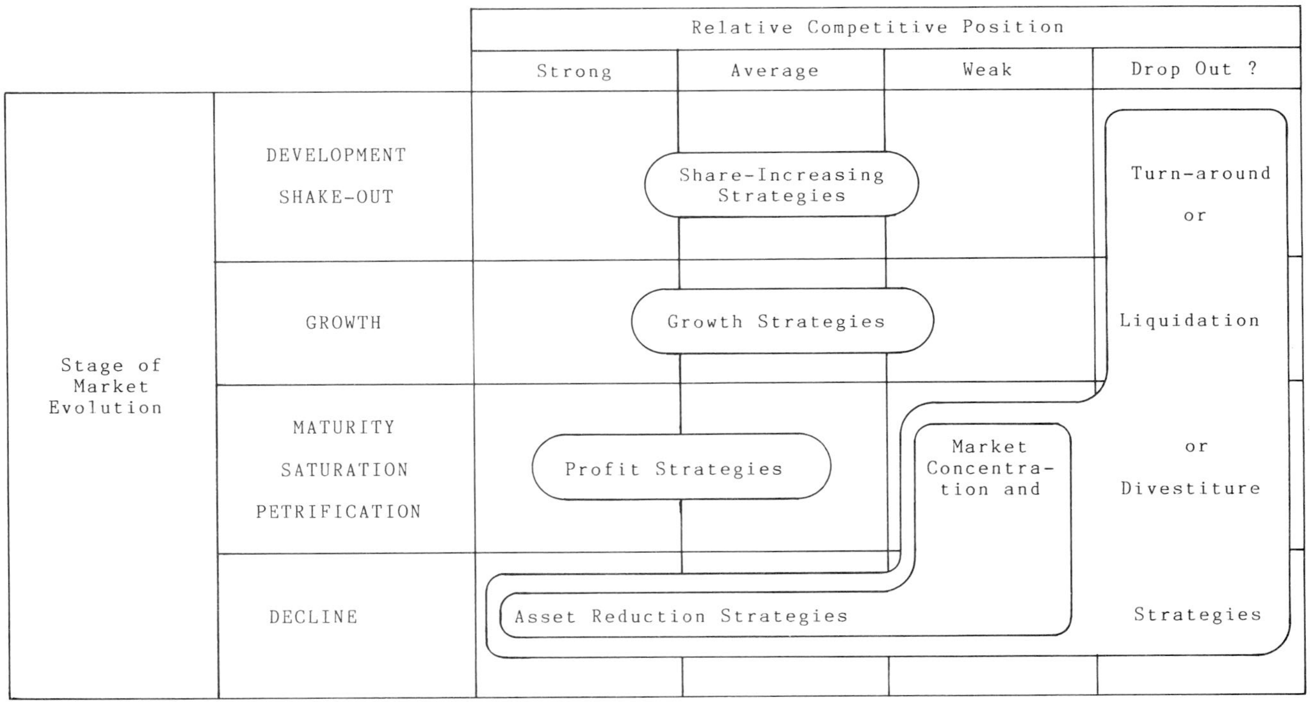
Relative Competitive Position
Strong
Average
Weak
Drop Out ?
Stage of Market Evolution
DEVELOPMENT
SHAKE-OUT
GROWTH
MATURITY
SATURATION
PETRIFICATION
DECLINE
Share-Increasing Strategies
Growth Strategies
Profit Strategies
Market Concentra-tion and
Asset Reduction Strategies
Turn-around or Liquidation or Divestiture Strategies

Figure 14 (Hofer and Schendel, 1978f). The axes are the relative competitive position and the growth rate of the industry. Businesses which lie in the upper left quadrant are called "Stars" because of the high growth rate; such businesses are roughly self-sufficient in terms of cash-flow. Businesses in the lower left quadrant are known as "Cash Cows" because of the combination of low growth rate and high market share which results in an excess of cash. Businesses in the lower right quadrant are called "Dogs" because of the poor market position in a low growth market which means the profits are generally low. These businesses should generally be liquidated. In the upper right quadrant are the "Question Marks" which are the businesses with the highest cash needs due to the poor market position in a high growth market. A Question Mark should grow into a Star or it should be divested.

Criticisms of the BCG matrix have led to the development of somewhat more sophisticated matrices such as the General Electric Business Screen (Hofer and Schendel, 1978g). The competi-

FIGURE 14

THE BCG BUSINESS PORTFOLIO MATRIX

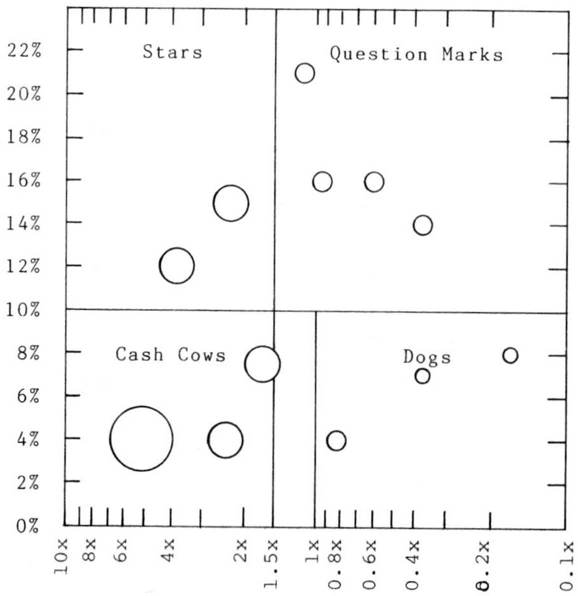

tive position is indicated as strong, medium or weak. High, medium or low are used to indicate industry attractiveness. The area of the circles in the BCG matrix represented the size of the business while in the GE matrix the area represents the size of the industry and the size of the company's market share is indicated as a "pie slice" within the circle. A modification of the GE matrix to give a 15 cell matrix has been made by expanding the industry dimension to specifically give five dimensions of product/market evolution, namely development, growth, shake out, maturity and decline. The latter modification is valuable if businesses consist of individual or small groups of related product/market segments. Otherwise the original GE matrix is superior. The BCG matrix may be used for initial screening to indicate which businesses need closer attention.

The idea of a finite lifetime is implicit in the concept of the BCG and GE matrices. A balanced portfolio consists of companies at different stages of development. Organizations have a finite lifetime. This is in contrast to the strategy-structure formulations of organizational change that assume growth or at least that the company stands still. Some regression could occur but the disappearance of a company is not covered. There should be a strategy-structure approach for optimizing the return from a company on the way down. Adizes has considered the stages of growth and decay of organizations (Adizes, 1979). According to Adizes an organization must do four things to be effective. It must Produce, Administer, be Entrepreneurial and Integrate. All four roles PAEI must be performed well but there is a different weighting on the roles depending on the position of the company on the life cycle curve. Table 12 indicates the stages of growth and decay along with the cause of premature mortality during growth. Also indicated is the relative emphasis on the four roles for each stage. Capital letters indicate high emphasis and lower case letters indicate lower emphasis. Different treatments for the various crises which occur during growth and decay are also indicated in the table. Up to maturity such treatments are preventative; after maturity the treatments are curative.

According to Adizes the correct handling of a Prime organization can lead to an organization distinguished by equal emphasis on PAEI. Such an organization can continue to grow and develop

TABLE 12

Life cycle of an organization

Stage	Role	Type of premature mortality	Treatment
Courtship	paEi	Aborted idea	Reality testing
Infancy	Paei	Infant mortality	Inexpensive support
Go-Go	PaEi	Founder's trap	Directive board
Adolescence	pAEi	Divorce	Rekindle the fire
Prime	PAEi	PAei	Recognize and Decentralize
Maturity	PAeI		
Aristocracy	pAeI		A'S/M therapy
Early bureaucracy	-A-I		Surgery
Bureaucracy	-A--		Euthanasia
Death	----		"Caretakers"

by utilizing a portfolio approach similar to the BCG matrix. Failure to properly handle the Prime organization can lead to a transition to the Mature organization. This train of events leads to decay of the organization via Aristocracy and Bureaucracy forms of organization.

An interesting feature of the Ardizes treatment is that the political and career structures of the organization are considered. The climate of the organization is also discussed, especially the receptiveness of the organization to innovation. This receptiveness undergoes a reversal at the Mature stage thus reinforcing the decline of the organization as a viable entity. Up to maturity organizational change can be done from within. After maturity the natural resistance of the organization to change means that effective change can only come from without.

Products, product areas, companies, businesses and industries all have life cycles. Stages occur during growth, maturity and decline which serve to categorize the relationships between product and market, between company units and processes and between companies. Use of the product life cycle concept allows the best fit of strategy, structure and process to be attempted.

At the very least the recognition of a finite lifetime indicates that at some stage it is appropriate to discontinue a product, divest or liquidate a company. On the positive side actions to reverse a declining trend may be available.

2.7 STRATEGY-STRUCTURE IMPLICATIONS FOR R & D

Strategy and structure formulations have concentrated on the macroscopic business aspects of the firm. In principle, R & D strategy should fit with the overall strategy of the company and the structure of the R & D operation should fit with the R & D strategy. The continuity of the means-ends chain should ensure that there is a coherent fit between R & D and the rest of the organization. The logical approach would be to consider the goals, to decide if the strategies involved a technological component and, if so, to decide on the extent of R & D involvement. Unfortunately, in the management literature the fit of R & D into the strategy-structure formulation of the firm has only been treated peripherally. One reason for this is that it is possible for R & D to have a poor fit and yet there will be no immediate effect on the organization; it may take a period of several years for the effect of poor R & D fit to become apparent. Thus there is no immediate pressure on the company to optimize R & D fit.

There are many intriguing questions concerning R & D and the structure-strategy relation. For example, the effects of an R & D strategy not in accord with the company strategy has not been examined. R & D interacts with the external environment to forecast and produce change. There is, thus, the question of the external fit of R & D to the environment and the internal fit to the strategies of the organization. If R & D strategy cannot fit both the external environment and the internal environment, is it better to concentrate on the internal fit or vice versa? These questions cannot be answered at the present time but from a consideration of strategy-structure relationships, the role of technology and the concept of the product life cycle, some indicators can be gleaned as to the fit of R & D.

Descriptions of the stages of growth and development provide a brief mention of R & D. Scott considers three stages of development (Scott, 1971). In Stage I, R & D is not institutionalized and is oriented by the owner manager. Stage II provides for an increasingly institutionalized search for product or process improvements while Stage III has R & D as an institutionalized search for new products as well as for improvements. The model shown in Table 7 has five organizational types and extends the Scott approach to the global multi-national. As far as R & D is concerned the Simple and Function types are identical to Stage I and Stage II of Scott. The Holding, Multi-Divisional and Global forms have the same overall mission as for the Stage III organization of Scott but there are differences of structure. In the Holding form, R & D is decentralized to divisions. For the Multi-Divisional form there is centralized guidance while for the Global form R & D is carried out in a centralized form and also decentralized at centers of expertise. It will be noted that although there is mention of R & D in the models of growth and development, mention is only in the most general of terms and is only of use as an overall guide to R & D.

Empirical studies on overseas R & D have been carried out (Behrman and Fischer, 1980). The studies have concentrated on the reasons for setting up R & D abroad, possible structures and management styles adopted. The results are of considerable benefit in the case of an overseas laboratory of an American corporation since most of the data were obtained from case studies of this situation. Thus critical size of the R & D operation, the integration of the laboratory into the worldwide structure and the appropriate management control style of the laboratory can all be addressed from the results of this work. However, the subject of R & D and its relation to strategy and structure as a whole is not discussed.

As mentioned earlier the work of Thompson provides a general statement of the strategy-structure problem with indications of solutions. In the area of structure R & D must be a boundary spanning function and must therefore lie at the edge of the organization domain. R & D is to be organized according to the heterogeneity and degree of dynamicism of the environment. Since organizations attempt to remove uncertainty R & D is in a powerful position if it can demonstrate the ability to remove uncertainty. R & D can do this be enabling the organization to obtain more control over the environment. It must also be realized that R & D often is a producer of uncertainty; a breakthrough by R & D presents the organization with many choices of action. If R & D can reduce uncertainty, however, a dependency is generated which will give R & D a privileged position in the organization.

The model of Miles and Snow enables several statements to be made regarding R & D. A Defender organization will place its emphasis in a narrow domain and will aim for continuous improvements in technology to maintain efficiency. Financial and product functions are the most powerful. R & D is subservient to these functions and may not even be explicitly identified as R & D. The company might get by with process improvement work located at the manufacturing locations and some design engineering to produce limited product development. From the study of distinctive competence, Defender organizations have a distinctive competence in applied engineering in the semiconductor and automotive industries but not in the plastics or air transportation industries; R & D is not mentioned (Snow and Hrebiniak, 1980).

Prospector organizations rely on high technology for growth and survival. The most powerful functions are marketing and research. Growth is by product and market development and so the thrust is in innovation. The organization must be flexible and so the tendency will be towards a product orientation. Technical competence is not localized but is spread throughout the company. Distinctive competence studies show that engineering is a distinctive competence in all four industries covered and that R & D is a distinctive competence in the semiconductor and plastics areas.

R & D in the Analyzer organization reflects the dual nature of the business. Miles and Snow predict a low investment in R & D since imitation of the successful products of others requires speed of action in the engineering sphere. However, marketing and applied research are the most influential functions followed closely by production. Applied engineering appears as a distinctive competence in the semiconductor and automotive areas but not in the other two areas.

Since the Reactor does not pursue a distinct consistent strategy there is no pattern to the organization or to the distinctive competence. Applied engineering, however, appeared as a distinctive competence in three of the four industrial areas for those Reactor organizations studied.

Although the model of Miles and Snow does give an indication of the general orientation of the organization in terms of strategy pursued, it does little more than to outline the part that R & D plays in the strategy and how R & D is structured. Except for the Prospector organization R & D does not explicitly appear as a distinctive competence although the descriptions of the dominant coalitions would indicate that applied research would be an important function in the Analyzer organization.

Consideration of the environment of the firm can give an idea of the importance of R & D and the emphasis to be placed on research as opposed to development. In a turbulent environment there will be a reliance on R & D to meet competitive challenge through new product development (Emery and Trist, 1963). Lawrence and Lorsch divided organizations into three main subsystems of marketing, economic-technical and scientific and supposed that the structure of each subsystem would vary depending on the predictability of the environment. Ansoff and Stewart related the technological profile to the rate of change of the environment and the distance of the technology from the state of the art. Conclusions may be drawn about the ratio of research effort to development effort. In general, environmental considerations may be used to indicate the type of R & D effort but specific predictions on the strategy and structure of the R & D function are not forthcoming.

Product life cycle considerations give an idea of the emphasis on R & D as a function of product lifetime (Hofer and Schendel, 1978e). Technological change and the emphasis on R & D occurs in the development stage as the product is introduced. Thereafter R & D is not involved but the emphasis on engineering during the growth stage and on cost reduction during shakeout and after suggests that applied research and development are involved along with process improvements as one way to achieve the desired cost reduction.

Steele has considered the role of technology in business strategy (Steele, 1975a). This is done using a matrix approach. The business strategies possible are hold/harvest, grow the present business or extend the present business. Technology inputs are to apply the state of the art, to extend the state of the art, to use competing technology or to use an alternative technology to supplant the old. This matrix is shown in Figure 15. Here the business strategies have been subdivided to give added focus to the strategy employed. The examples indicate different levels of strategy and technology. Thus option A is simply involved with cost reduction using the existing technology. Option B aims at an increased market by improving the life of a bearing; this involves an extension of the state of the art in the technology. Option C proposes an extension of the present business by integration backwards; this involves adopting existing technology used by others. Option D is the most ambitious. It proposes an entry into a new market by developing a new linear motor, a technology that is still being developed. It may be noted that the different options involve technology but do not necessarily involve R & D. Option A definitely does not involve R & D; option B probably is at the level of development or, at most applied research. Option C might involve some R & D input but probably would only call on R & D as trouble-shooters. Option D definitely would involve R & D. As stated before, the fact that technology is involved does not mean that R & D is involved. Other alternatives are available. The approach of Steele combines the product/mission matrix and the diversification matrix of Ansoff. The diversification aspects are treated by Steele as an extension to the business. Steele does not focus on the mission aspects; the emphasis is on product development rather than customer development. Steele explicitly discusses the strategy of hold/harvest while Ansoff does not. Both approaches should be employed. In addition, the three dimensional representation of Abell is useful since it enables R & D to examine the technology and its relation to customers and customer functions.

Recent work has examined the changes in the functions of the firm over the product life cycle (Moore and Tushman, 1980). In this study the product class life cycle was considered rather than the product form or the brand life cycle. Thus the emphasis

FIGURE 15

BUSINESS STRATEGY-TECHNOLOGY MATRIX FOR MOTOR DEVELOPMENT

BUSINESS OBJECTIVES / TECHNOLOGY INPUTS	HOLD/HARVEST			GROW PRESENT BUSINESS			EXTEND PRESENT BUSINESS		
	Improved Performance	Improved Cost	New Features	Improved Performance	Improved Cost	New Features	New Level of Value Added	Associated Markets	Associated Technologies
Apply the State-of-the-Art		A							
Extend the State-of-the-Art				B					
Competing Technology Used by Others							C		
New Alternative Technology to Supplement Old									D

A - Reduce shop cost;

B - Redesign bearing to improve life;

C - Produce own magnet wire;

D - Develop linear motor.

was on significant increments of innovation in the marketplace. Three stages of the life cycle were considered; introduction, growth and maturity. Statements were made regarding the relative importance of R & D, marketing, manufacturing and other functions during these three life cycle stages. The remarks on R & D will be discussed here.

In the introductory stage of the product life cycle the primary form of innovation is major product innovation. The primary functions are marketing and R & D. In the early stages of introduction the market needs are ill-defined and relevant technologies are not clear. There is, therefore, little incentive for organized R & D but as a dominant design is approached, market needs become clearer and as relevant technologies clarify organized R & D becomes very important; the function of R & D at this stage is to develop new and use existing knowledge to improve product performance. More successful laboratories will have extensive contact with suppliers, vendors and customers as well as with the scientific community. Laboratories will be small and informally organized with strong links to marketing. Key roles are filled by technological innovators and gatekeepers. Moore and Tushman remark that large companies seldom generate radical innovations in their own industry; most frequently the innovators are totally new entrants.

The decision to standardize the product and to rationalize the process signals the start of the growth phase. During this phase major process innovations begin to increase the interdependence between the product and manufacturing process. As the technical core becomes clear the company can divert its investment into the R & D area with greater certainty. Typically during this stage there is specialization with a small effort (5-10 percent) on basic research in disciplines which are or may be important to the firm. The remainder of the effort is focussed on applied development, engineering and technical service work. The drive is to new improved products and processes. Depending on the rate of change of the technology and the average program length the laboratory will be organized by discipline or by program. When technologies are substantially different and there are multiple programs, a matrix organization is appropriate. Strong links to marketing and manufacturing must be maintained by the development activity while research should be well linked to sources outside the organization.

In a mature stage there is little or no sales growth. Product development is incremental but this does not mean that substantial changes cannot be made; cumulatively the effects of many incremental changes can be as important as the initial innovation. Strategies do not have to be deficient in R & D. Good returns associated with medium and high levels of R & D/Sales and with high product quality in low growth industries have been demonstrated (Hammermesh and Silk, 1979). Generally, however, R & D is less important and is more focussed during the mature stage. Emphasis is on efficiency and concentration of effort on process changes and other incremental changes is evident. Contacts with the outside are reduced and the interaction with other company units is more formalized. During this phase the company is most resistant to new ideas and new technology from R & D.

As Moore and Tushman point out the difficulty is to manage the necessary transitions during the life cycle. In the healthy company there must be a balance between stability and change, consistency with inconsistency, centralization with decentralization, order with disorder, commitment with tolerance for ambiguity and entrepreneurship with stewardship. Only in this way can the organization remain innovative.

The BCG and GE matrices provide a convenient framework for analyzing the expenditure on R & D. Product policy and development in a multi-national company have been discussed in terms of the business matrix (Bergen, 1980). However, it is Bolton who has explicitly examined the use of the BCG matrix in R & D allocation (Bolton, 1980). The use of the BCG matrix focuses on the position of a company, business or product line in the life cycle and also indicates the market share. In general terms it would appear to be of little value to fund R & D programs for Dogs. In the case of a Cash Cow there should be the minimum expenditure necessary to sustain the business. A Cash Cow is probably at the mature stage of the life cycle and thus R & D is to be focussed on incremental improvements. In the case of Question Marks a close examination of the business is needed. If it appears that a Question Mark can become a Star then the necessary action should be taken. This action does not necessarily involve R & D. R & D is most likely to be involved with Stars. Presumably such companies are on the growth part of the life cycle and need to commit a large effort to institutionalized R & D.

It has been suggested that a company consist of a balanced portfolio of businesses. This does not give any indication as to the way in which R & D is to be organized. If R & D is centralized then a differentiated effort attuned to the different needs of the companies in the portfolio is required. The alternative is to have the R & D effort physically located according to the need of the business with perhaps a small long-range R & D program at the Corporate level.

Nystrom has examined the manner in which companies choose new markets and new areas of technology and how the research effort is focussed (Nystrom, 1979a). Companies are considered to be either positional or innovative in character. Positional companies resemble the Defenders of Miles and Snow while innovative companies resemble Prospectors. A distinction is made between intended and realized R & D strategies. Intended strategies are expressed in explicit policies relating to R & D activities while realized strategies refer to consistent patterns of behavior which may or may not be the result of implementing policy decisions. In the Nystrom study R & D policy was assessed from interview data while realized strategies were inferred from product data. The empirical analysis was based on 11 Swedish companies, three in pharmaceuticals, two in steel, four in electronics and two in the chemical area.

Based on the interviews three main policy dimensions evolved. The first was concentrated versus diversified R & D which referred to the extent that a company wanted to move into new product and technological areas. The second was technological or market orientation and the third dimension was offensive versus defensive. Patterns emerged from the data. Thus diversified companies tended to be technologically oriented and also offensive in strategy. Relating the dimensions to the classifications of positional and innovative companies a generalization may be made. Diversification, technological orientation and offensiveness are indications of an open innovative R & D policy while concentration, market orientation and a defensive policy are indications of a closed positional policy.

In a similar fashion to R & D policies, realized R & D strategies are measured along three dimensions. The first dimension is internal versus external orientation. Internally oriented companies emphasize internal competence and their own experts in new product development while externally oriented companies rely to a large extent on external consultants in carrying out R & D activities. This orientation is measured both with regard to idea generation and to technical product development. The second dimension concerns technology use. Some companies emphasize the isolated use of technology while others emphasize the synergistic use of technology. The third dimension concerns fixed versus responsive organization of R & D. The aspects of organization are considered in three areas. The first is the organization of the company's external information and contact network. This has been classified as wide or narrow depending on how extensively the company searches in its external environment for ideas and cooperation in R & D. The second aspect of organization is idea and project evaluation. This activity may be centralized or decentralized. The third aspect refers to how internal project work is carried out and whether project groups are fixed or flexible.

Nystrom found that the variation between the companies studied was related to concentrated versus diversified R & D. Highly or relatively fixed organization of R & D was related to a concentrated policy. Responsive organization of R & D was related to a diversified R & D policy. This finding is consistent with what would be expected from structure-strategy formulations and was reinforced by the results from the examination of realized strategies. External orientation, synergistic technology use and responsive organization of R & D were found to be indications of open R & D strategies. Internal orientation, isolated technology use and fixed organization of R & D were indicative of closed R & D strategies.

Relationships between the type of R & D strategy and the level of innovation were found. External orientation was associated with a higher level of technological innovation than internal orientation. A high level of technological innovation was correlated with synergistic use of technology.

From the results it would be expected that a consistent combination of policy and realized strategy would give a more successful result in R & D than an inconsistent combination. This means that a more innovative policy should be accompanied by a more open realized strategy and a more positional policy by a more closed strategy. The approach of Nystrom gives interesting insights into the effect of strategies on R & D. It is an approach that is within the strategy-structure sphere. Analysis of a company along the dimensions suggested should allow predictions to be made as to the adequacy of the strategy, the organization of R & D and the consistency of the fit. With the strategy-structure approaches heretofore the difficulty has been to use the generalized models for specific cases. For the Nystrom approach the need is to move from the empirical to a more general formulation. The full benefit of the treatment may then be realized.

All of the above treatments on strategy-structure, the environment and on the product life cycle provide clues as to the organization of R & D and the strategies to be employed. Throughout the whole discussion the central theme has been that R & D strategy must reflect corporate strategy even if the conclusion in the extreme is that there should be no R & D carried out. This connection between R & D and the corporate hierarchy is so important that it will be discussed separately. The relationship in effect specifies the position of R & D in the company.

2.8 CORPORATE STRATEGY AND R & D

To be effective a company must decide how to divide resources between current and future needs. However, corporate plans frequently make no reference to products for the future (Parker, 1980). Technological resource allocation in general and R & D resource allocation in particular must be made with reference to the corporate plan.

The means-end chain for R & D has been summarized and is shown in Figure 16 (Steele, 1975b). Starting with the corporate strategy and business objectives, the role of technology in the strategy is outlined. This determines the R & D strategy bearing in mind the corporate mission for R & D. The procedure should be an iterative one since the resources and capabilities of R & D may influence the goals which are achievable for the corporation. The whole decision making process for R & D is shown in Figure 17 (Twiss, 1980a). The selection of new projects and the continuation of some existing projects follow the definition of R & D strategy.

FIGURE 16

THE RELATIONSHIP OF R&D TO CORPORATE STRATEGY

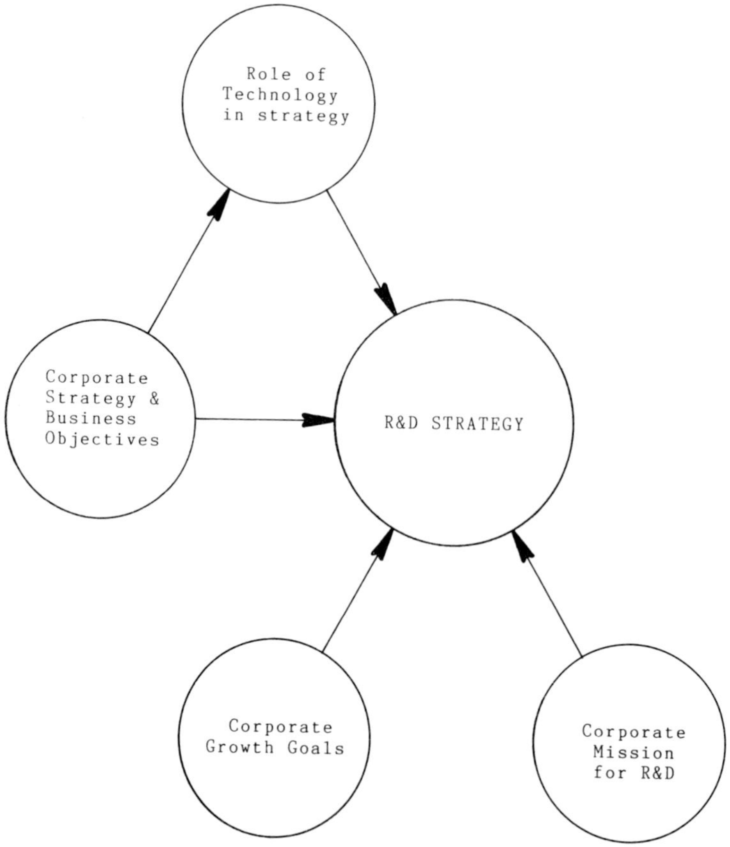

FIGURE 17 : THE R&D DECISION PROCESS

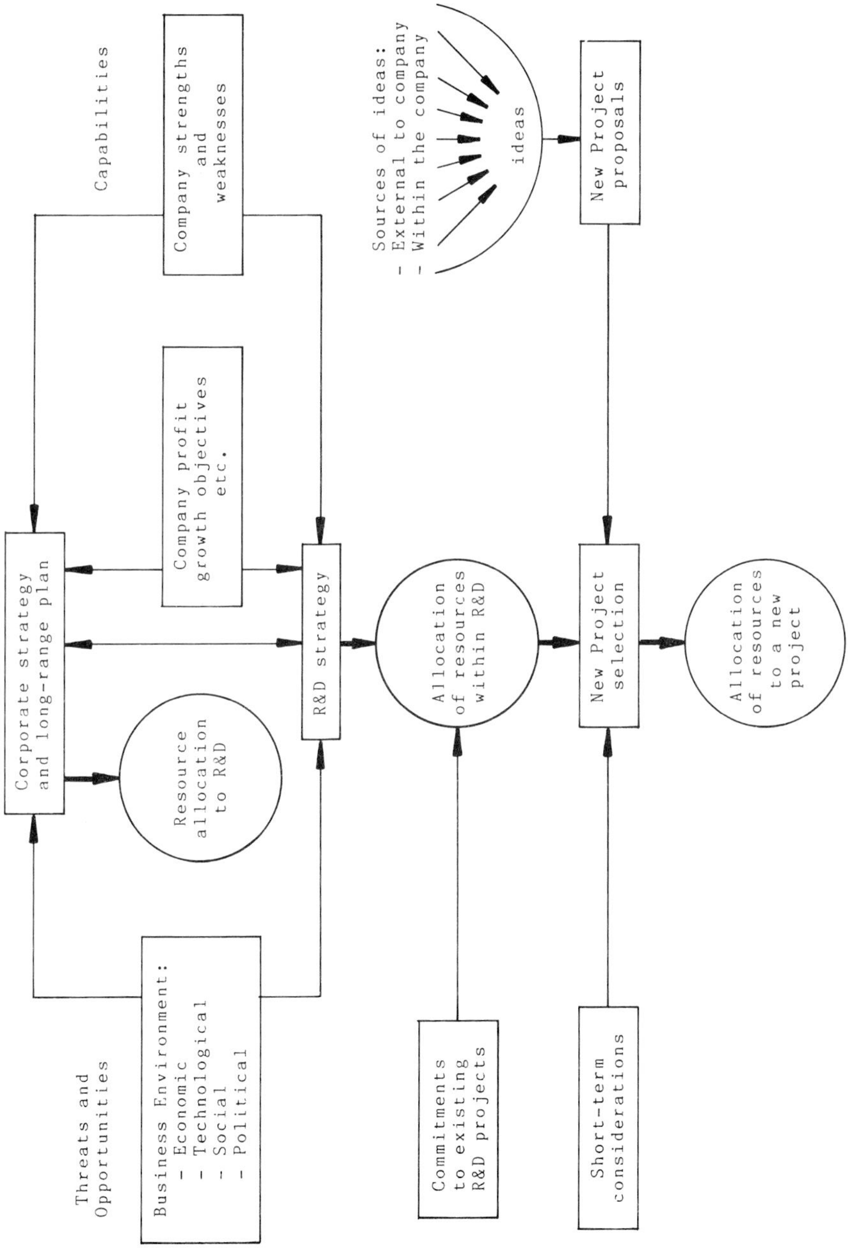

*Reprinted by permission from "Managing Technological Innovation" 2nd ed., by Brian Twiss, Longman Group Limited, publisher.

It is difficult to directly correlate corporate involvement with the setting of R & D strategy and the success of the corporation. However, there are several indicators that such involvement is important. The empirical work of Nystrom has been extended and shows that it is important for companies to systematically consider product development strategies as an important part of product planning (Nystrom, 1980). R & D strategies need to be integrated into corporate planning. The studies have shown that more wide and open searching R & D strategies in general were associated with greater technological, and probably with commercial, success than more narrow and closed restricted strategies. Thus R & D strategies greatly influence success and should not be decided in a parochial local fashion.

There is another reason for tying in R & D to corporate strategies and this is the great effect that corporate strategies can have on R & D (Conley, 1972). Recent changes in the business environment have led to vigorously contested market shares and highly competitive product technology. There has been a shift from an environment in which a high R & D capability was uncommon to one where R & D capability is well distributed. There is increasingly rapid entry of companies into new business and more rapid duplication of successful products and processes. Price competition is increasing. For all these reasons Conley believes that R & D will come under increasing scrutiny with an emphasis on product cost reduction, project selectivity and budget stability. It is important in such a climate that R & D strategy be closely linked to corporate strategy to most efficiently use the available resources.

It is of interest to examine corporate attitudes to R & D. Collier has stated that R & D faces two conflicting attitudes from management. The first is brought on by declining profit and the second by the growing social consciousness of top management (Collier, 1973). The approach of management has see-sawed from unquestioned support and optimism in the 1960's to withdrawn support and discouragement with investments in technology in the 1970's (Frohman, 1980). According to Frohman a new attitude is developing in which it is recognized that the technology is a strategic asset which needs to be managed. Management now no longer decides how much to spend on technology but decides how and where the funds should be invested. The technical assets and functional competencies must be fitted to the business strategy. A survey has indicated that a major reason for R & D failures is the isolation of R & D from other corporate functions and the lack of top management involvement in the direction of the R & D effort (Gruber et al, 1973).

Communication between R & D and the corporate level is hindered by a lack of understanding on both sides. Often R & D objectives do not match corporate goals for this reason (Gee, 1970). Based on this finding a method for relating research to corporate goals was suggested (Gee, 1975). Bobbe has considered the options that are available when there are no corporate research goals (Bobbe, 1970).

It is now widely recognized that R & D should be a component of corporate planning and several schemes for the process have been suggested (Domsch, 1978; Bemelmans, 1979; Holt, 1980). These approaches do not, however, consider the communication aspects of the process but merely concentrate on the mechanics. Wilkinson has demonstrated how research policy formulation is carried out in one company (Wilkinson, 1975).

Although the specific pattern of the strategy making process will depend on the company involved, there are generalized approaches that can be used (Wissema, 1980). Wissema indicates how the R & D effort is tied in to corporate strategy and to the other functional areas of the company.

R & D must maintain good relations with all parts of the company but nowhere is it more critical to have a good relationship than with the corporate management. Goals and strategies at the corporate level must be translated into goals and strategies for R & D. Management must understand the function of R & D if it is to be effectively used as a company resource. R & D personnel must be aware of the significance of corporate strategies if the R & D effort is to coherently match the overall thrust of the company. The actual mechanism employed in the planning process will be specific to the company itself but the interactive communication aspects are all important.

2.9 THE ORGANIZATIONAL STRUCTURE OF RESEARCH AND DEVELOPMENT

2.9.1 Organizational Structures

The organization of any given company is often complex but in general the structure can be related to model structures either in a pure form or as a mixed mode of organization. The subject of organizational structure was briefly discussed earlier but before going on to describe the ways of organizing R & D, it is appropriate to consider these model structures in more detail.

(i) Functional Forms

The typical functional organization is shown in Figure 18a. In this structure each function reports vertically upwards. The figure shows a general manager as supervising the functions. The advantage of the functional structure is that each function is staffed by experts in that particular discipline. This makes for greater efficiency within each function but may not be effective due to the lack of communication between the various functions (lateral relationships). Decisions must be made by the general manager and in a changing environment requiring many decisions there will be overload of the decision-making process. It is then very possible that different parts of the organization will be working at cross-purposes.

Functional organizations are generally adopted for production facilities where there is day-to-day stability in the tasks. However, some entire companies adopt a functional organization. Such companies would only expect to prosper in a stable environment.

(ii) Product Forms

Diversity of product lines leads to organizational structures arranged by product. Figure 18b illustrates a typical organization arranged so that each part of the organization can concentrate on its own product or product line. The advantage of the organization is that segmentation deals directly with product diversity; the disadvantage is that resource duplication may be required, for example each product structure may require its own R & D. In addition innovative new products may require synergism between the different product activities. Lateral relationships would improve communication.

(iii) Area Forms

In some situations a company may be selling a line of products which have to be modified or sold in a different fashion depending on the area of sales. Under these circumstances it is logical to structure the organization by area. This is entirely analogous to the product organization and in the figure the boxes referring to Products A, B and so on may be regarded as referring to Areas A, B, C, D and E respectively.

(iv) Matrix Forms

The matrix organization attempts to take the strengths of the functional and product (or area) organizations and to blend these strengths into one organization. Figure 18c gives a view of a matrix organization. Each product line (profit center) has its own complement of functional units (service centers) but each of the service centers participates in more than one product activity. Resource allocation can be arranged and modified according to the degree of activity in the different product areas. Communication is improved between the activities

FIGURE 18

EXAMPLES OF ORGANIZATIONAL STRUCTURES

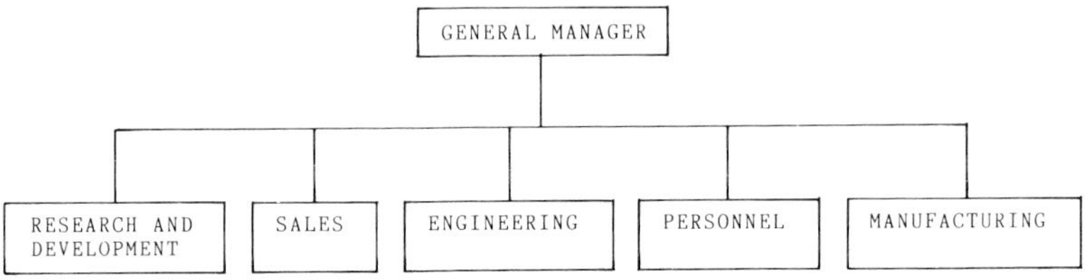

(a)

FUNCTIONAL ORGANIZATION

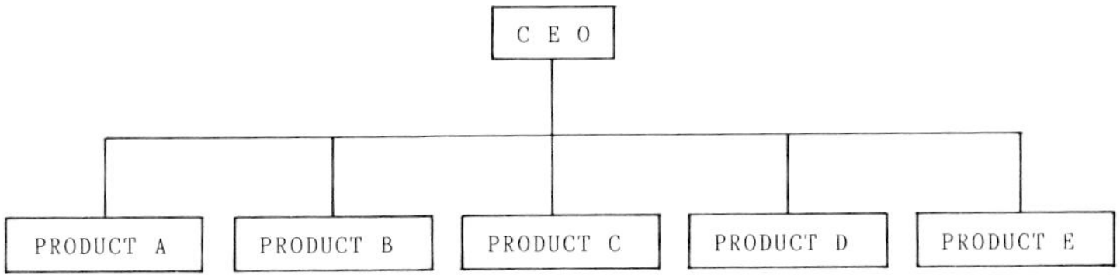

(b)

PRODUCT ORGANIZATION

FIGURE 18 (Cont'd)

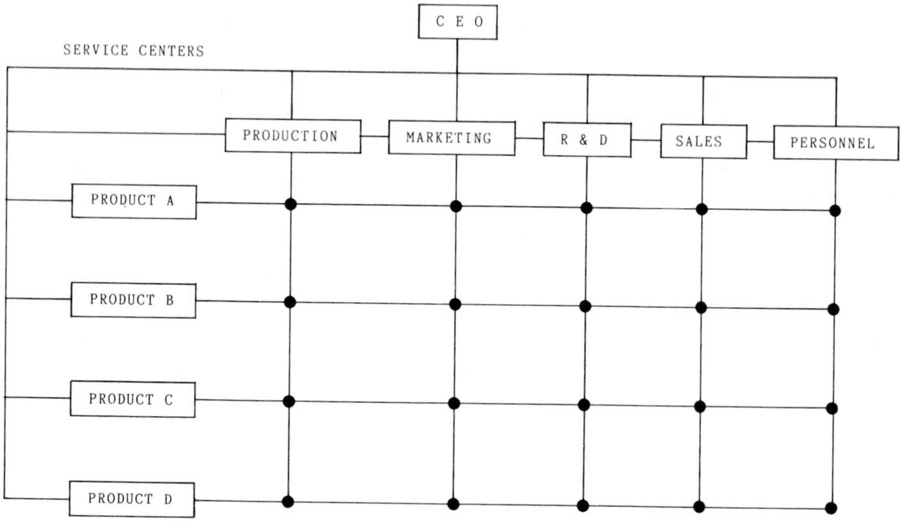

(C)

MATRIX ORGANIZATION

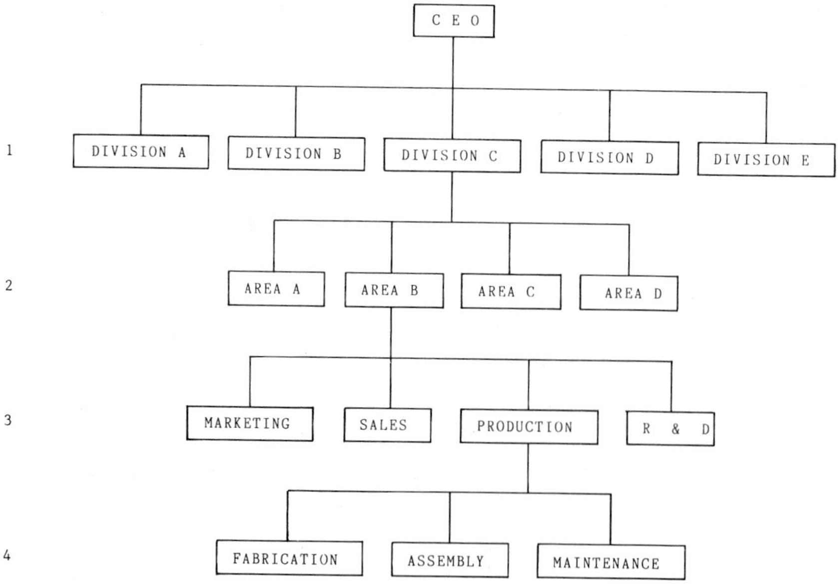

(d)

MIXED ORGANIZATIONAL STRUCTURE

in the different product areas because the service centers are engaged in more than one product. This type of organization is often adopted for projects such as the development of aerospace systems. As projects are cancelled or come to fruition the project team can be disbanded and the team members assigned to other projects. It should be noted that the price paid for the advantages of the system is the role ambiguity of the units or team members. At each nodal point in the diagram there is a dual reporting relationship to the functional supervisor and to the product program supervisor. Since the requirements of these supervisors may not always (or ever) be congruent there can be considerable conflict in the organization.

The matrix organization is difficult to set up and to maintain. The structure is suitable only for those companies facing great uncertainty often with high rates of technological change and diversity of environmental pressures. Few companies have adopted the matrix form and some companies which did organize around this structure in the 1970's have since moved to more conventional structures.

(v) Mixed Forms

Very few organizations adopt a single form especially if the organization is of any size. Figure 18d illustrates one example. Here the structure is broadened to show that the organization is first segmented into divisions. Usually each division would be concerned with a particular business which could, but not necessarily, be a homogeneous product line. In some very large corporations the divisions may themselves consist of several different businesses, which would interpose a further level of hierarchy in the structure. Level 2 indicates subdivision by area while level 3 shows that within an area the organization is by function. Level 4 indicates a further functional subdivision. It should be pointed out that the organization has some of the disadvantages of the earlier structures in that there may be resource duplication and little lateral communication. Resource duplication in a diversified corporation is often handled by central financial decision making so that the overlapping demands of the divisions are minimized and by locating shared functions at headquarters as corporate level staff. Corporate staff do not have line (profit) responsibilities but engage in company-wide activities of various types. Already mentioned is the placement of shared functions at the corporate level.

Examples of such functions are the legal department, licensing and acquisition and regulatory affairs. Functions such as finance are located at the corporate level for control purposes. Research and development may also be a corporate function either to cover overlapping research requirements of the divisions or to take care of long-range basic research.

2.9.2 Differentiation and Integration

Studies of strategy and structure have shown the different ways in which organizational structures change and adapt to changes in strategy. A common factor has been the differentiation of organizations in the face of non-homogeneous environments. Differentiation, of course, has a biological basis; in the body different cells are formed by differentiation from lines of progenitor cells. In this way there is great efficiency since different cells perform specific tasks (functional organization) and cells of the required type can be created by differentiation as needed, for example, as in response to injury. The body also demonstrates differentiation by product and by area. Specific cells turn out specific products such as insulin and many cells are confined to specific areas of the body. If the body is regarded as an example of differentiation then the body also illustrates the principle of integration. Ultimately the whole organization is under the control of the brain. This represents a hierarchical chain of command. There are also many mechanisms for lateral communication so that the activities of one group of cells influences the activities of another without involving the hierarchy. In the body it is easy to see that integration is necessary to ensure that the activities of the different parts are for the greater good of the organism as a whole. It may also be pointed out that the higher organisms are those that are more differentiated in function. The biological analogy with organizations appears to be a good one.

Just as integration is necessary in the body so integration is required in organizations. Only in the case of conglomerates, where the only link between businesses is financial, is there a minimum of integration. Studies on the success of organizations have focused on the balance between differentiation and integration. The most effective firms are those that differentiated functions to the extent needed to adapt to functional sub-environments and at the same time had developed the mechanisms to integrate the differentiated functions in order to deal with the competitive issue of the overall corporate environment (Lawrence and Lorsch, 1967).

Table 13 gives an ordered list of integration mechanisms (Galbraith and Nathanson, 1978d). Organizations may select from this list of integration mechanisms but the list is arranged in order of increasing complexity and increasing cost. In general an organization should start with implementation of those mechanisms at the head of the list and should go down the list only as far as is necessary to give sufficient integration of activities.

TABLE 13

List of integration mechanisms

Hierarchy
Rules
Goal setting (planning)
Direct contact
Interdepartmental liaison roles
Temporary task forces
Permanent teams
Integrating roles
Integrating departments

The hierarchy of authority is the principal mechanism for the resolution of interdepartmental problems. If the same type of problem continually arises then rules and procedures may be employed to avoid the difficulty. The setting of goals employing methods such as budgeting, scheduling, milestones and endpoint definition may be used to integrate efforts from various company functions. Problems which are not easily dealt with by goal setting or rules are handled by direct contact; if this fails then resort is made to the hierarchy.

The first four integration mechanisms are used extremely widely and may suffice in those companies which do not face great uncertainty and change. Under conditions of change and diversity, however, it is necessary to expend more resources on integration and to employ other mechanisms on the list. For example, product managers responsible for a single product line may be used for interdepartmental liaison. The product manager integrates the efforts of departments such as design, manufacturing and marketing to ensure the smooth launch of the right product, on time and with the necessary level of informational backup for sales.

Temporary task forces are used where there is a problem that must immediately be tackled without the addition of extra resources (it is assumed that sufficient slack is present to allow the team members to work on the problem). Permanent teams are often used for high technology projects with a high time pressure. At the end of the program team members must be reassigned to other programs. Integrating roles and integrating departments represent the highest form of integration.

The matrix organization already discussed provides integration of activities via the project team approach. This highlights the point that integration often must be paid for in role ambiguity of the integrators. The integrator often has difficulty in identifying a core set of activities since the activities are of an interactive nature with many different departments. The integrator often has low authority and must accomplish tasks by persuasive techniques. From an organizational viewpoint integration is extremely important but it takes special types of individuals to accomplish the demanding activities.

2.9.3 Process

The subjects of differentiation and integration are part of the subject known as process. Structure may be viewed as the manner in which the work is segmented into roles such as production, sales, administration and so on and the manner in which these activities are combined into departments or divisions using any of the grouping methods so far discussed. Processes concern the direction, nature and frequency of work and information flow linking the different parts of the organization. Thus structure is the way the task is divided and process is the way the task is carried out. Processes between differentiated parts of the organization for the purpose of more effective task accomplishment amount to the integration mechanisms dealt with above.

One interesting and important type of process is decision-making and studies have shown that the decision process depends on the organizational structure (Gailbraith and Nathanson, 1978a). Especially interesting is resource allocation. Here three phases, definition, impetus and approval have been identified and three levels of hierarchy, department, division and corporate. The definition phase takes place at the level of the division; a need is recognized and a proposal formulated to satisfy that need. There is impetus when a divisional manager takes ownership of the proposal and in effect becomes the champion. Finally, approval by corporate management is given for the allocation of scarce resources (often money). However, there are variations in the allocation process depending on the degree of diversification of the company. For integrated companies all three levels are active in the definition stage while corporate and divisional levels act in the impetus stage. For diversified corporations definition occurred at the departmental level and impetus at the divisional level. For both types of firms the approval phase took place at the corporate level.

It appears that decision-making will be pushed down the organization where there is diversity and uncertainty but that for very important decisions central control is exercised. Authorization limits for the expenditure of funds are often used as a means of control. Small amounts can be spent at the departmental level, larger amounts with the approval of the division and still larger amounts with the approval of corporate headquarters. Unfortunately two effects must be noted, First, the approval amounts for R & D are usually unrealistically low. Second, it is simple for over-control by corporate headquarters by arbitrary reduction of approval levels or by keeping levels fixed despite inflation.

2.9.4 Organizational Choice

The analysis of the fit of an organization leads to the study of strategy, structure and process aspects of the firm. In a recent review of organizational structures a decision tree analysis of the means of choosing the "right" organization has

FIGURE 19

DECISION TREE ANALYSES TO GIVE ORGANIZATION DESIGN

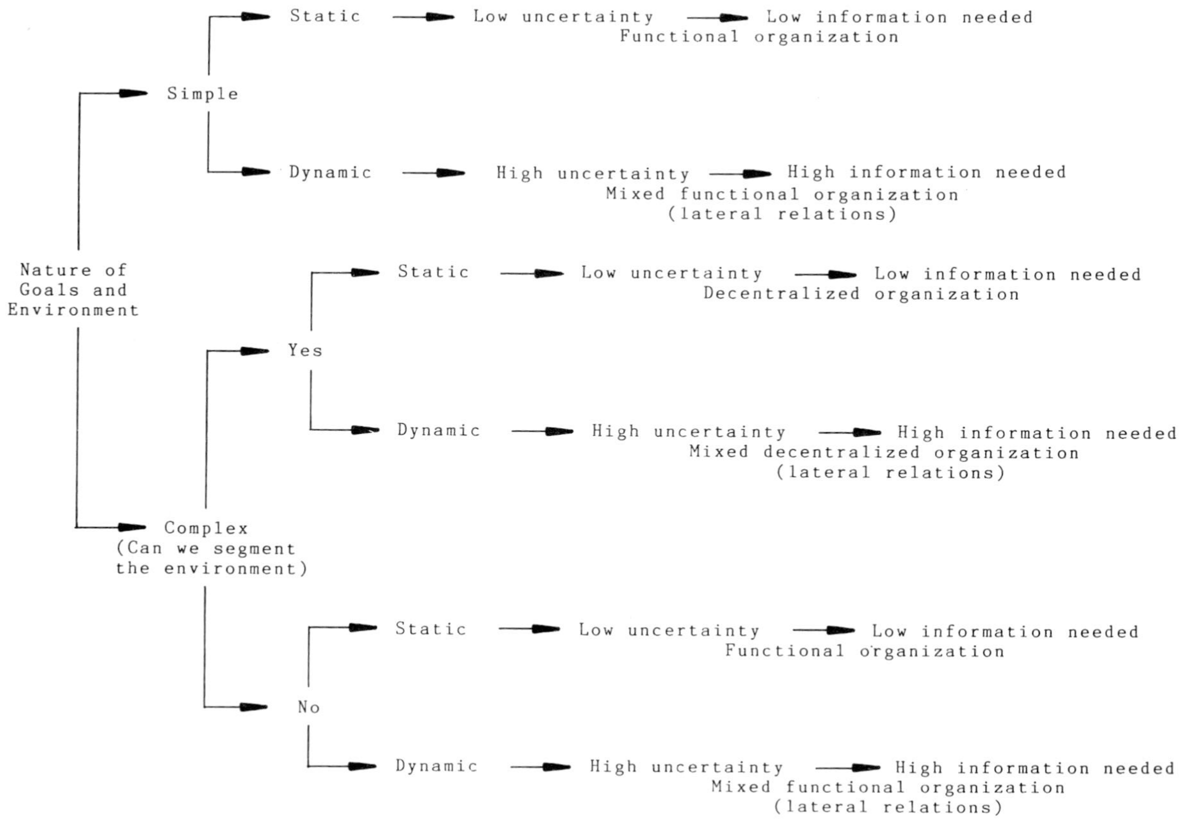

been given (Duncan, 1979). The approach taken is to use environmental demands to derive the structure giving the best fit (Figure 19).

In the analysis the environment is classified along two dimensions: simple-complex and static-dynamic. A simple, static environment can be handled by a functional organization. The highest degree of uncertainty is found for a complex environment which is segmented and which is dynamic. A mixed decentralized organization with emphasis on lateral relationships (integration) is required. It should be noted that the decentralized structure can handle complex environments with low uncertainty. It is the provision of lateral relationships that provides the basis for handling dynamicism in the environment. The provision of lateral relationships has other benefits. Decentralized organizations give difficulty in resource allocation especially if there is diversity in terms of technology, products and markets since it is necessary to provide corporate staff with a wide range of expertise. Radical innovation is rare because each division is specialized and does not have the range of information to be innovative. In the limit, as the strength of lateral relations increases until lateral relationships have an equal weight to vertical (hierarchical) relationships there is a matrix organization.

The decision tree approach is an interesting way of examining the elements of strategy, structure and process already discussed. Presumably the measure of organizational fit adopted, short term or long term, profit or sales volume will also determine the degree of satisfaction with the organization.

2.9.5 Factors Influencing the Structure of R & D

There are many factors that can influence the way in which the R & D effort of a company is organized. Some of the factors are as follows:

- Size
- Position in life cycle
- Product diversity
- Market diversity
- Geography
- Environmental factors
- Management philosophy

The size of the firm influences R & D directly. Small companies are usually organized on functional lines which gives rise to the functional organization of R & D. Growth of the company usually leads to decentralization. Thus different products lines may require different R & D organizations and perhaps different emphasis on research or on development depending on the rate of change of the particular technology. Market diversity may also give rise to the need for different R & D structures. Geographical considerations may require the decentralization of R & D to address the particular needs of different areas such as the United States, Europe and the Far East. It may even be politically wise to locate R & D units in different geographical areas because of favorable grants, tax treatment or favored treatment of the company.

The location of R & D also depends on the nature of the business. Process research requires that the laboratories be placed adjacent to manufacturing facilities or the expenditure of sizable amounts for pilot plant facilities.

Environmental factors that influence R & D are often those that influence the company as a whole. These factors include diversity, dynamicism and other elements such as the influence of government regulation. The impact of a new technology, however, may necessitate the restructuring of R & D well before its effects are felt by the whole organization.

The position in the life cycle is very important in determining the organization of R & D. Partly this is a direct effect of size, product and marketing diversity but is also due to the attitudes within the company to R & D. In the young, vigorously growing company the R & D activity is centralized but is loosely structured. The activity is central to the growth and direction of the company. As the company becomes successful the R & D activity is more structured and, although it is important, the activity is not crucial to the survival of the firm. Further increase in the size of the company, which is usually accompanied by diversification, will lead to a complete separation of the R & D organization. As attitudes harden towards change and bureaucratic procedures become institutionalized it becomes ever harder for the R & D organization to take part in the introduction of new products.

The subject of the organizational life cycle raises two issues. As the company diversifies by product and market so it is necessary to have differentiated R & D. At the same time lateral relationships must be implemented to ensure the integration of the R & D effort into company activities, to reduce resource duplication and to provide contrasting inputs for innovating. Management philosophy is of overriding importance. Reliance on acquisition and licensing will minimize the reliance on R & D. Companies which are "dogs" will do little R & D. "Cash cows" may do varying amounts of R & D depending on whether there is a need for new product introductions to maintain profitability. An emphasis on long-range basic research by management will probably lead to the establishment of a central laboratory while emphasis on new product introductions on a frequent basis with incremental improvements requires decentralized R & D. Since management philosophy also dictates the amount to spend on R & D, it is true to say that the philosophy is the most important of the factors considered.

2.9.6 The Organization of R & D in Multinationals

The organization of R & D in multinational corporations provides an excellent example of the variety of structural forms possible (Behrman and Fischer, 1980). The study cited also gives information on overseas R & D.

The level of overseas R & D activity is higher than has been realized. Within 31 out of 34 American multinational firms that had any overseas R & D activities there were 106 active foreign groups. Of these 106 activities some 62 had missions that included a substantial commitment to new product research.

In order to understand why a firm pursues R & D abroad it is necessary to examine the orientation of the company to the international market. Companies with a primary orientation to their home market would be expected to have little overseas commitments. Host-market firms have an orientation towards the national markets where they are located. A third category of companies is that in which the firms have an international market orientation. The data of Behrman and Fischer suggest that world market firms, those with an international orientation, are more likely to set up overseas R & D to look after new product research.

The most powerful reason for the location of R & D in a foreign country was the presence of a powerful affiliate in that country and a growing and sophisticated market with an adequate scientific and technical infrastructure. The primary obstacles to setting up R & D abroad was the perception of difficulties in assembling an adequate R & D staff and the economies of scale with centralized R & D. In many cases it was found to be more expensive to carry out R & D abroad than in the United States and so cost savings was not a reason for overseas R & D. Other reasons cited for setting up an overseas laboratory are the need to examine overseas technology, the need to follow developments in markets becoming more sophisticated and the need to be responsive to the overseas markets. In addition technical resources may be available overseas which can be tapped for R & D.

The Behrman and Fischer study found that there was no one generally accepted pattern of organizing foreign R & D activities. Five different styles of management could be identified: absolute centralization, participative centralization, cooperation, supervised freedom and total freedom. However, these styles were developed for the relationship between a parent and a subsidiary company. Companies which practice absolute centralization separate R & D from the line divisions and exercise substantial control over the contacts with the divisions. R & D appears to be isolated from the rest of the corporation.

Companies which adopt different styles have different patterns of innovation. The participative centralization firms are much more tightly coordinated and pay attention to detail on projects. Global allocation of resources by such a company is also undertaken. Invention and innovation are influenced by much more than the style of management in R & D; the overall company climate and the system of rewards and punishments are among factors of importance in influencing invention and innovation. Of importance is the concept of "critical mass" or minimum effective size. This concept was subscribed to by most of the managers in the Behrman and Fischer study. There are considerable variations in the estimates given for the critical size but in pharmaceuticals the minimum group in any given area is 8-9 Ph.D.'s. For the most part, foreign R & D activities are smaller in size and more restricted in scope than R & D activities pursued at home.

It is of interest to examine the organizational structures that have been employed in multinational firms with overseas R & D. These structures follow from the study of Behrman and Fischer where it was pointed out that there was no one right way to structure foreign R & D activities. Nearly every firm studied had a structure unique in itself but the structures could be viewed on the continuum of centralized-decentralized management styles. These structures will be briefly described.

Figure 20(a) shows the structure for a company with centralized management of R & D. All of the R & D is conducted at the corporate headquarters and is the responsibility of an Executive Vice President who is at the same level in the hierarchy as the Executive Vice President in charge of manufacturing and sales worldwide. The link between the divisions and corporate R & D is the research committee consisting of the President, the two Executive Vice Presidents, the Director of Research and a marketing representative.

A global product manager system is used in the organization shown in Figure 20(b). This organization has a central research facility with each product group (division) having support units in R, D and E. For each product, a worldwide responsibility is held by a particular staff member who determines the R & D portfolio for that product.

Figure 20(c) illustrates a firm with a "home market" orientation. Overseas R & D is managed centrally but most of the research is done at home where there is a matrix arrangement for R & D. Foreign activities are chiefly of a technical service nature.

The company shown in Figure 20(d) has R & D run by a committee. The Central Application Technology unit is responsible for new processes and new product lines not already in existing divisions. Each of the operating divisions has a research facility which produces new processes and product adaptations. In such a situation the facility is included in the total R & D budget otherwise the activity is charged to production. An R & D coordination group handles administration matters only. A Central Committee made up of the heads of Central Research,

FIGURE 20 (a)

CENTRALIZED MANAGEMENT OF R&D

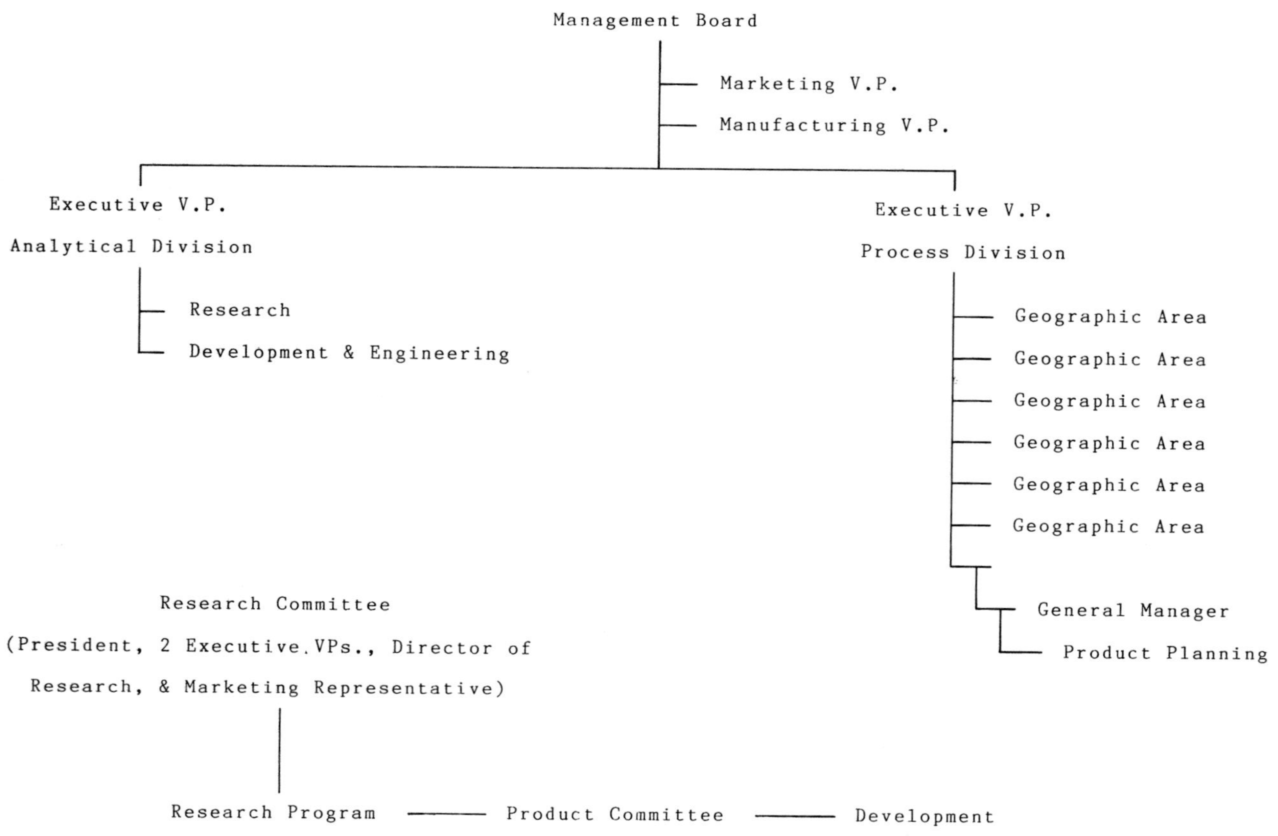

FIGURE 20 (b)

CENTRALIZED R&D COORDINATION THROUGH PRODUCT MANAGERS

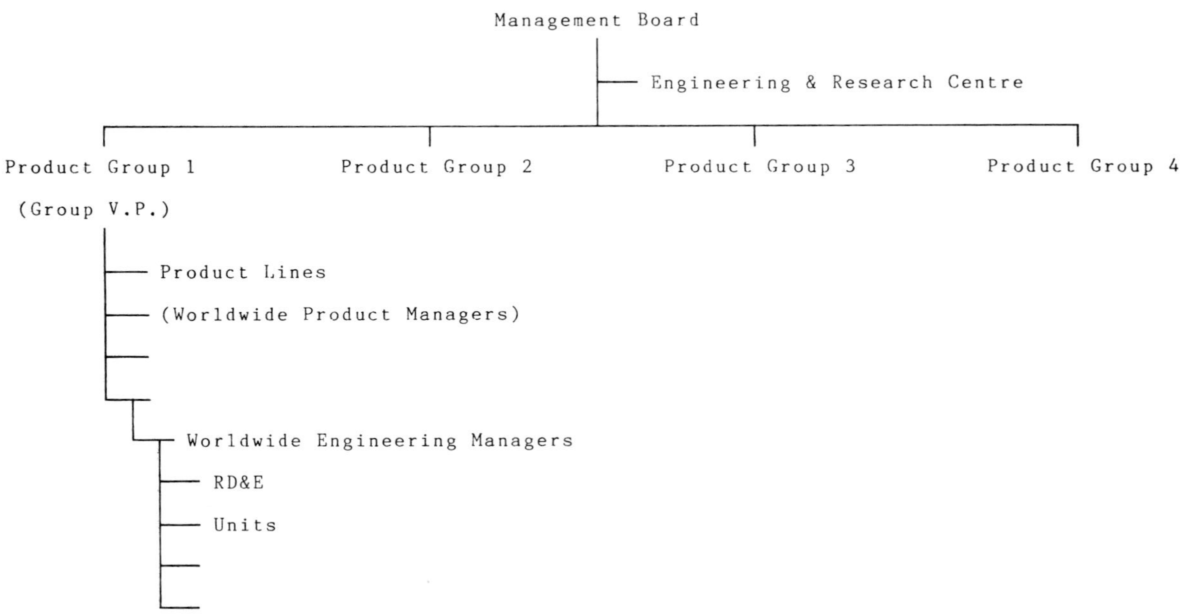

FIGURE 20 (c)

CENTRALIZED MANAGEMENT OF FOREIGN R&D WITH MATRIX MANAGEMENT OF DOMESTIC R&D

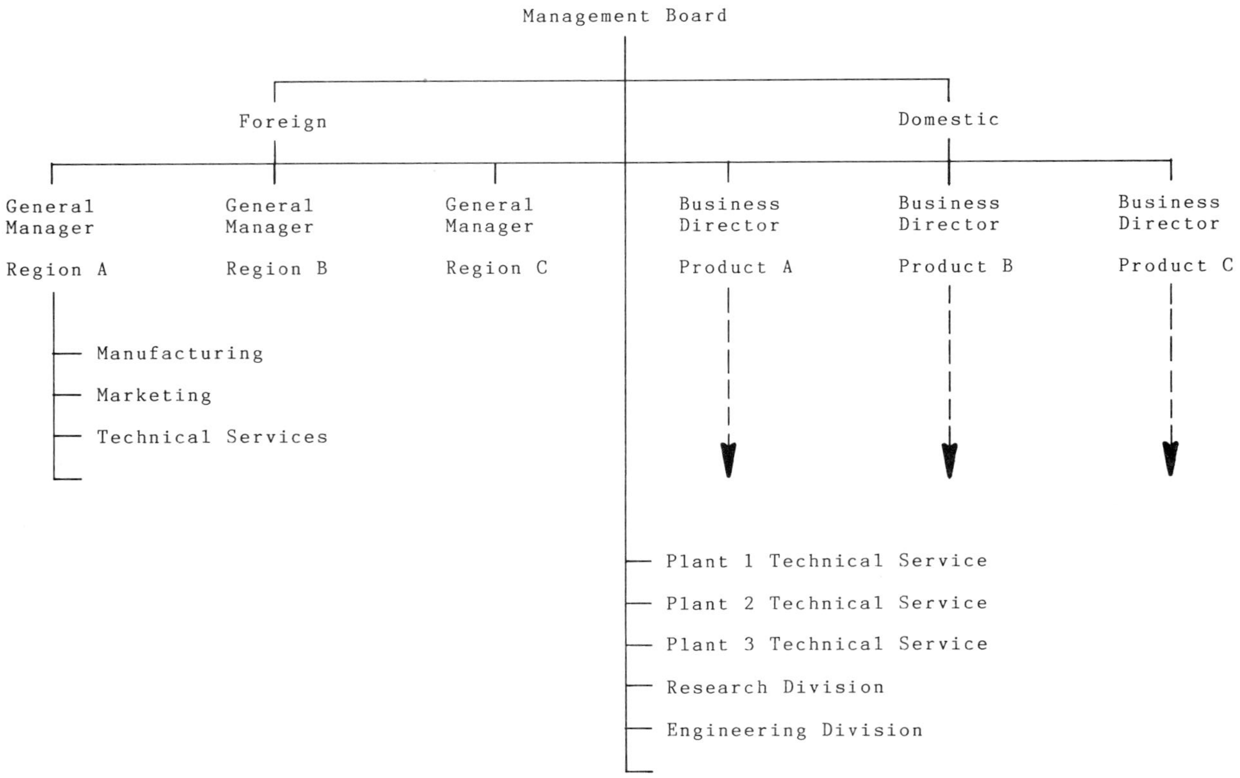

FIGURE 20 (d)

COMMITTEE MANAGEMENT OF R&D

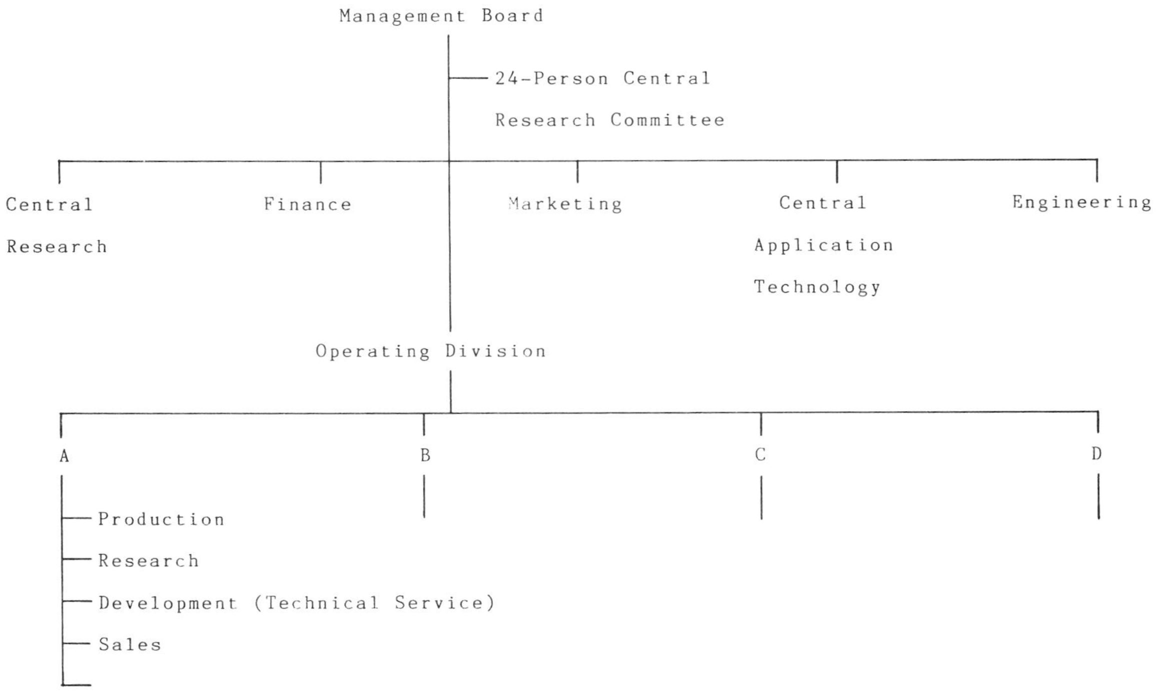

Central Applied Technology and the Engineering Division Research Unit, the Research Heads in each of the Operating Divisions, the Divisional Technology Application Directors and some of the corporate staff, meets four times a year for a thorough review of budgets, programs and policy matters. This committee resolves matters between the central research unit and the other divisions.

Figure 20(e) shows a system of centralized control and coordination. Here R & D for all of the affiliates is carried out by a central laboratory. Each functional unit makes inputs to the R & D program but the R & D group has the final say on whether or not the program goes ahead. The operating divisions also have R & D capabilities that primarily perform R & D in support of the group objectives. Laboratories can perform projects for other groups and bid for projects in order to reduce overheads. The Research Coordinating Committee is responsible for the allocation of projects among the competing laboratories.

The company with the organization shown in Figure 20(f) has a complete separation of the research and development functions. Each of the operating divisions has its own development group. Research is supported by an automatic levy against each division with no accounting to the divisions for the R & D expenditures.

A hybrid management type is shown in Figure 20(g). The R & D Division has no relationship with the existing businesses and concentrates on trying to take the firm into new fields. Each division, however, has its own product-directed R & D activities. Foreign and domestic R & D activities are not well coordinated.

The example shown in Figure 20(h) represents the result of a merger situation with the placement of R & D laboratories originally established by the formally independent companies. There are laboratories in the divisions and also in departments. Branches also have laboratories. The laboratories report to the operating management at that level. Basic research is sponsored by the Central Research Division which has no laboratory of its own. Coordination and control is accomplished by persuasion and

FIGURE 20 (e)

CENTRALIZED CONTROL AND COORDINATION

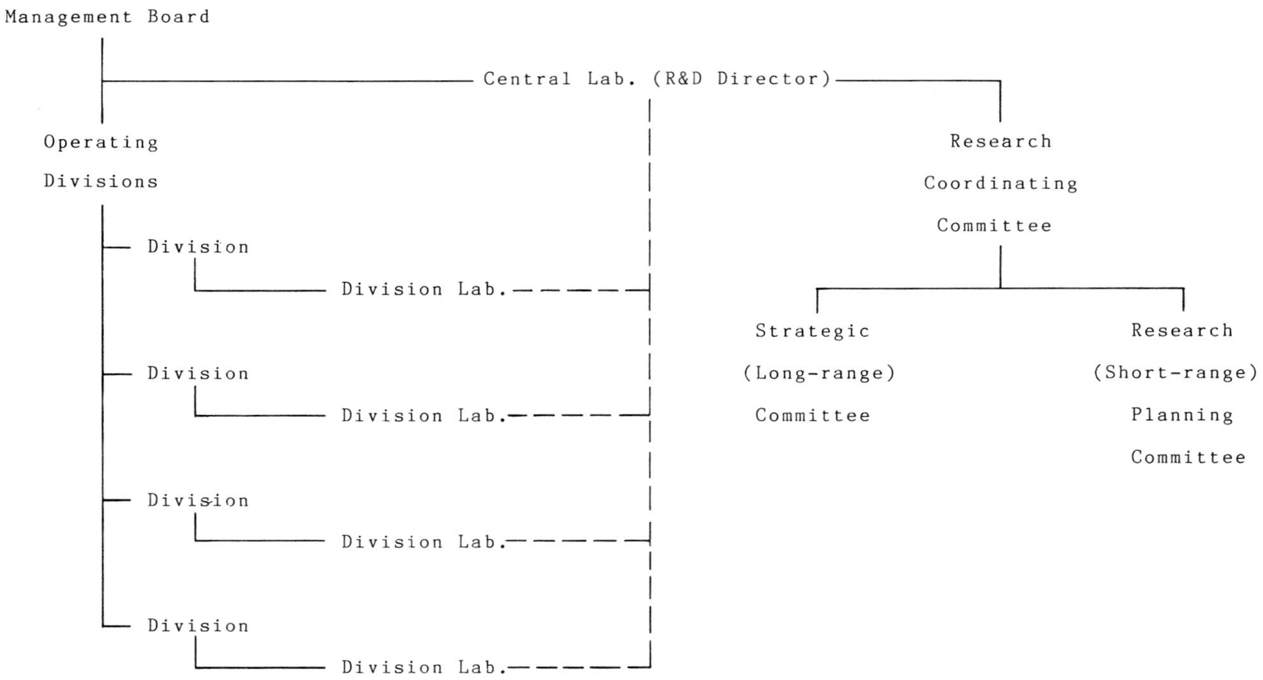

FIGURE 20 (f)

INDEPENDENT R&D UNITS (LINE RESPONSIBILITY)

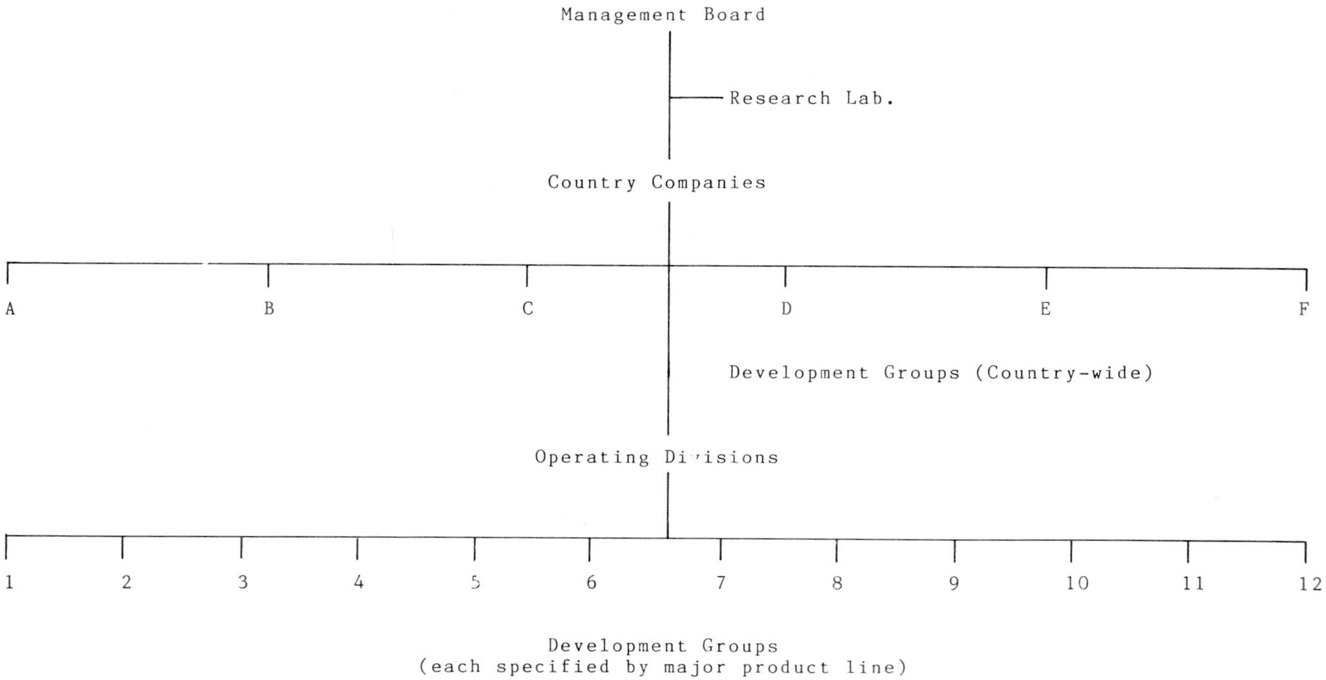

FIGURE 20 (g)

DOMESTIC PRODUCT LINE MANAGEMENT

AND FOREIGN GEOGRAPHIC MANAGEMENT

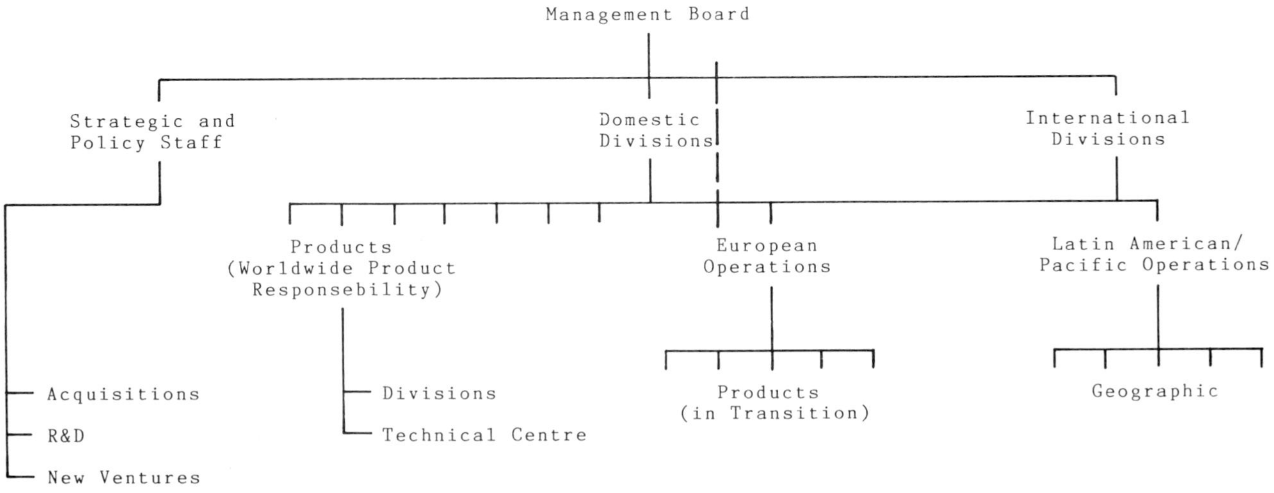

FIGURE 20 (h)

THREE-LEVEL LINE RESPONSEBILITY

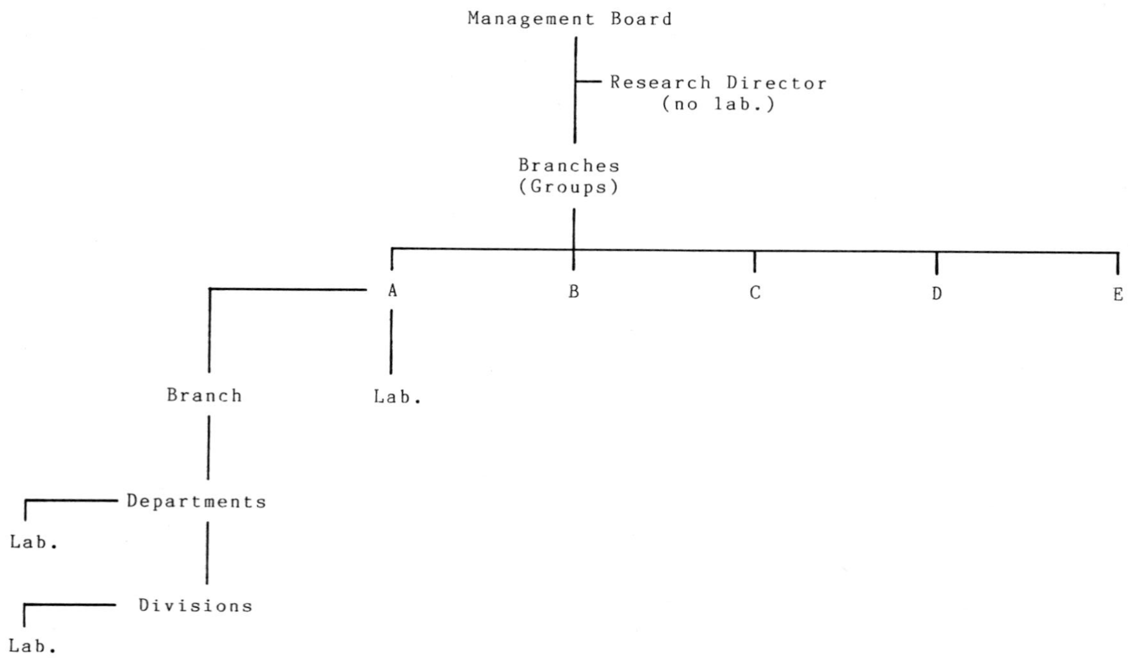

discussion stressing the corporate philosophy of R & D but each business unit has the final say as to what R & D will be undertaken.

Figure 20(i) illustrates the situation of decentralized R & D with corporate supervision. A position of Vice President for Futures was created to coordinate new ventures, corporate R & D and the central corporate laboratories. This individual has "dotted-line" responsibility for the general technical direction of the R & D activities in the firm. However, R & D program determination is the responsibility of each business unit. Interaction with the R & D efforts of the business unit is effected through a Corporate Technology Committee with members drawn from each corporate laboratory and the business unit R & D group. This committee governs the sponsoring of work by the business units at the corporate laboratories.

Figure 20(j) shows a firm in which the responsibility for technical direction in particular fields is assigned to Lead Divisions. These units also supply technology to foreign operations. Coordination and day-to-day technical activities occur directly between the Lead Divisions and the foreign affiliates. The technology center is involved where new processes or new materials are involved. The purpose of the technology center is to work on high risk/long term programs for the corporation as a whole plus those short term projects supported by the operating divisions.

R & D is totally decentralized within the business divisions in the organization shown in Figure 20(k). The role of the technical director is to review the plans and capabilities of the business divisions and to manage their performance. The technical director has no direct responsibility for the company's R & D activities but serves as a link between laboratories. The corporate R & D budget is split among the divisions with the technical director having a small budget with which to sponsor R & D corporate interest in the business unit laboratories.

FIGURE 20 (i)

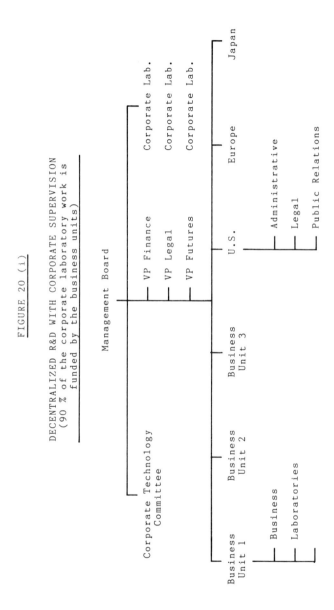

DECENTRALIZED R&D WITH CORPORATE SUPERVISION
(90 % of the corporate laboratory work is
funded by the business units)

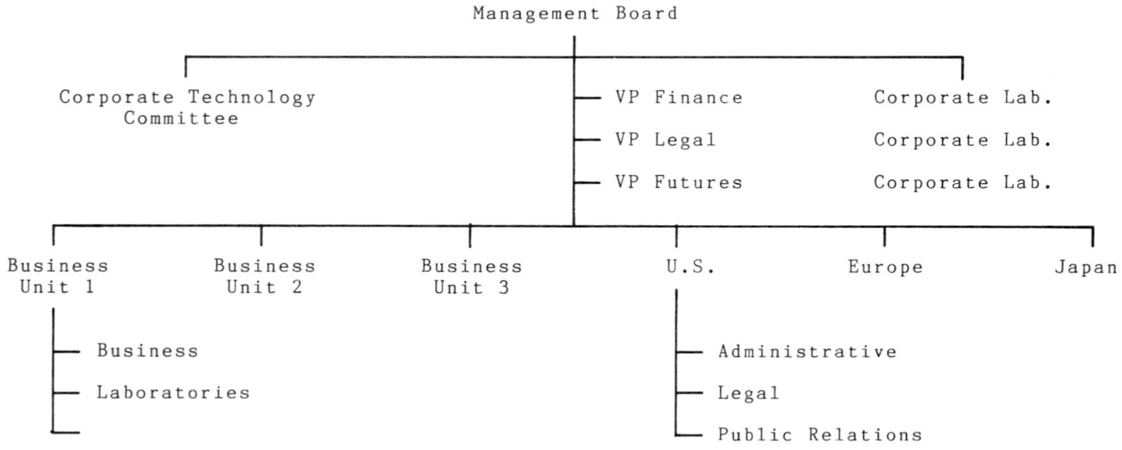

FIGURE 20 (j)

LEAD DIVISION

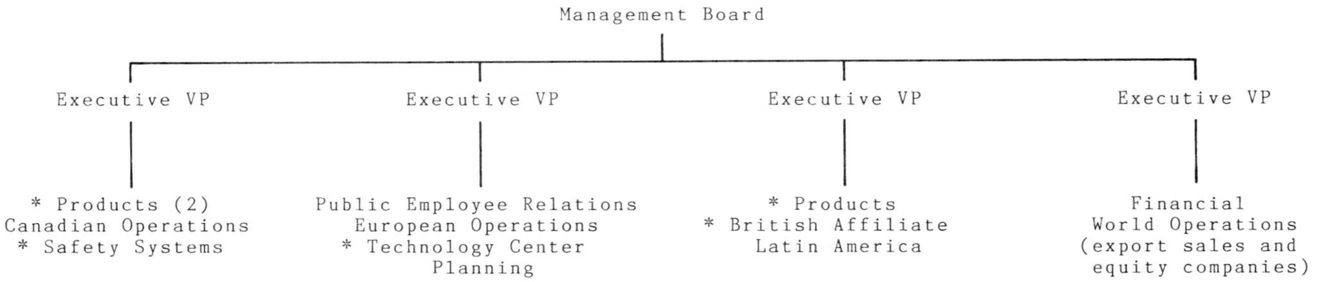

* Lead Divisions

FIGURE 20(k)

DECENTRALIZED R&D

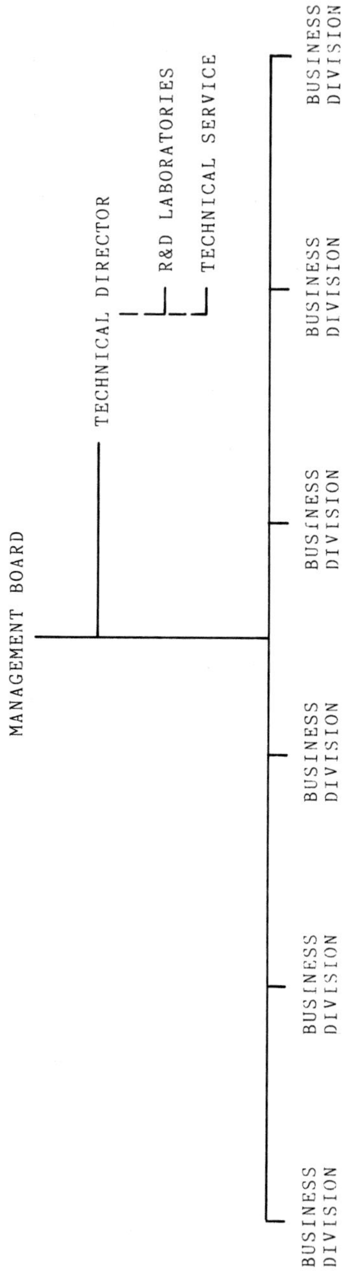

The examples of R & D structures from the study of Behrman and Fischer are not exhaustive and there will be other ways in which R & D can be and is organized. However, the range of options given is large enough that it can be employed to examine the structure of R & D in any multinational organization with overseas interests.

2.10 SUMMARY

It is now recognized that technology management in a company is a key to success. As part of the technological capability of the firm, R & D management is highly desirable. In looking to the literature in the management area it becomes clear that only recently has the importance of R & D as an integrated function been realized. There is no complete theory of R & D in the firm. Rather it is necessary to consider the implications for R & D from the studies in the strategy-structure and other areas.

The models of strategy, structure and process indicate how an organization grows and develops but there is only the broadest mention of the effect on R & D during this process. The work of Thompson allows the implications of uncertainty reduction to show how R & D can gain influence in the corporation. The classifications of Miles and Snow indicate the typical dominant coalition and distinctive competence; once a company has been classified certain statements may be made about the form and importance of R & D.

The influence of technology on the business may be used to give the technological profile of the firm. The implications of the rate of change of technology on the type of R & D and on the likelihood and the location of a breakthrough can be predicted. Such factors as homogeneity and dynamicism of the environment also play a role in deciding on the structure and type of R & D to be contemplated.

Life cycle concepts appear to be especially rich in predictive power for R & D. The fact that many life cycles are linked to change in technology immediately involves the R & D process. Generalization of the life cycle idea to businesses has led to the BCG and GE matrices which may be used as a guide in the allocation of R & D effort. Further extension to the life cycle of the company has direct implications for the receptiveness of the firm to new ideas; general aspects of change and the likelihood of R & D outputs being well- or ill-received may be predicted.

The indications are that it is of crucial importance to ensure the coherence of R & D and corporate strategy. Studies have shown that success, at least in the technological area, depends upon the setting of R & D strategy with reference to the corporate strategy. Several different schemes have been put forward but the integration of R & D into corporate planning is not yet widely practiced.

The approach adopted here has been to consider the implications for R & D of different strategy-structure options. There are several frameworks available which will allow the strategy of R & D to be determined. Structures to best fit the adopted strategy may then be considered. It would appear that there is much work still to be done on the fit of R & D into the strategy-structure-process of the firm. Even so it is likely that such studies would only generate general guidelines and that specific cases would have to be considered separately taking into consideration those special features that apply. At the present time the best approach in the case of a specific firm is to utilize the frameworks presented taking into account any features specific to the company.

REFERENCES

Abell, F.D.	1980	"Defining the Business", Prentice-Hall, Englewood Cliffs, N.J. Chapters 3 and 9.
Adizes, I.	1979	"Organizational Passages - Diagnosing and Treating Life-Cycle Problems of Organizations:, Org. Dynamics, Summer, 3-25.
Ansoff, H.I.	1965	"Corporate Strategy", McGraw-Hill, New York.
Ansoff, H.I. and Stewart, J.M.	1967	"Strategies for a Technology Based Business", HBR, 45, #6, 71-83.

Behrman, J.N. and Fischer, W.A.	1980	"Overseas R and D Activities of Transnational Companies", Oelgeschlager, Gunn and Hain, Inc., Cambridge, Mass.
Bemelmans, Th.	1979	"Strategic Planning for Research and Development", Long Range Planning <u>12</u>, #2, 33-44.
Bergen, S.A.	1980	"Product Policy and Development in a Multi-National Company", Manchester Business School, Communication from A. Pearson.
Bobbe, R.A.	1970	"What To Do When There Are No Corporate Research Goals", Res. Mgmt., <u>XIII</u>, #4, 251-263.
Bolton, W.K.	1980	"Industrial Product Strategy", R and D. Mgmt., 10th Anniversary Conference on "Industrial R and D Strategy and Management", Manchester Business School, 30 June-2 July.
Burns, T. and Stalker, G.M.	1966a	"The Management of Innovation", Tavistock Publications (2nd ed.), Chapter 5.
Burns, T. and Stalker, G.M.	1966b	Ibid. pp. xi - xii.
Chandler, A.D.	1962	"Strategy and Structure", M.I.T. Press, Cambridge, Mass.
Collier, D.W.	1973	"How Can Industrial Research Meet the Changing Expectations of Corporate Management", Res. Mgmt., <u>XVI</u>, #3, 56-63.
Conley, P.	1973	"How Corporate Strategies Are Affecting R and D Today", Res. Mgmt., <u>XVI</u>, #3, 18-20.
de Kluyver, C.A.	1977	"Innovation and Industrial Product Life Cycles", Cal. Mgmt. Rev., <u>20</u>, 21-33.
Dessler, G.	1976a	"Organization and Management", Prentice-Hall, Inc., Englewood Cliffs, N.J. p. 2.
Dessler, G.	1976b	Ibid, Chapters 4 and 5.
Domsch, M.	1978	"The Organization of Corporate R and D Planning", Long Range Planning, <u>11</u>, #3, 67-74.

Downey, H.K., Hellriegel, D. and Slocum, W. (Jr.)	1975	"Environmental Uncertainty: The Construct and Its Application", Admin. Sci. Quart., 20, 613-629.
Duncan, R.B.	1972	"Characteristics of Organization Environments and Perceived Environmental Uncertainty", Admin. Sci. Quart., 17, 313-327.
Duncan, R.B.	1979	"What is the Right Organizational Structure", Organizational Dynamics, Winter, 59-80.
Emery, F.E. and Trist, E.L.	1963	"The Causal Texture of Organizational Environments", Human Relations 18, 21-32.
Franko, L.	1974	"The Move Towards a Multi-Divisional Structure in European Organization", Admin. Sci. Quart., 19, 493-506.
Frohman, A.L.	1980	"Managing the Company's Technological Assets", Res. Mgmt. XXIII, #5, 20-24.
Galbraith, J.R.	1977	"Organization Design", Addison-Wesley, Reading, Mass., Chapter 14.
Galbraith, J.R. and Nathanson, D.A.	1978a	"Strategy Implementation: The Role of Structure and Process" West Publishing Co., St. Paul, Minn., Chapter 8.
Galbraith, J.R. and Nathanson, D.A.	1978b	Ibid, 121-123.
Galbraith, J.R. and Nathanson, D.A.	1978c	Ibid, p. 22.
Galbraith, J.R. and Nathanson, D.A.	1978d	Ibid, Chapter 4.
Galbraith, J.R. and Nathanson, D.A.	1978e	Ibid, Chapter 6.
Gee, R.E.	1970	"How Often Do Research Objectives Match Corporate Goals", Res. Mgmt., XIII, #6, 451-459.
Gee, R.E.	1975	"A Method for Relating Research to Corporate Goals", Res. Mgmt., XVIII, #6, 11-16.
Greiner, G.	1967	"Patterns of Organizational Change", HBR, July-August, 121-138.

Gruber, W.H., Poensgen, O.H. and Prakke, F.	1973	"The Isolation of R and D from Corporate Management", Res. Mgmt., XVI, #6, 27-32.
Hammermesh, R.G. and Silk, S.B.	1979	"How to Compete in Stagnant Industries", HBR, 57, Sept.-Oct., 161-168.
Hayes, R.H. and Wheelwright, S.C.	1979a	"Link Manufacturing Process and Product Life Cycles", HBR, 57, #1, 133-140.
Hayes, R.H. and Wheelwright, S.C.	1979b	"The Dynamics of Process-Product Life Cycles", HBR, 57, #2, 127-136. 127-136.
Hofer, C.W. and Schendel, D.	1978a	"Strategy Formulation: Analytical Concepts", West Publishing Co., St. Paul, Minn., Chapter 1.
Hofer, C.W. and Schendel, D.	1978b	Ibid, Chapter 2.
Hofer, C.W. and Schendel, D.	1978c	Ibid, pp. 111, 149.
Hofer, C.W. and Schendel, D.	1978d	Ibid, 135-137.
Hofer, C.W. and Schendel, D.	1978e	Ibid, 107-109.
Hofer, C.W. and Schendel, D.	1978f	Ibid, 30-31.
Hofer, C.W. and Schendel, D.	1978g	Ibid, 32-34.
Holt, K.	1980	"Integrating R and D Planning with Corporate Strategy", R and D Mgmt. 10th Anniversary Conference on "Industrial R and D Strategy and Management", Manchester Business School, 30 June - 2 July.
Lawrence, P.R. and Lorsch, J.W.	1967	"Organization and Environment: Managing Differentiation and Integration", Graduate School of Business, Harvard University, Boston, Mass.
Lawrence, P.R. and Lorsch, J.W.	1969	"Organization and Environment", Richard D. Irwin, Homewood, Ill.
Miles, R.E. and Snow, C.C.	1978a	"Organizational Strategy, Structure and Process", McGraw-Hill, New York, Part 1.

Miles, R.E. and Snow, C.C.	1978b	Ibid, 164-165.
Moore, W.L. and Tushman, M.L.	1980	"Managing Innovation Over the Product Life Cycle", Graduate School of Business, Columbia University, New York. Research Working Paper #380A. #380A.
Nystrom, H.	1979	"Creativity and Innovation", John Wiley and Sons, Chichester, England, Chapter 6.
Nystrom, H.	1980	"Company Strategies for Finding and Developing New Products", R and D Mgmt. 10th Anniversary Conference on "Industrial R and D Strategy and Management", Manchester Business School, 30 June - 2 July.
Parker, R.C.	1980	"Innovation and the Corporate Plan", R and D Mgmt. 10th Anniversary Conference on "Industrial R and D Strategy and Management", Manchester Business School, 30 June - 2 July.
Richards, M.D.	1978a	"Organizational Goal Structures", West Publishing Co., St. Paul, Minn. Chapter 1.
Richards, M.D.	1978b	Ibid, Chapter 2.
Rink, D.R. and Swan, J.E.	1979	"Product Life Cycle Research: A Literature Review", J. Bus. Res., 7, 219-242.
Salter, M.	1970	"Stages of Corporate Development", J. Bus. Pol., 1, 40-57.
Scott, B.R.	1971	"Stages of Corporate Development", Intercollegiate Clearing House, Harvard Business School, Boston, Mass.
Smith, W. and Charmoz, R.	1975	"Coordinate Line Management Working Paper, Searle International, Chicago, Ill.
Snow, C.C. and Hrebinak, L.G.	1980	"Strategy, Distinctive Competence and Organizational Performance", Admin. Sci. Quart., 25, 317-336.
SPI	-	"Market Position: Build, Hold or Harvest", Pimsletter, Cambridge, Mass.

SPI	1979	"Entry of New Competitors: How Safe Is Your Industry", Pimsletter, Cambridge, Mass.
Steele, L.W.	1975a	"Innovation in Big Business", Elsevier, New York, Chapter 8.
Steele, L.W.	1975b	Ibid, Chapter 9.
Stopford, J. and Wells, L.	1972	"Managing the Multinational Enterprise", Longman's, London.
Thompson, J.D.	1967	"Organizations in Action", McGraw-Hill, New York, Part 1.
Tosi, H., Aldag, R. and Storey, R.	1973	"On the Measurement of the Environment: An Assessment of the Lawrence and Lorsch Environmental Uncertainty Subscale", Admin. Sci. Quart., 18, 27-36.
Twiss, B.C.	1980	"Managing Technological Innovation", Longman, London, Chapter 6.
Weber, J.A.	1976	"Planning Corporate Growth with Inverted Product Life Cycles", Long Range Planning, 9, #5, 12-29.
Weick, K.E.	1969	"Enactment Processes in Organizations" in "New Directions in Organization Behavior", Addison-Wesley, Reading, Mass.
Wilkinson, J.B.	1975	"Research Policy Formulation", J. Gen. Mgmt., 3, #1, 32-41.
Wissema, J.G.	1980	"Strategic R and D Management", R and D Mgmt. 10th Anniversary Conference on "Industrial R and D Strategy and Management", Manchester Business School, 30 June - 2 July.

PART 3

CREATIVITY

3.1 INTRODUCTION

The process of producing a new product is long and costly especially if a significant degree of innovation is involved. Ideas must be generated, feasibility demonstrated and scale-up accomplished. Factors external to the firm must be taken into account as well as internal factors such as the fit of the project to functional, business and corporate strategies. Resources must be provided. Lateral relationships must be nurtured with groups such as marketing, engineering, development and production. Time schedules should be maintained as well as keeping to limits on project costs. The market for the new product must be prepared. The miracle is not that so many projects fail but that so many succeed.

The activities described above can be grouped into several categories. Strategy-structure relationships are important. Also of importance are process factors that involve information flow and interactions between people within the R & D organization and between R & D personnel and personnel in other company units. These areas have been discussed earlier and further attention will be given to people and group interactions later. What has only briefly been touched on is the subject of idea generation. This will be discussed here under the heading of creativity.

The creation of a work of art or an outstanding scientific theory is usually, but not always, the object of public acclaim. Creativity is highly prized in most cultures and the work of genius, if not immediately recognized, is usually accorded its true place in time. There is no generally accepted definition of creativity but the characteristics of the creative product include novelty and social usefulness. Although the subject of creativity has been of interest for well over one hundred years it is only in the past thirty years that intensive efforts have been made to investigate the nature of creativity and to identify creative individuals. In addition there has been a growing interest in enhancing creativity.

It might be thought that creativity is not a subject of prime importance in the management area and in the strict sense of creativity and genius perhaps this is so. But if the conditions are relaxed so that creativity includes products which are novel, but are not of the highest degree of novelty, then it is clear that industry has a need for a creative resource. Reference to the literature will show that creativity is desirable even if the creativity referred to is not of the highest order.

As discussed in Part 1 there is a spectrum of activities which ranges from the Artist to the Trader with limiting categories of the Dreamer and Mandarin (Kingston, 1977a). The characteristics of the Artist in this classification would identify that person with the creative. Thus there is a clear link between creative acts and the sale of goods in the market-place although the link may not be one of immediate cause and effect. Also discussed is the subject of discontinuous versus continuous change (Kingston, 1977b). Schumpeter's A-phase is identified with the "big leap" or discontinuity which would occur as the result of a creative act. On the other hand Schumpeter's B-phase is concerned with many small changes each arising from its predecessor and determining its successor. The A-phase requires high creativity but the B-phase does not require creativity to be high. It is implicit from the above that the number of creative acts is much smaller than the number of incremental improvements.

It is widely accepted that industry must change to prosper and grow. Companies which have been very successful have often been based on creative products such as the Xerox process and Polaroid "instant" photography. The need for creativity and innovation in the development of the firm has been recently emphasized and it has been pointed out that the economic theory of the firm does not pay much attention to company creativity and innovation (Nystrom, 1979a). According to Nystrom creativity is the precursor to innovation; discontinuous change is often required in the firm.

Recent descriptions of change in organizations have emphasized the requirement of creativity. Thus it has been stated, "It is increasingly necessary in today's complex organizations to have a planned, managed-from-the-top, organization-wide effort to create a set of conditions and a state that will allow the organization to "creatively" cope with the changing outside demands on it..." (Beckhard and Harris, 1977).

The need for more creative individuals in organizations due to the emergence of areas such as advertising and R & D has been mentioned (Dessler, 1976). Dessler points out that the survival and growth of an organization depends increasingly on the capacity for changing creatively and for solving new and unique problems. In a discussion of product planning, creativity is emphasized and it is stated that rigid statements of product strategy inhibit or restrict creativity (Day, 1975). The use of creative product line extensions to prolong the maturity phase of a product has been discussed (Kijewski, 1978). The need for creative market segmentation has been emphasized (Abell, 1980). Evidently there is a need for creativity in various industrial activities even if the intent of the authors is only to imply novelty rather than the true creative act.

It is in the R & D area that there is a ready identification between the ongoing activities and the need for creativity. Studies of scientific creativity are widespread and although industrial R & D has not been investigated in depth there have been many publications regarding the need for creative individuals in the R & D laboratory. In fact the concept of creativity is implicitly used in organizations which have a "dual ladder" which allows creative individuals to follow a scientific reward path whilst others follow an administrative route. It is also commonly accepted, but not proven, that the creativity of an individual decreases after a certain age and this again influences the way in which older scientists are treated. Some companies arrange for scientists to automatically transfer to non-R & D functions at a certain age.

Referring back to the model of R & D (Figure 6), the R & D process requires creativity. According to Steele an R & D organization is almost unique in its dependence on the creative dedication of its staff (Steele, 1975). The need for widespread creativity in the organization but especially in the R & D laboratory is discussed in detail by Twiss (Twiss, 1980). A recent discussion of the impact of R & D on profits states that one way to improve the return on R & D expenditures is to increase the creativity and productivity of R & D scientists and engineers but adds that, "This is as difficult as it is important" (Ravenscraft and Scherer, 1981). Support of the universities is urged to prevent the well of ideas and creative personnel from running dry.

The above discussion indicates at least the need for novelty if not for outright creativity throughout the various activities of industry but especially in the R & D area. A brief survey of the creativity literature will be given and subsequently a description of the application of some of these findings will be outlined within an R & D context. In looking at the work on creativity a rigorous approach will not be taken because the interest is in whether the findings can be used in an R & D area to increase the flow of novel products. This involves the extrapolation of findings from the level of genius to that of more modestly endowed individuals but there are extensive studies which cover individuals with a spectrum of abilities. To be pragmatic it may be stated that it is only necessary for an organization to be relatively more creative than its competitors to hold an advantage and that this need not imply a high absolute creative level at all.

3.2 APPROACHES TO CREATIVITY

The literature on creativity is voluminous and no attempt will be made to present a balanced summary. However, Vernon has summarized many important contributions in this area and a brief description will be made following his approach (References not specifically cited in this section will be found in Vernon, 1970).

The approaches to exploring creativity described include empirical studies, introspective materials, theoretical contributions, psychometric approaches and personality studies.

Stimulating creativity is additionally considered. Also to be discussed is a recent provocative appraisal of creativity (Perkins, 1981).

Galton, himself a genius, considered the hereditary aspects of creativity by studying the ancestry and family relationships of eminent individuals. Characteristic differences were found in those relationships between individuals eminent in different fields. Terman and Cox utilized biographies to study the characteristics of eminent men. Several interesting facts emerged from this study. It was noted, for example, that the direction of later achievement was foreshadowed by the interests of childhood. There are difficulties in such a retrospective study and so the authors engaged in a prospective study in which children were tested for I.Q. (Intelligence Quotient) and were followed through life to see whether predictions could have been made on the basis of the earlier testing. In general, children identified as gifted did achieve far more than those of lower I.Q. but the influence of personality factors was shown to be important. In the gifted group success was associated with emotional stability and beyond a certain level of intellectual ability success was largely determined by non-intellectual factors. Retrospective studies were carried out on eminent scientists by Roe who investigated the background, interests, achievements and personality of these men. Again interesting facts emerged. For example, theoretical physicists had suffered a high incidence of serious illness in childhood. Biologists and physicists had a characteristic pattern of a shy, lonely, over-intellectualized child. On the other hand, social scientists were highly social from an early age and were deeply involved in organizations. But all scientists had in common a driving absorption in their work. These pioneering studies present interesting facets which have been followed up but the picture is still incomplete. What is clear is that eminent people are different intellectually and in personality from each other and from the general population.

Further examination of creativity may be done by reading the descriptions of the creative act given by eminent individuals. There are at once similarities and differences in experience. Thus Mozart stated that a composition would come to him all at once and that it was then only necessary to write it down.

Tchaikovsky stated that the germ of a composition would come to him which would require further conscious effort to produce the finished work. Spender has emphasized the extreme effort needed to take an idea to the finished poem. In the scientific area Poincare emphasized the hard work and frustration which occurred before the inspiration which suddenly allowed the answer to be revealed. Detailed examination of the experiences of creative individuals has largely been responsible for the formulation of descriptions of the stages of creativity but it is clear that there are significant differences between individual experiences.

Theoretical studies of creativity have been concerned with various aspects. Rogers has discussed creativity in terms of the need for self-realization whilst Freud was particularly interested that creativity might be a kind of sublimation of repressed complexes. The cognitive aspects of creativity have been especially studied. Guilford has given a theory of the intellect in which intellectual functioning is described in terms of three factors: operations, contents and products. In this model it is suggested that the most important operation for creativity is divergent thinking. The operation of convergent thinking implies the selection of a narrow region which is examined in great detail. Convergent thinking has been erroneously regarded as the antithesis of creativity. Theoretical studies include description of the stages of creativity mentioned above.

A large amount of effort has been expended on the development of tests by which creative individuals may be identified. A wide variety of tests are available ranging from simple tests designed to identify word fluence and divergent thinking abilities to more complex evaluations such as the Barron-Welsh Art Scale. Often such tests have been administered along with I.Q. tests and attempts made to divide individuals into groups, it being assumed that intelligence and creativity are orthogonal. There is great confusion at the present in this area of psychometric evaluation.

Considerable effort has also been given to personality studies of individuals of high and lesser creativity. Cattell has shown the typical personality profile of eminent researchers. Barron has also studied the personalities of individuals eminent in different fields. The personality approach has not been as popular as the psychometric approach largely because it is easier to ask an individual to take a series of simple tests, such as giving as many uses as possible for a paper clip, that appear non-threatening. Personality approaches require more extensive testing and require the individual to open up to the investigator.

The last area considered by Vernon is that of the stimulation of creativity. This is the most recent area of creativity investigation and most studies have been carried out in the United States. Subjects at school and at university have been involved in such studies and further investigations have been done in industry. There are perhaps two main themes to this work. In the first, subjects are tested, involved in a series of exercises intended to enhance the usage of creativity, and then retested. In the second approach attempts are made to stimulate the production of creative ideas via techniques such as Brainstorming and Synectics. According to Vernon it is difficult to find sober psychological assessments of the effectiveness of work in this area.

Based on the review of Vernon there are several different approaches to the investigation of creativity but there is no indication that any one approach by itself will be entirely fruitful. Undoubtedly a combination of approaches is necessary. At the present time, the use of retrospective studies is not an active area and there is not a high degree of activity in the theoretical area. There is little activity in the area of introspective studies. Psychometric testing still appears to be of interest and there is renewed interest in the psychological testing area. Programs to enhance creativity usage are available at universities in Europe and the United States. In the last few years there has been great interest in the development of techniques to stimulate creative problem solving.

The description given is the classical view of creativity. Some of the different aspects have been reexamined (Perkins, 1981) including descriptions of creative acts by creators. The conclusion would seem to be that many of the accepted "principles" about creativity are on shaky ground and may not be correct. With this in mind the next few sections will explore in more detail many of the aspects covered by Vernon.

3.3 THE NATURE OF CREATIVITY

Creativity itself may be regarded in different ways (Whitfield, 1975a). The special quality concerned with anything creative is implied in words such as novelty, inspiration, ingenuity, originality, invention, genius and imagination. All these descriptions have the quality of creativity in common but separately describe an idea, the mental activity which forms it or the ability of the person to carry out the activity.

Creativity applied to the product of mental activity may be defined in terms of the association of ideas. The creative idea is the result of the remoteness of the association which gives the novelty of the end result. The implication is that whatever is created is derived from what is already present.

As a mental activity creativity appears to involve not only the conscious mind but also the unconscious. Creativity may be regarded as an action that produces a new idea or insight. The effect spills over into the emotional area since the creative act gives rise to anticipation before the event and pleasure following the event. The moment of creation appears to be largely outside conscious control but the process is aided by relaxation or turning away from the problem. It is believed that at such times the unconscious continues to create patterns or associations, without hindrance from the conscious, which assists in the attainment of insight. It may also be remarked that creativity has both constructive and destructive aspects since rigid patterns of mental organization must be disrupted to achieve the new synthesis. The more remarkable the creative product the greater must be the modification of vast complexes of associated ideas, attitudes and habits of thinking.

The consideration of creativity as an ability comes back to the consideration of creativity as a process since ability can only be measured in terms of the power of an individual to perform a mental or a physical task. Intelligent behavior has been described as the forming of associations, the integration and synthesis of ideas, showing good judgment and reasoning, adjusting to new situations, abstract thinking and constructive thinking directed to the attainment of some end. These attributes may be necessary but are not sufficient to describe creative ability for there is the requirement of novelty which cannot be ignored. As mentioned earlier divergent thinking has been given as an ability necessary for creativity. Guilford has cited those aspects of the human intellect which contribute to creative ability. This list includes: sensitivity to problems, fluency of thinking, flexibility of thinking, originality, redefinition and elaboration. In addition to these factors, which indicate the potential for creativity, it is also necessary to consider personal temperament and the motivation to create.

Up to this point a definition of creativity has not been given and in fact the difficulty is not in the giving of a definition but in deciding which definition to give. Parnes and Brunelle have stated, "Let us define creative behaviour as the production and use of ideas that are both new and useful to the creator" (Whitfield, 1975a). On the other hand Stein states "Creativity is a process that results in a novel work that is accepted as useful, tenable or satisfying by a significant group of people at some point in time" (Stein, 1974a). The former definition employs an internal criterion whilst the latter employs external evaluation to judge the novelty and usefulness of a work. According to Welsh, "There is, however, a dismaying lack of concensus in definition...," and he refers to over 100 definitions analyzed by Taylor (Welsh, 1975a).

For the purposes here the definition proposed by Stein will be taken. Although for the individual it may or may not be important whether there is external appreciation of a novel work, it is important in the industrial sense that novel products not only have a social usefulness but that the products are perceived by the marketplace to have a usefulness. For industry the purpose is to sell goods, although some companies are happy to be regarded as the place where novel products, not necessarily useful, are produced. This high regard can be translated into sales of other products and the attraction of high calibre personnel.

It may be noted that the definition of Stein does not demand that a novel work be immediately recognized and indeed there are many cases where a work was only recognized as novel long after the event. Furthermore, there is no constraint on the size of the group making the evaluation. In industry the interest would be in recognition of the novel work in a short time frame by a large enough group to ensure an adequate ROI.

The nature of creativity is seen rather differently by Perkins (Perkins, 1981). The work of creating is viewed as the work of selection that leads toward a creative result. Most of the processes of creativity relate to planning, abstracting, undoing and making means into ends. Creativity is thus regarded as a more deliberate process depending on inborn and learned abilities. Although the approach would say that creativity is less mystical than often proposed, there is little clue as to why, of two persons of equal ability, one is creative and the other not. Personality factors obviously play a large role in the process.

3.4 THE CREATIVE STAGES

Studies of the creative process in different fields have led to the description of the various stages in the process. From three to seven stages have been given by different workers (Haeffle, 1962). In historical order these stages are:

By Helmholtz:
(i) Preparation
(ii) Incubation
(iii) Illumination

By Wallas:
(i) Preparation
(ii) Incubation
(iii) Illumination
(iv) Verification

By Young:
(i) Assembly of material
(ii) Assimilation of material in our minds
(iii) Incubation

(iv) Birth of an idea
(v) Development to practical usefulness

By Rossman:
(i) Observation of a need or difficulty
(ii) Analysis of the need
(iii) Survey of the available information
(iv) Formulation of objective solutions
(v) Critical analysis of the proposed solutions for advantages and disadvantages
(vi) Birth of the new idea, the invention
(vii) Experimentation to test out the most promising solution; perfection of the embodiment by repeating some or all of the previous steps.

By Osborn:
(i) Orientation: pointing up the problem
(ii) Preparation: gathering pertinent data
(iii) Analysis: breaking down the relevant material
(iv) Hypothesis: piling up alternatives by way of ideas
(v) Incubation: letting up, to invite illumination
(vi) Synthesis: putting the pieces together
(vii) Verification: judging the resultant ideas

All the descriptions of the stages give essentially the same process as does Wallas. It may be noted that the stages overlap and are not intended to be discrete steps in the process. From a consideration of the stages proposed, Haeffle has decided that the creative process is best described in terms of four stages:

(i) Preparation
(ii) Incubation
(iii) Insight
(iv) Verification

The stage of preparation involves familiarization with the subject area or problem. There is an intense desire to develop the idea or to solve the problem but all attempts are frustrated. The individual then turns away from the problem either to another problem or to relaxation. During this time of incubation the unconscious still works on the problem and the moment of insight occurs when the solution rises from the unconscious into the conscious. The solution must then be verified. Further work may need to be carried out to develop the idea further.

The process described is not a guarantee for success. Insight may never come. There are instances of insight leading to the wrong or inappropriate solution. However, from the introspective descriptions by creative individuals it does appear that the four stages generally describe the creative process.

The identification of the stages of the creative process is important because different factors play a role at each stage. Haeffle and Stein give extensive detail on the manner in which the approach to these stages may be optimized. The preparation stage requires hard dedicated work. Flexibility of thinking is required in the definition of the problem and in the formulation of proposed solutions. In contrast, the stage of incubation requires distancing from the problem. Once insight has been achieved more convergent thought processes are required to reach and to verify the final solution.

Although Haeffle has well summarized the stages of the creative process, the problem of evaluation as to novelty and usefulness is not really considered. Thus the verification stage implies an examination of the appropriateness of the creative product by the individual but does not imply any external evaluation. An interesting view of the creative process has been given by Stein (Stein, 1974b) who regards the process as consisting of the following stages:

(i) Hypothesis formation
(ii) Hypothesis testing
(iii) Communication of results

Stein regards the preparatory or educational stage as not uniquely a part of the creative process and so it is excluded. Incubation and insight are to be included in the stage of hypothesis formation. Hypothesis testing is akin to the verification stage. Whilst other workers regard the creative process as a uniquely individual one, Stein states that intrapersonal and interpersonal factors play a role at each stage. Creativity occurs in a social context and depends upon the transactions between an individual and the environment. The creative process is not ended with the verification stage but extends to include the communication of results. This is an extremely important concept since there is a large body of literature concerned with communication. It is noted that the communication is not usually a neutral flow of information since it is often necessary to influence the external evaluating group where approval is necessary for the legitimacy of the creative work to be recognized. As is well recognized the more the creative work deviates from the accepted norm the greater will be the resistance to change. Thus the subject of creativity leads into the consideration of change in organizations and in other social systems.

3.5 INTELLIGENCE AND CREATIVITY

The pioneering prospective study of Terman already referred to showed that eminence could be predicted on the basis of I.Q. tests carried out in the formative years. The implication here is that creativity might be predicted from tests of intelligence. The connection between intelligence and creativity also appears from considerations of theories of intelligence (Whitfield, 1975a). Cattell suggests that general intelligence is a function of two separate but correlated factors: a crystallized intelligence which is employed in those situations requiring learned habits of thinking and a fluid intelligence which is needed in situations where existing patterns of cognitive skill are inadequate. Thurstone obtained seven factors of intelligence: verbal comprehension, word fluency, numeracy, spacial ability, memory, perception and reasoning. In the study of how different tasks are performed, Spearman concluded that intelligence was made up of a general ability "g" which contributed to all the tests and a number of specialized abilities. The model of Guilford gives a theory of the intellect in which there are 120 categories linking an operation, a content and a product. From these models it will be seen that intellectual ability is multi-faceted. In addition some of the factors considered in theories of intelligence likely play a major role in the determination of creativity. It is clear that intelligence and creativity must be considered together. Whether or not intelligence and creativity are directly connected has been, and is, the subject of continuing study.

Before considering possible connections between intelligence and creativity it is in order to examine intelligence itself and this entails a discussion of the I.Q. test. It is readily accepted that creativity is a nebulous concept but is intelligence fully understood? The answer would appear to be in the negative. Thus Welsh states, "There is a widespread belief held by most laymen that intelligence is fairly well understood by "science" and that there are many tests available to measure intelligence precisely" (Welsh, 1975b). Welsh emphasizes that intelligence is an attribute not an entity and can only be recognized through intelligent behavior (the same can be said for creativity). He cites the findings of a symposium which concluded that, "Intelligence is whatever is measured by intelligence tests".

The subject of intelligence testing is not straightforward. The conclusion at the end of the last paragraph provides a working definition. However, there are other definitions which might be used (Welsh, 1975b). In practical or pragmatic definitions common sense behavior in everyday life is employed. Thus some people "catch on" more quickly to new problems. Welsh proposes that although I.Q. and behavior in life may be correlated at lower levels of I.Q. it is likely that the relationship breaks down at higher levels because success depends on personality characteristics such as attitudes, interests and motivation.

The disagreement over the measurement of intelligence is reflected in the statements of other authors. There has been heated argument over whether intelligent ability is inherited or whether it is largely shaped by the environment. Eysenck believes strongly that intelligence is largely determined genetically while Kamin holds that environmental factors play a large role in the determination of intelligent behavior (Eysenck and Kamin, 1981). A balanced view would hold that both genetic and hereditary factors would play a role in the determination of intelligent ability but what is extremely interesting in the evaluation of the arguments of the proponents of each view is that in many cases the same studies have been taken and presented so as to support each opposing viewpoint. At the very least these studies are inadequate and the conclusion to be drawn is that the field of intelligence is in some disarray.

Accepting that intelligence is whatever is measured on an intelligence test the question remains as to whether such a test has predictive power. It will be remembered that Terman was able to forecast likely eminence on the basis of I.Q. tests. Eysenck states that school success and college success can be predicted on the basis of I.Q. results and that the "higher social occupations" are filled by people with higher I.Q. values (Eysenck and Kamin, 1981). Hudson would modify this view and refers to the results of MacKinnon who found little or no connection between adult I.Q. and adult achievement above a minimum level which lay somewhere in the region of I.Q. 120 (Hudson, 1967). However, Hudson believes that the intelligence test should not be discarded because, "Tests of this kind perform perfectly well the function for which they were originally conceived: the rapid and impersonal assessment of intellectual ability in the population as a whole". Difficulties arise only when the I.Q. test is regarded as a precise indicator of mental power. Hudson points out that the relationship of I.Q. to distinction in the intellectual fields is very complex. At the low levels of I.Q. there is a definite relation but this peters out at higher I.Q. levels and above a certain point a high I.Q. is of little advantage. The point at which the relationship dwindles depends, however, on the occupation. In the arts the relationship breaks down at a lower level than in science. Hudson estimates that the I.Q. lower limit in science lies in the region of 115 and believes that a high I.Q. is of little advantage above 125. For the arts the lower limit probably lies in the range 95-100 with the upper limit around 115. Two possibilities have been suggested as to why the relationship between I.Q. and accomplishment breaks down. First, it is possible that above a certain level of I.Q. motivation is of overriding importance. Second, the intellectual skills measured in an I.Q. test are too simple. It is possible that if tests were devised to assess more complex skills the predictive power might be extended.

Since the description of the intellect is extremely difficult, multifaceted models will require complex tests to separately elucidate each facet if such is possible. It is appealing to use a simple two-parameter model and this was first done by Getzels and Jackson (Vernon, 1970). These authors took children at a private school and used tests to select two groups, a high I.Q. group and a high creative group. These groups were then studied for similarities and differences. The high I.Q. group were in the top 20 percent in I.Q. but were below the top 20 percent on the creativity measures; the high creativity group were similarly defined. Differences were found between the two groups. For example, the high I.Q. subjects tended to converge on stereotyped solutions whilst the high creative group tended to diverge from the customary. The high I.Q. students were preferred by teachers over the high creative group presumably because the high I.Q. students were more conventional and not as questioning. It was found that both groups performed equally well academically.

Although the results from Getzels and Jackson are intriguing there are several methodological problems. The groups chosen in actuality were both high I.Q. since the minimum I.Q. in the study was 140. As discussed above the relationship between I.Q. and achievement peters out at these higher I.Q. levels. There is other evidence for this. MacKinnon in a study of creative and uncreative architects could not find any difference in I.Q. as measured on the Wechsler Adult Intelligence Scale, each group had an I.Q. of about 130 (Barron, 1969a).

A more vigorous methodology was developed by Wallach and Kogan (Vernon, 1970) who employed different tests especially to determine creativity. Four groups of children were obtained by dichotomizing the distributions of intelligence index and creativity: high creativity/high intelligence; high creativity/low intelligence; low creativity/ high intelligence and low creativity/low intelligence. Differences between these groups were found. Thus the high creativity/high intelligence group could exercise both control and freedom. The high creativity/low intelligence group were in angry conflict with themselves but could blossom in a stress-free environment. The low creativity/high intelligence group were addicted to academic achievement whilst the low creativity/low intelligence group engaged in manoeuvers such as passivity or intense social activity as a defense mechanism.

Welsh has employed a four-fold characterization recently but there are differences to the above analyses (Welsh, 1975c). A practical approach was taken to the definition of the two dimensions and different types of test were used since the intent was to investigate personality rather than cognitive factors. The Concept Mastery Test (CMT) was employed to measure "intellectence" a term suggested for the personality dimension related to performance on intellectual measures such as the CMT. Similarly "origence" was defined as the personality dimension measured by the Barron-Welsh Art Scale and related measures. This dimension is related to originality or creativity. Differences between the groups were found in attitudes and beliefs, direction of activity, inter-personal conduct, preference for occupation and so on. For example, low origence/high intellectence subjects had a preference for the sciences, high origence/high intellectence subjects for the arts and humanities and low origence/low intellectence subjects for the business area. If, however, subjects of low origence tend to go into the sciences it is interesting to speculate as to the origin of the creative scientist.

The subject of the differing orientation of individuals has been studied in the context of convergers and divergers (Hudson, 1967). The tendency in the literature has been to regard divergers as potentially creative whilst convergers are not. Hudson remarks on a conflict in the literature. The studies of Roe (Vernon, 1970) show that eminent research workers in the physical sciences strongly resemble the converger; the work of MacKinnon shows that creative persons in all fields are more divergent than their non-creative colleagues. According to Hudson these findings can only be reconciled provided the divergent attributes to which MacKinnon refers exist within a relatively narrow range. Thus all scientists are inhibited but the creative ones less so. Artists are uninhibited but the creative ones have some of the rigour and dedicated single-mindedness of the scientist. It is pointed out that this idea accords with the concept of a tension between tradition and revolution being necessary for creativity. The approach of Hudson is not one in which individuals are classed directly in terms of intelligence and creativity but even so the approach showed that subjects could be categorized on the basis of intelligence tests, creativity tests and autobiographical data and that predictions could be made as to the occupation which would be chosen by subjects in each category. The results indicated that there was an intellectual spectrum in which each occupation attracted individuals of a certain type. Convergers were attracted to one end of the spectrum and divergers to the other.

Recent work on the structure and function of the brain suggests that there may be a physical basis for the idea of convergent and divergent thinking (Furst, 1979). These findings have been based on work with patients in which it was necessary to sever the corpus callosum, the major connection between the left and right hemispheres of the brain. It has long been known that speech and language are associated with the left hemisphere of the brain but the function of the right hemisphere was little understood. It is now believed that verbal, logical analytical thinking is the domain of the left hemisphere and that this hemisphere excels in human speech, abstract conceptualization, logic and mathematics. The domain of the right hemisphere is visuo-spatial or pictorial. Whereas the left hemisphere is involved in events which occur in a logical sequence the right hemisphere is involved with configurations in space where relationships occur simultaneously. The suggestion has been made that the left hemisphere is involved with convergent thought processes whilst the right hemisphere is involved with divergent thought processes. The work on the functioning of the brain has recently been taken up in the management area where "left-brain thinking" has been regarded as involved in planning and "right brain thinking" is managing (Mintzberg, 1976; Brown, 1979).

3.6 PERSONALITY FACTORS

The investigation of personality factors in creative persons has proceeded at the same time as the work in the cognitive area. It appears that personality plays a very important role in creative behavior. Hudson quotes Cox as stating that "....high but not the highest intelligence, combined with the greatest degree of persistence, will achieve greater eminence than the highest degree of intelligence with somewhat less persistence", (Hudson, 1967). Again quoting MacKinnon, "Our data suggest, rather that if a person has a minimum of intelligence required for mastery of a field of knowledge, whether he performs creatively or banally in that field will be crucially determined by nonintellective factors" (Hudson, 1967).

Creative aptitudes seem to need many strengths in addition to special talents in a particular field. One way of representing such features is given by the Field Theory of Lewin (Whitfield, 1975b). According to this theory a person is subjected to forces due to the attraction or repulsion of aspects of the environment which arouse internal tensions. These vectors fall into three classes; knowledge, intellectual abilities and temperament. Knowledge refers to specific and general information which acts as a store for novel recombinations. Intellectual abilities refer to those mental skills which enable knowledge to be used effectively. Temperament refers to those aspects of personality which make a person more likely to use skills and knowledge and to take risks in developing novel ideas.

Stein has examined personality theories including the fulfillment model and the conflict model and has reviewed the literature on the personality characteristics found to be associated with the creative individual (Stein, 1974b). These characteristics are given in Table 14. The findings do not characterize any single individual. No single creative individual has all these characteristics but a creative individual probably has more of them than does a less creative person.

The emphasis on personality characteristics has also been made by Welsh in an examination of the literature on creativity (Welsh, 1975d). It is suggested that the creative person is likely to possess, in addition to superior intelligence and cognitive skills, a distinctive motivational structure and personality. Welsh continues on to criticize the use of cognitive tests. Welsh believes that creativity has been conceptualized too narrowly in cognitive terms and that the tasks explored have been too close to those tapped by conventional I.Q. tests. This subject will be considered again in the area of testing.

The approach of classifying individuals on scales of intellectence and origence resulted in the hypothesis of psychological similarity related to each category (Welsh, 1975c). Thus

TABLE 14

Characteristics of Creative Individuals

Characteristics
1. An achieving person.
2. Has curiosity and a need to satisfy this curiosity.
3. Need for order.
4. Self-assertive, self-sufficient, has initiative.
5. High power need.
6. Not inhibited, less formal, less conventional, low on measures of authoritarianism.
7. Persistence, liking and capacity for hard work, dedicated.
8. Independent.
9. Has constructive criticism, not contented, dissatisfied with what is.
10. Widely informed, wide range of interests, versatile.
11. Open to feelings and emotions. Subjective. Enthusiastic.
12. Low on economic values.
13. Aesthetic in values.
14. Freer expression.
15. Introverted. Low on need for interpersonal relationships.
16. Emotionally unstable but well adjusted.
17. High self-image and low on self-criticism.
18. Intuitive.
19. Can influence others.
20. Has empathy.

all persons falling at the same position on origence and intellectence should be of the same general personality type. In addition a consistency corollary predicts that for two persons at the same position the one who is more consistent in all of the characteristics of that particular location will be more effective in performance and better satisfied. The characteristics associated with each origence-intellectence type are given in Table 15. It will be noted that the descriptions accord with the characteristics of the creative person given earlier and that given for convergers and divergers.

TABLE 15

Summary of characteristics associated with origence-intellectence dimensional types *

	INTRAPERSONAL ORIENTATION		
I.	Extroversive	II.	Introverted
	exhibitionistic acting out		withdrawing ruminative
III.	Extroverted	IV.	Introversive
	outward directed responsive		inward directed speculative
	INTERPERSONAL CONDUCT		
I.	Sociable	II.	Asocial
	outgoing many acquaintances amicable		isolative few friends impersonal
III.	Social	IV.	Unsocial
	friendly indiscriminate sociality benevolent		shy guarded sociality humanitarian
	DIRECTION OF ACTIVITY		
I.	Interactive	II.	Proactive
	interdependent responder		autonomous detached viewer
III.	Reactive	IV.	Active
	dependent follower		independent leader
	NATURE OF SELF CONCEPT		
I.	Self-seeking	II.	Self-centered
	egocentric		egoistic
III.	Self-effacing	IV.	Self-confident
	allocentric		egoistic
	ATTITUDES AND BELIEFS		
I.	Irregular	II.	Unorthodox
	uncommon "don't conform"		unconventional "take risks"

* I	High orig/low int	II	High orig/high int
III	Low orig/low int	IV	Low orig/high int

Table 15 (Continued)

III.	Orthodox common "play safe"	IV.	Regular conventional "follow rules"
	COGNITIVE STYLE		
I.	Imaginative fantasy improvisation simile	II.	Intuitive insight meditation metaphor
III.	Customary industry persistence allegory	IV.	Rational logic deliberation analogy
	COGNITIVE DEVELOPMENT		
I.	Proto-integration without differentiation diffuse global imprecise	IV.	Integration with differentiation synthesis organization composition
III.	Proto-differentiation without integration fragmented detailed unrelated	IV.	Differentiation with integration analysis specification resolution
	VOCATIONAL PREFERENCE		
I.	Histrionic action performing and dramatic arts sales occupations	II.	Intellectual ideas arts and humanities related occupations (e.g., journalist)
III.	Pragmatic practical problems commerce and business service occupations	IV.	Scientific concepts sciences and mathematics related occupations (e.g., statistician)

Personality studies have been carried out on scientists (Barron, 1969b). Barron refers to the work of Cattell on eminent researchers using the 16PF inventory. Figure 21 gives the mean profile of eminent physicists, biologists and psychologists.

FIGURE 21

CATTELL PROFILE OF THE CREATIVE PERSONALITY

Mean 16 P.F. Profile of Eminent Researchers (N=140)
in Physics, Biology, and Psychology

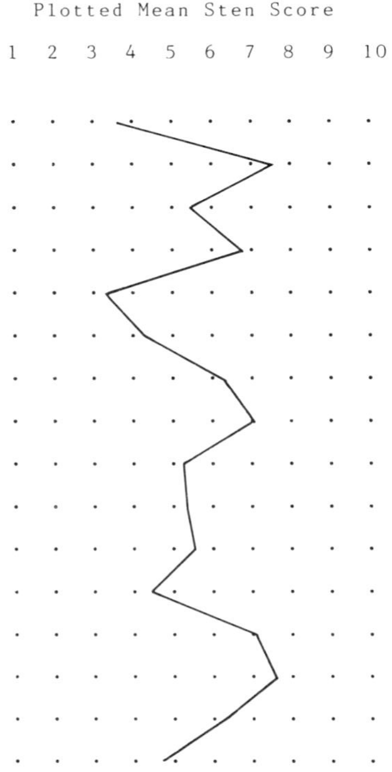

Personality Dimension Label at Lower Pole	Mean Stens	Personality Dimension Label at Upper Pole	
A-Schitzothymia	3.36	Cyclothymia	A +
B-Low Intelligence	7.64	High Intelligence	B +
C-Low Ego Strength	5.44	High Ego Strength	C +
E-Low Dominance	6.62	High Dominance	E +
F-Desurgency	3.15	Surgency	F +
G-Low Group Super Ego	4.10	High Group Super Ego	G +
H-Threctia	6.01	Parmia	H +
I-Harria	7.05	Premsia	I +
L-Low Protension	5.36	High Protension	L +
M-Praxernia	5.36	Autia	M +
N-Simplicity	5.50	Shrewdness	N +
O-Low Guilt Proneness	4.38	High Guilt Proneness	O +
Q1-Conservatism	7.00	Radicalism	Q1 +
Q2-Low Selfsufficiency	7.52	High Selfsufficiency	Q2 +
Q3-Low Self Sentiment	6.44	High Self Sentiment	Q3 +
Q4-Low Ergic Tension	4.91	High Ergic Tension	Q4 +

Especially to be noted are two unusual combinations: high schizothymia with high ego strength and high dominance with high desurgency. The picture given is of a rather cold, introspective, solemn, strong-willed, unconventional and highly intelligent person. Later studies have reinforced this picture. Studies at the institute for Personality Asssessment and Research generally agree with these findings. On the Barron-Welsh Art Scale the more creative scientist differs from the less original one by preference for more complex asymmetrical figures and in this respect is similar to the artist.

The work of Gough is cited in which scientists were classified into different categories (Barron, 1969b). Eight types were noted: the zealot, initiator, diagnostician, scholar, artificer, aesthetician, methodologist and independent. The point is made that there are many different methods of functioning and different ways of making a contribution. It would be a mistake to rely on a group profile.

Summarizing the literature Barron has given the profile of the creative scientist: high ego strength and emotional stability, strong need for independence, high degree of control, superior general intelligence, a liking for abstract thinking and a drive towards elegance in explanation, high personal dominance, rejection of conformity in thinking, a preference for dealing with things rather than with people, a liking for order and controlled risk-taking. The question is whether methods are available to identify such persons. In addition, if the characteristics of the creative person are identified it may be possible to modify the behavior of the ordinary individual to make him more creative.

3.7 CREATIVE PROBLEM SOLVING

The subject of problem solving is related to that of creativity. In creative problem solving the object is to come up with a novel solution. For problems in industry the intent is also to produce a product, in the broadest sense, that is socially useful and recognized as such. This is the justification for regarding this approach to problem solving as creative.

The question arises as to whether all creativity is problem solving. For example, even a work of art may be the result of the resolution of creative tension in an individual and so may be the "solution" to the presence of an internal problem. The same may be said of the reason for the production of a new scientific theory. The difference between creativity in general and creative problem solving in particular appears to often lie in the formulation of the problem. In creative problem solving the origin of the problem usually lies outside the individual and is not a direct consequence of personality factors although the perception of the problem may be. For the creative individual the internal component to the problem resolution which results in a creative product is dependent critically on personality factors. According to Stein, novelty can be achieved in a variety of ways including trial and error, serendipity and problem solving (Stein, 1974c). Stein states, ".... it would be an error to think of my presentation of the creative process as the same as problem solving" and ".... the creative process is distinguished from problem solving in that the former results in greater leaps, in giving things more of a twist, and the final result is regarded as much more novel than the result of problem solving". This may be due to the fact that the really creative advances are made by persons of genius whereas problem solving is usually the province of individuals of lesser talent. Furthermore the creative process may be largely unconscious whilst problem solving involves a series of steps more or less consciously chosen. There is, therefore, not a one-to-one correspondence between creativity and problem solving but there are areas in common and in creative problem solving there is an effort made to utilize some of the elements of creativity in order that a more novel solution may be achieved.

In the following, many different techniques of problem solving will be given but discussions in detail will not be made. However, several points will be covered. Two independent paths to problem solving appear in the literature (Davis, 1973). In the first approach, studies on animals and later computer studies have focused on the psychological aspects of problem solving involving the cognitive and Gestalt views of problem solving and problem solving from the point of view of stimulus-response theory. The second approach taken is that dictated by the constant needs of industry for new solutions. It is the latter experience which will be detailed here.

Problem solving methods may be divided broadly into individual and into group techniques. However, some techniques may be employed at either level. It may be noted that once a group technique is introduced the whole theory of group dynamics becomes relevant. Individual and group techniques have been discussed in detail (Stein, 1974d; Stein, 1975). No one technique is suitable for all stages of the problem solving process and so it is necessary to employ a selection of techniques on the road from problem definition to problem solution. There is again a wide variety of techniques and it is not necessary to learn every one. It is preferable to become conversant and proficient with a small number of techniques that can be employed effectively. Listings of techniques are widely available (Whitfield, 1975c; Rickards, 1974; Schlicksupp, 1977; Rickards, 1979; Rickards, 1980; Parnes et al, 1977; de Bono, 1977; de Bono, 1971; de Bono, 1980). Some of these discussions are specifically directed to the industrial situation as in new product development and marketing and others specifically to management aside from the general discussions on problem solving.

The problem-solving process may be regarded as an alternating series of convergent and divergent steps. Figure 22 shows the process (Whitfield, 1975c). The problem-solving sequence begins with a convergent process in which the problem is first perceived. A divergent process then follows in which more data are gathered and other viewpoints chosen. There is then a convergent process in which the problem is finally defined. It is emphasized that problem definition is extremely important as the selection of the "right" problem determines all subsequent steps. Once the problem is defined a divergent stage follows in which there is a search for solutions, existing and novel. A convergent stage involving selection of a solution and evaluation of that solution follows. The solution must then be implemented. Although not shown in the diagram it may be necessary to repeat the cycle at any stage or even to return again to the beginning of the problem. It may not be realized until a solution has been reached and implemented that the "wrong" problem

FIGURE 22

PROBLEM-SOLVING SEQUENCES, HELPFUL TECHNIQUES AND DEVELOPMENT METHODS

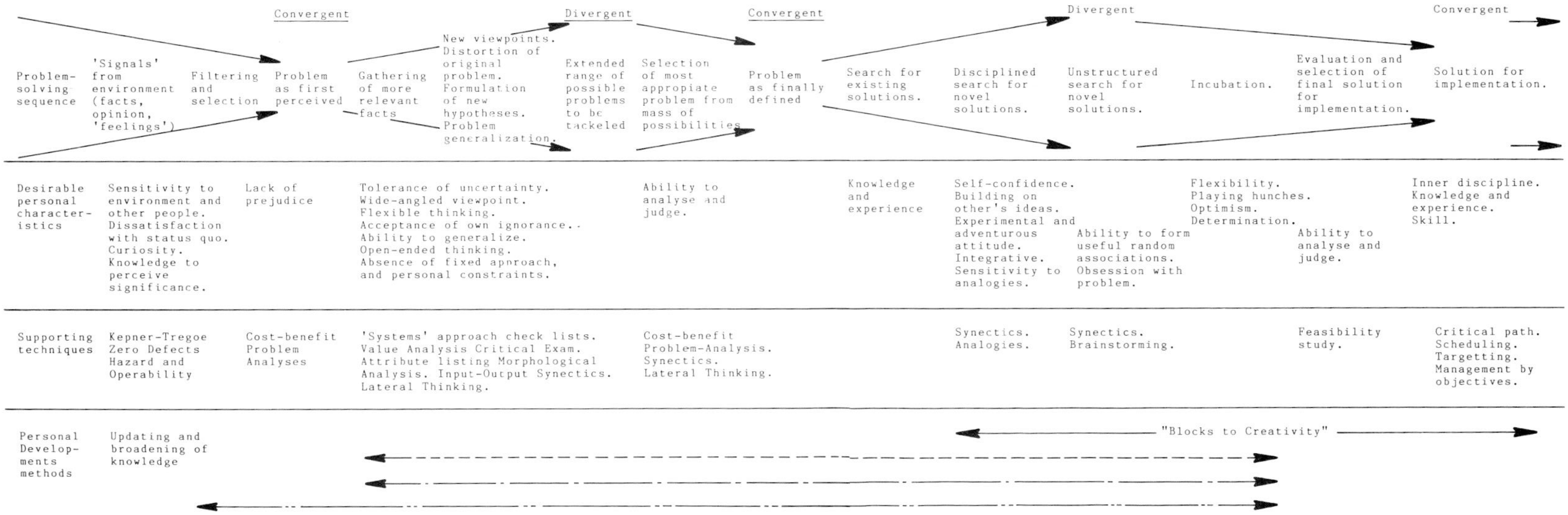
Convergent
Divergent
Convergent
Divergent
Convergent
Problem-solving sequence
'Signals' from environment (facts, opinion, 'feelings')
Filtering and selection
Problem as first perceived
Gathering of more relevant facts
New viewpoints. Distortion of original problem. Formulation of new hypotheses. Problem generalization.
Extended range of possible problems to be tackeled
Selection of most appropiate problem from mass of possibilities
Problem as finally defined
Search for existing solutions.
Disciplined search for novel solutions.
Unstructured search for novel solutions.
Incubation.
Evaluation and selection of final solution for implementation.
Solution for implementation.
Desirable personal character-istics
Sensitivity to environment and other people. Dissatisfaction with status quo. Curiosity. Knowledge to perceive significance.
Lack of prejudice
Tolerance of uncertainty. Wide-angled viewpoint. Flexible thinking. Acceptance of own ignorance. Ability to generalize. Open-ended thinking. Absence of fixed approach, and personal constraints.
Ability to analyse and judge.
Knowledge and experience
Self-confidence. Building on other's ideas. Experimental and adventurous attitude. Integrative. Sensitivity to analogies.
Ability to form useful random associations. Obsession with problem.
Flexibility. Playing hunches. Optimism. Determination.
Ability to analyse and judge.
Inner discipline. Knowledge and experience. Skill.
Supporting techniques
Kepner-Tregoe Zero Defects Hazard and Operability
Cost-benefit Problem Analyses
'Systems' approach check lists. Value Analysis Critical Exam. Attribute listing Morphological Analysis. Input-Output Synectics. Lateral Thinking.
Cost-benefit Problem-Analysis. Synectics. Lateral Thinking.
Synectics. Analogies.
Synectics. Brainstorming.
Feasibility study.
Critical path. Scheduling. Targetting. Management by objectives.
Personal Develop-ments methods
Updating and broadening of knowledge
"Blocks to Creativity"

*Reprinted by permission of Penguin Books Ltd.

was solved. Evaluation of a solution may indicate it is incorrect or inappropriate and it is then necessary to consider another solution for evaluation.

It will be noted that the problem-solving sequence requires different personal characteristics at different stages. No one individual will be equally strong at all the stages. This is the attraction of using group techniques in which different individuals can supply needed expertise. Thus one individual may provide knowledge and experience, another the ability to judge and analyze and yet another the divergent ability to suggest novel approaches for a solution. However, this points up the need for a leader who can ensure that the proper contributions can be made at the correct time and who balances the input from the group members so that, for example, the group is not overly evaluative.

Also indicated on the diagram are personal development methods. These methods may be employed to provide more desirable personal characteristics and also to enhance group interaction. There are many supporting techniques which may be used at the appropriate stages of the problem-solving process. However it will be noted that technique is not enough; personality characteristics and knowledge must also be available. What the technique accomplishes is the optimum structuring and utilization of personality characteristics and knowledge to provide the best solution available with the resources to hand.

A particularly good description of problem-solving techniques has been given with examples of use (Rickards, 1974). Table 16 gives examples of techniques along with subroutines. Creative problem solving is only to be used in open problem situations as distinct from closed problem situations. A distinction between the two classes of problems is given in Table 17. A flow diagram indicating the place of each class of techniques is given in Figure 23. It may be pointed out that restructuring techniques, decision aids and redefinitional procedures are often used by the individual problem-solver whilst brainstorming and synectics procedures are used in a group context.

TABLE 16

Problem-solving techniques and their subroutines

Techniques for individual problem-solving	Techniques for group problem-solving
Class 1 Restructuring techniques	Class 4 Brainstorming
1.1 Morphological analysis	4.1 Osborn's methods
1.2 Relevance systems	4.2 Trigger sessions
1.3 Attribute lists	4.3 Recorded round robin
1.4 Research planning diagrams	4.4 Wildest idea
	4.5 Reverse brainstorming
	4.6 Individual brainstorming
Class 2 Decision aids	Class 5 Synectics
2.1 Weighting procedures	5.1 Active listening/constructive group behavior
2.2 Checklists	5.2 Goal orientation
Class 3 Redefinitional aids	5.3 Itemization
3.1 Goal orientation	5.4 Changed meeting roles
3.2 Successive abstractions	5.5 Excursion procedures (speculation and analogy)
3.3 Analogy procedures	5.6 Individual synectics
3.4 Wishful thinking	
3.5 Nonlogical stimuli	
3.6 Boundary examinations	
3.7 Reversals	

TABLE 17

Characteristics of open and closed problem situations

Open	Closed
Boundaries may change during problem-solving	Boundaries are fixed during problem-solving
Process of solving often involves production of novel and unexpected ideas	Process marked by predictability of final solution
Process may involve creative thinking of an uncontrollable kind	Process usually conscious, controllable and logically reconstructable
Solutions often outside the bounds of logic - can neither be proved nor disproved	Solutions often provable, logically correct
Direct (conscious) efforts at stimulation of creative process to solve problems is difficult	Procedures are known which directly aid problem-solving (algorithms or heuristics)

FIGURE 23

DECISIONS INVOLVED IN THE SELECTION OF SUITABLE TECHNIQUES FOR OPEN-ENDED PROBLEM-SOLVING

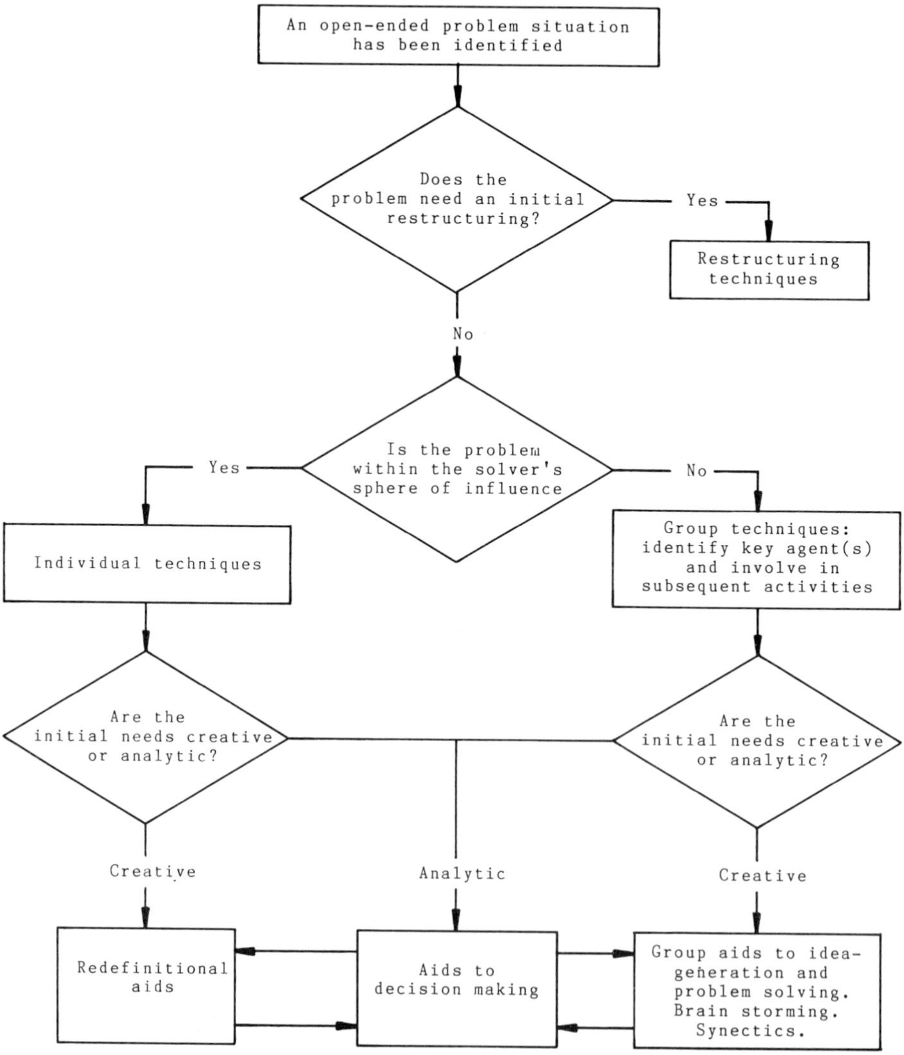

Based on an assessment of techniques Rickards states that brainstorming and synectics are most widely used followed closely by morphological analysis and then analogies/bionics and checklists (Rickards, 1980). Although it is not the intent to give detailed accounts of these techniques the connection with creativity studies will be given.

The function of brainstorming is to generate many ideas while suspending judgment and to encourage building on ideas. The precept is that with a large flow of ideas more novel associations will be obtained. Synectics relies on the stimulation of new ways of looking at a situation through the use of metaphors, similes and analogies. A metaphoric excursion may compress the incubation period prior to insight. Morphological techniques involve the structured search along all possible dimensions and allows the search to be widened. Judgment is suspended. The systematic nature of the search avoids overlooking obvious possibilities and may bring to light promising and unusual combinations. Lateral thinking techniques allow the restructuring of perceptions away from preferred patterns, assist in the suspension of judgment and encourage the imagination. Rather than being a single technique lateral thinking is a whole array of techniques.

One question to be asked is whether these techniques allow the solution of problems otherwise insoluble or for which only mundane solutions would be found. Some indication of the utility of these techniques has been given but in general the validity of these techniques remains unproven (Osborn, 1957; Rickards, 1974; Gordon, 1961).

3.8 CREATIVITY AND TESTING

In connection with creativity it is often desirable to evaluate individuals as to creative potential and for creative abilities. This may be necessary so that factors such as personality may be studied in groups differing in creativity. In a program designed to increase creativity usage it may be desirable to measure the status of individuals before and after the event. As has been discussed earlier there is disagreement over the definition of creativity and as to the weighting of cognitive and personality factors. It is, therefore, hardly surprising that there is a divergence of opinion as to whether creativity can be measured and, if it can be measured, there is disagreement as to the best method of measurement.

In the study of subjects in the area of creativity it is, in fact, essential to consider more than tests of creativity. Tests which have been used fall into four classes: intelligence tests, personality evaluation, tests of "creativity" and other tests. These will be discussed. Details on many of the tests in the first three categories are to be found elsewhere (Anastasi, 1976a). Examples of tests which have been employed in connection with creativity are widely available (Hudson, 1967; Parnes et al, 1977; Welsh, 1975d; Vernon, 1970; Haeffle, 1962; Barron, 1969a).

In the area of intelligence the variety of tests is limited since the subject will be restricted to adults. Thus the Wechsler Adult Intelligence Scale provides a satisfactory means of testing general intelligence.. This test comprises eleven sub-tests with six of these grouped on a verbal scale and five grouped into a performance scale (Anastasi, 1976b). For special populations other types of tests may be employed. Thus Barron and Welsh have used the Concept Mastery Test (Anastasi, 1976c). This test provides sufficient ceiling for the examination of highly superior adults and uses both analogies and synonym-antonym items. The Watson-Glaser Critical Thinking Appraisal, although not classed as a measure of general intelligence but rather one of a measure of separate abilities, in this case critical thinking, may also be useful (Anastasi, 1976d).

Personality tests have often been used in connection with creativity programs. Examples of such tests are the 16PF, Minnesota Multiphasic Personality Inventory (MMPI). California Psychological Inventory (CPI), Myers-Briggs Indicator (MBI), Strong Vocational Interest Blank (SVIB), Adjective Check List (ACL) and the Rorschach Test (Anastasi, 1976e). The 16PF test is based on work by Cattell who assembled personality traits from factor analysis of a heterogeneous group of 100 adults. According to Anastasi, due to the shortness of the scales reliability of scores for any single form of the 16PF is low. The MMPI is a very widely used personality inventory and covers such areas as health, psychosomatic symptoms, neurological disorders and many other areas including neurotic or psychotic behavior manifestations. The MMPI has been used widely by Barron in the study of creative individuals. The MMPI served as the basis for the CPI which Anastasi regards as one of the best personality inventories available. Again the CPI has been used by Barron in the evaluation of creative and other individuals. The MBI is based on Jungian psychology and is of the forced-choice type. The SVIB examines interests, attitudes and values and indicates preferred occupations. It may be noted that occupations that score high on one theme tend to score low on its direct opposite. Thus a high score on the Artistic dimension would go with a low score on its direct opposite, the Conventional dimension. In this regard similarities to the work of Hudson on convergers and divergers may be seen since these types preferred quite different careers. Again there is a connection with the work of Welsh who showed preferences for different types of occupation by different individuals. The ACL is employed to evaluate self-concepts such as the Need for Achievement, Dominance,Affiliation, Order and Autonomy. Other scales refer to Self-Confidence, Self-Control, Flexibility and Personal Adjustment. The connection between some of those attributes and creative ability has been given earlier in the list of personality characteristics of creative individuals. The Rorschach test is different from the above in that it is a projective test used most widely in the clinical setting. It has been employed, however, on a limited basis in the evaluation of creative subjects. Also to be included under personality testing are tests of aesthetic preference such as the Barron Welsh Art Scale. This test asks the subject to state a "like" or "dislike" for designs and figures. There are some indications that liking for less symmetrical figures is related to creative tendencies. Recently Welsh proposed a Revised Art Scale (RA) which was used to categorize subjects along a scale of origence (Welsh, 1975b).

While intelligence tests do not directly indicate the proclivity to creative behavior there is an indication that some personality tests may be used to identify creative individuals. The "creative profile" of eminent researchers on the 16PF has already been mentioned. The RA test has been used to directly identify creative individuals. However, the main use of personality evaluation has been to determine the personality characteristics of subjects identified as creative by means of other types of tests often referred to as creativity tests. It is in the area of these so-called creativity tests that there is most controversy.

TABLE 18

Guilford tests and results

Test	Task required for item	Factor Related to Test
Sentence Analysis	List all facts or assumptions contained in simple sentences.	Analysis
Paragraph Analysis	Analyze paragraph into five basic ideas.	Analysis
Figure Analysis	Pick out objects jumbled together in drawing with lines in common.	Analysis
Figure Concepts (uncommonness)	Find features in common in picture of objects. (Score as the number of uncommon responses).	Originality
Impossibilities	List things that are impossible.	Fluency
Plot Titles (low quality)	Write titles for story plots. (Score is the number of low-quality titles written).	Fluency
Plot Titles (cleverness)	Aim at cleverness of titles. (Score is the number of clever titles written).	Originality
Brick Uses (fluency)	List different uses for a brick. (Score is number listed).	Fluency
Brick Uses	Develop as many classes of uses as possible. (Score is the number of classes of uses listed).	Flexibility
Number Associations (uncommonness)	List associations for given numbers. (Score is the number of statistically uncommon responses).	Originality
Consequences Test (low quality)	List consequences of certain changes. (Score is the number of more obvious consequences).	Fluency
Consequences Test (remoteness)	(Score is the number of indirect or remote consequences listed).	Penetration
Match Problems	Take away matches and leave certain number of squares or triangles.	Flexibility
Quick Response (uncommonness)	Word associations. (Score is the number of uncommon responses).	Originality
Word Transformation	Regroup letters in series of words, without changing order, to form new set.	Redefinition
Sentence Synthesis	Make sentence out of words in scrambled order.	Synthesis

Table 18 gives the test battery proposed by Guilford (Haefle, 1962). These tests are to measure divergent ability. It is here that disagreement arises for there is little real evidence to suggest that divergency is synonymous with creativity. Hudson has emphasized that the two subjects must be clearly separated (Hudson, 1967). A similar position is taken by Nicholls after a careful study of the literature (Nicholls, 1972). Be this as it may it is clear that open-ended tests do measure something not tapped by conventional intelligence tests and as such may be useful. The link between divergent stages of problem solving and divergency in the individual may be mentioned. Furthermore the tests or variations on the tests of Guilford have been widely used in the evaluation of subjects. Thus Hudson was able to employ tests of divergency to separate subjects into convergers and divergers.

While the Guilford tests are the product of factor analysis research on the nature of the intellect, the Torrance Tests were developed on the basis of classroom experience (Anastasi, 1976d). These tests comprise three groups: verbal, pictorial and an auditory battery. Twelve separate tests in all are employed. The battery has been mostly used with school children.

In many of the open-ended or divergent tests quantity is emphasized but it is clear that quality might also be important; a smaller quantity of high quality responses (in terms of novelty and usefulness) could well be a better indicator of creative tendencies than a large quantity of mundane responses. Unfortunately this introduces a subjective element into the test scoring. It may be further remarked that the calculation of correlation coefficients has often shown a higher correlation between divergent test results and intelligence test results than between the different types of divergent tests themselves. This has led to the criticism that divergent tests are simply another type of intelligence test.

It may be noted that one of the best methods of forecasting creative behavior is on the track record of the individual. An individual who has been creative in the past will probably be creative in the future. This suggests that a non-test evaluation for example by peer reviews would be valuable in the identification of creative individuals. Incidentally this approach has something in common with the evaluation of the contribution of R & D in the firm described in Part 1. Discussions of peer and superior evaluation are available (Haeffle, 1962; Taylor and Barron, 1963). Of special interest is the procedure adopted to identify creative architects (Vernon, 1970).

With regard to evaluation of scientists and engineers in organizations formal review forms have been produced (Stein, 1982). The purpose of the techniques are twofold: to provide management with a systematic and integrated method for studying the salient and critical features of industrial research organizations that are related to the effectiveness and creativity of the people employed and to aid in the recruitment and hiring of new scientists and engineers. The techniques are as follows:

(i) The Research Environment Survey - this survey gathers facts and opinions from R & D personnel on a variety of factors related to their jobs.

(ii) The Survey for Administrators - the purpose of this survey is to collate the views and opinions of the top administrators of the R & D organization on those factors that are critical to the effective management of the organization. The results may be compared with those of the Environment Survey to determine the degree of mismatch.

(iii) Individual Qualification Form - this form allows a systematic description of a job position to be made and is an aid in recruitment.

(iv) Personal Data Form - on this form the individual is given ample opportunity to give background information, accomplishments and job preferences. The form may be used with the preceding form to fit the right person to the job.

(v) Research Personnel Review Form - the form provides the basis for the periodic review of performance of the individual scientist or engineer. On this form an attempt is made to evaluate the creativity of the individual via number of ideas, quality and novelty. However, there is also great emphasis on the aspects of idea communication, interpersonal relationships, persistence and planning ability.

The need for five techniques demonstrate that a creative individual is not enough. The individual must be able to communicate findings to others and if necessary persuade them of the worth of the approach. The organization must encourage creative behavior and must be in a position to make use of creative outputs.

In the context of testing to evaluate personality characteristics or to examine abilities, it is convenient to discuss learning styles. This area has been popularized by work on individual experiential learning (Kolb, 1976). It is theorized that learning is a four stage process which includes concrete experience, reflective observation, abstract conceptualization and active experimentation. People are not likely to place equal emphasis on each stage and the degree of emphasis can be measured using a questionnaire approach, the Learning Style Inventory (LSI). The claim is that individuals can be categorized along orthogonal learning dimensions based on continua of active-reflective orientations and concrete-abstract orientations. This leads to a four cell matrix that shows different learning styles (Kolb, 1976). Figure 24 indicates the learning model. The four types of individual have different strengths. Assimilators have the ability to create theoretical models and excel in inductive reasoning and the integrating of disparate observations. Assimilators are more interested in abstract ideas than with people and prize theoretical elegance over practicality. On the other hand, accommodators have the opposite characteristics i.e. accommodators are interested in the execution of a program. The converger is abstract and active and is the planner while the diverger is rooted in concrete experience and is reflective. The implication is that the diverger dreams up potential solutions and that the converger homes in on a solution.

It must be stated that there is considerable discussion of the Kolb model and studies on the validity and uses of the model are under way both in Europe and in the United States. Some workers have flatly rejected the model (Freedman and Stumpf, 1980; Kolb, 1981, Freedman and Stumpf, 1981). However, others have accepted the basic premise of the model but seek to determine improvements. There are certainly several areas for examination. The LSI is very simple and non-specific. It is quite possible that individuals adopt different styles when faced with

FIGURE 24

KOLB LEARNING STYLES

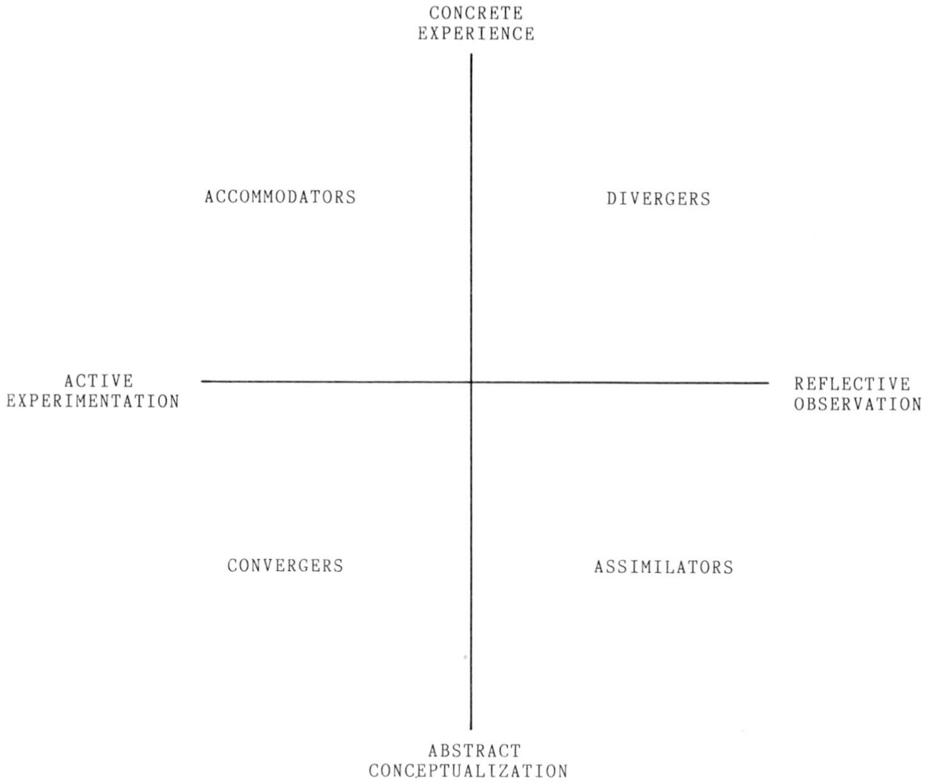

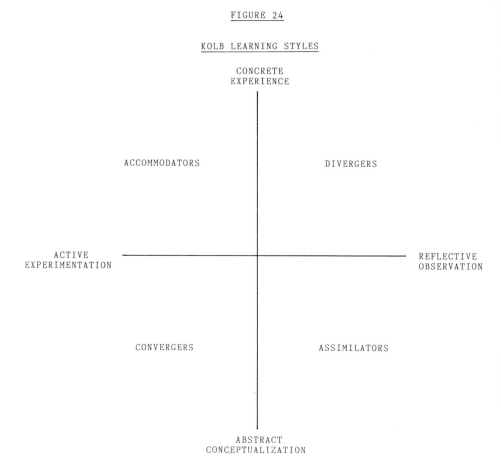

different problems, either different in type or different in context. This would not be demonstrated on the LSI. The situation is akin to that with simple tests that do not differentiate between individuals on the basis of I.Q. regarding performance. Other studies have examined the orthogonality of the axes of the model and have concluded that six dimensions rather than four are required. Despite the above-mentioned difficulties, the Kolb model is being actively used and its validity explored. The model appears useful not only for individuals but in a group context at least for stimulating discussion on learning behavior and on problem solving.

Besides the tests of intelligence, personality and "creativity", it is often necessary to employ other tests as a background. As has been remarked creativity does not take place in a vacuum and the environment can play an important role in determining whether creative potential is realized. Especially in R & D individuals work in a group or team. Thus the group interactions can determine the local climate for creativity. The overall company climate is also important especially the value that the company places on creativity and innovation. Perhaps it is not the actual climate but the perception of that climate by the individual in the R & D group that is important. In any event it is important to consider other than the individual factors in making an evaluation of creativity..

3.9 INDIVIDUAL, GROUP AND ORGANIZATION

There is a growing body of information on the influence of company climate, in general and at the local level, on individual creativity.

Organizational climate has been defined as the quality of the organizational environment subjectively perceived or experienced by the members of the organization (Litwin and Stringer, 1968). The different aspects of climate are described in terms of different dimensions. Different numbers of dimensions have been used for models depending upon the interest of the investigator. Typically the dimensions include: structure, responsibility, risk, rewards, support, conflict, warmth, and standards. In an organization measurement of the climate is usually based on questionnaire response with the questionnaire structured so that the answers reflect the status against the number of dimensions chosen. Nystrom states that, "Successful innovations therefore depend on creativity", (Nystrom, 1979b) and sees creativity as a cause and innovation as an effect. If the basic conditions for individual creativity are lacking there is nothing that the organization can do by way of compensation. However, organizational and economic factors may well act as constraints to individual creativity. Specific organizational factors are not given although it is indicated that company flexibility and decentralization are desirable.

Organizational characteristics and creativity have been extensively discussed (Steiner, 1965). Thus creative organizations encourage open channels of communication, experiment with new ideas, are more decentralized, evaluate ideas on their merits, expect risk-taking and so on. It is recommended that creativity can be encouraged by suitable reward systems in pay, recognition and channels for advancement. The individual should be given freedom of action.

Table 19 shows further examples of those factors that can lead to a release of creativity (Taylor, 1972). These factors are to be distinguished from those which constrain creativity such as lack of trust, tight control and distortion of communication.

TABLE 19

Managing for the release of creativity

Basic organizational factors	Management behavior and attitudes	Typical effects of management behavior and attitudes
Emotional climate	High trust Low fear	More impulsive, uncensored behavior Greater risk taking, more error Greater creativity and range of response Trust in own impulse and unconscious life
Communication flow	Free flow of communication Clarity Open strategy and planning	Greater spontaneity in response and feeling Greater expression of foolish, irrelevant, and seemingly meaningless ideas and behavior Greater emotionality More feedback up and down Interaction and "piggy-backing" of ideas
Goal formation	Allowing self-determination Allowing self-assessment	Reward for risk-taking Sharing and mutual stimulation of ideas Greater diversity and non-conformity More sustained creativity
Control	Interdependent, emergent and intrinsic controls residual in life processes	Experimentation with work and structure Open expression of conflict and disagreement Greater innovation Priority of diversity and creativity over conformity

*Reprinted with permission from "Climate for Creativity", C.W. Taylor, editor, Copyright © 1972, Pergamon Press.

There have been many publications especially in the R & D literature that relate similar factors to those above to performance in R & D. Many of these publications simply give lists of factors without proof of effectiveness. However, one approach did evolve a methodology capable of identifying and measuring some facilitatory and inhibitory influences on creativity in an R & D environment (Osbaldeston et al, 1978).

Osbaldeston et al employed nine dimensions for the definition of climate. These dimensions are as described below:

(i) Structure - the perceived limitations of the work environment i.e. the constraints perceived in existing rules, regulations and procedures.

(ii) Autonomy/Responsibility - the feeling of being "your own boss" and knowing that there is accountability for performance.

(iii) Reward/Recognition - the feeling of reward for a job well done, an emphasis on positive rewards and perception of fair pay and promotion policies.

(iv) Risk/Tolerance of Conflict - the position as regards conflict resolution, the feeling that different opinions may be permitted; there is a sense of calculated risk and challenge.

(v) Warmth/Support - the feeling of working in a supportive atmosphere with cooperative attitudes, mutual trust and confidence between all levels and friendly relationships within work groups.

(vi) Standards/Pressure - the perceived importance of implicit and explicit goals and performance standards; the emphasis placed on doing a good job and the challenge represented in personal and group goals.

(vii) Identity/Belonging - the feeling that you belong to a respected organization and are a valuable member of a team; individual identification with and commitment to group goals.

(viii) Information/Communication - perceptions of the adequacy of information provision and exchange throughout the organization.

(ix) Management Style - the extent to which management is thought to consult staff and involve them in decision-making; the confidence held by subordinates in the style of management practiced by their boss.

An instrument was designed based on the above dimensions and a questionnaire was administered to 175 R & D staff. The responses were used to identify that group of personnel who felt they had the best climate for creativity. It was pointed out that of all the dimensions Management Style probably played the most important role in defining the right motivational conditions for creative work. It may be noted that in response to questions on the barriers to the exercise of creative behavior, 73 per cent of all the respondents stated that such barriers existed citing workloads, time pressures and deadlines as the most frequent reasons.

It is not the intent here to investigate the area of climate in detail but rather to point out that climate can influence creativity and so must be examined at the same time that individual creativity is studied. It should be pointed out that exercises in the area of creative development can influence company climate either in perceptual changes of the participating individuals or because of outside perceptions due to the more creative working of the group.

3.10 THE INFLUENCING OF CREATIVITY

The examination of the behavior of creative individuals is interesting and the identification of creative persons is useful but what would be of great importance would be the improvement in creative behavior for persons not classed as creative. This subject has received some attention. At the outset it must be admitted that there is little evidence as to the effectiveness of work in this area. The persons advocating creativity training are enthusiasts who tend to employ anecdotal evidence or those whose experiments are poorly controlled (Vernon, 1970).

The view of Stein is interesting (Stein, 1974d). It is believed that creativity can be stimulated if individuals are helped to become more like those persons who are known to be creative. This view rests on the fact that there is much knowledge of the psychological characteristics associated with creativity and that there is potential for change in the individual. For example, it has been learned that creative individuals are self-confident and so an encouragement of self-confidence might be expected to facilitate creativity.

Apart from creativity stimulation by the personality approach there are other ways in which enhancement of creativity usage may be achieved. Some of these methods such as the use of alcohol, caffeine and the mind-expanding drugs will not be discussed here other than to note that the intent in these approaches has been the "raising of the level of consciousness" leading to freer association of ideas; the results have been equivocal (Stein, 1974d). As mentioned creativity usage may be blocked by environmental factors. Thus improvement in the environment can make it supportive of creativity.

Much is known about the creative process so it might be expected that an improvement in creative behavior would result by teaching the individual to follow a sequence likely to promote the optimum usage of creative potential. According to Parnes a considerable part of creative behavior is learned. Attention to perceptual blocks can improve creative behavior. Examples of blocks are difficulty in isolating problems, difficulty from narrowing the problem too much, inability to define or isolate attributes and failure to use all the senses in observing. The use of deferred judgment is emphasized (Vernon, 1970).

Many of the techniques suggested for use in problem-solving can be used to improve the usage of creativity by the individual and this effect may carry over away from the problem solving setting. It is conceivable that the use of techniques such as brainstorming, synectics, attribute listing and so on with group involvement could lead to the attainment of more novel and useful solutions than would have been achieved by more traditional routes. In the industrial setting, at least, the question of whether or not the creativity usage of the individuals has increased in an absolute sense is of little importance; what is important is whether or not more creative solutions are obtained than would be obtained without the techniques.

There are a few studies concerning the improvement in individual creativity through special training. Barron describes some of the approaches used and has given the results of a program designed to make teachers more creative (Barron, 1969c).

The approach taken was to address the psychodynamics of the individual. The study involved a retreat in a secluded spot and initially the subjects were tested using the MMPI and CPI.

Interviews were made concerning life history and career. To measure creativity the Barron-Welsh Art Scale and tests from the Guilford battery were given. Detailed description of the program was not given but it involved discussions of creativity, unconventionality, achievement motivation and later feedback and discussion of test scores. Participants were tested again some ten months later at the end of the program and "significant gains were registered on all the measures of performance hypothesized to be related to creativity". The program did not benefit all individuals equally; some individuals showed marked changes whilst others showed no change at all. The degree of change was related to the personality profile of the individual. An individual must want to change to benefit from such a program.

Other information on creativity programs is given by Parnes (Parnes et al, 1977). At the State University of New York at Buffalo there is now a highly structured program in the area of creativity and evidence has been presented as to the effectiveness of this program. These results and a detailed outline of the program itself are given in the "Guide to Creative Action" (Parnes et al, 1977). In general what is claimed is that creative problem-solving students show substantial gains in the quantitative production of ideas and superiority on the qualitative production of ideas over control groups. Students showed substantial improvements in dominance but not in self-control or need to achieve.

The time required to bring about changes in subjects in the creativity area has been examined. In general, it appears that six months is the minimum time period necessary. Some programs run for a year or even two years. It is important to examine the lasting effects of a program since it is possible that subjects will return to their original state unless there is follow-up.

3.11 ASPECTS OF CREATIVITY PROGRAMS

Many books and articles cover creative processes, testing

methods and other aspects of creativity. However, there are few accounts of how to apply the techniques. In a classical approach subjects may be tested for creativity, then involved in a creativity enhancing program and subsequently be retested to determine whether the usage of creativity had been influenced by the program. From a scientific viewpoint such a study is attractive but there are many difficulties in applying this approach in an R & D setting. There is no direct test of creative abilities and so a range of tests is necessary. Creativity testing alone is insufficient since the creative behavior of individuals is influenced by factors such as motivation and organizational climate.

Although there are conceptual and methodological problems in the carrying out of a test-program-retest approach in an R & D setting there are still reasons why a creativity program is attractive. The creative approach i.e. the search for novel solutions is obviously of importance in R & D projects leading to more innovative products. In addition, this approach may be employed to improve interpersonal relationships and to minimize interface problems between R & D and other company units such as marketing and manufacturing. In this context it must be remembered that not every problem requires a creative solution. Tried and trusted methods may be successful in the solution of many problems. Most organizations are not equipped to handle a constant flow of creative solutions. Thus, if the creative approach is developed, it should be used as a resource only in those situations where a novel solution is desired or where mundane solutions are inadequate. Even if the problem requires a creative solution it should not be tackled unless the problem is under control of the problem owner. An advantage of the creativity approach is that a more novel, and hopefully more successful, solution may be obtained and, furthermore, since the solution is reached by the group through a process of involvement, the commitment to adopting the solution is all the stronger.

A core program in the area of creativity may consist of "creativity" tests and retests following exercises to enhance flexibility of thinking.If trained personnel are available then the MMPI or the CPI may be used. Failing this the 16PF serves as an indicator of personality characteristics. Intelligence (IQ) may be measured using the WAIS. Divergent-thinking tests are employed to measure fluency; the results may be measured for quantity but an added dimension is obtained if the quality of the output is sought. Divergency tests measure fluency in the areas of words, associations, ideas, expressions and sentences (Guilford 1964; Wilson et al, 1954). Other tests examine the ability to solve difficult problems (Fixx, 1978), and the Watson-Glaser Critical Thinking Test is also useful (Watson and Glaser, 1964; Nyfield et al, -). As mentioned earlier the Kolb learning style questionnaire may be used to indicate preferred learning styles (Kolb, 1976). The Kolb test has been widely used to describe the R & D process. Since the literature indicates that motivation is important for the usage of creativity, it is recommended that a need-achievement (n-Ach) test be employed. A simple test is available in which ten items measure achievement motivation and seven items comprise a "carelessness" scale (Smith, 1973; Opolot, 1977). An organizational climate measure should also be carried out. For example, the survey developed at the University of Michigan at Ann Arbor is designed around the dimensions of organizational climate, supervisory leadership, peer leadership, group process and satisfaction; there are ample data indicating the norms for different organizational settings with this instrument.

Besides the above tests which are well established, there are other evaluations either under development or in limited use. For example, a questionnaire representing a first attempt to determine tendencies towards left brain (analytical) or right brain (divergent) thinking may be obtained from Dr. T. Rickards of the Manchester Business School. A creativity audit is also available (Rickards and Bessant, 1980). This is an organizational survey with an emphasis on the creativity element at the local level and at the organizational level. The instrument has been used in an R & D context and also in process research and production areas. A questionnaire, Past-Present-Future, is available from D. R. Talbot of the University of Manchester Institute for Science and Technology. The individual is given an opportunity to examine himself, his job, the company and the business environment at three times (one in the past, at present and one in the future) and to contrast his actual achievements and status with the ideal. The responses may be used to develop constructs for the group and also to discuss any individual problems.

Just as a whole spectrum of tests are needed so there is no one outcome from a creativity program. The series of events in a typical program is shown in Figure 25. Table 20 gives a list of tests and evaluations. The initial testing phase, which covers many aspects besides creativity, is used to classify individuals in the R & D group as to strengths and weaknesses. From the

FIGURE 25

CORE CREATIVITY PROGRAM WITH SPIN-OFF BENEFITS

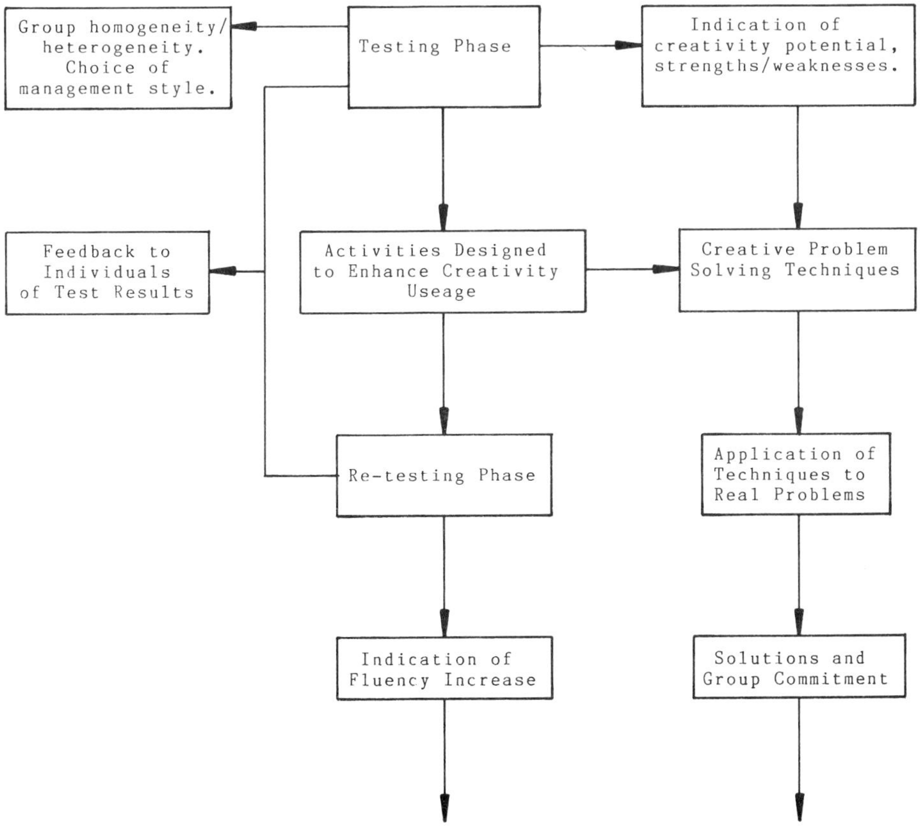

results the group may be classified on a homogeneous-heterogeneous dimension; this classification has been shown to have implications as far as group creativity and management style are concerned. Besides the identification of a group, individuals will benefit from feedback of the results from the testing phase. Sharing of group results, without necessarily giving the identification of individual results, can lead the group to appreciate the diversity of resources available and can lead to a fuller group interaction. Participation of the supervisor in the program can give rise to an improved group/leader relationship.

TABLE 20

Outline of the tests and evaluations for a core creativity program

Name of test of evaluation	Purpose
MMPI, CPI or 16PF (A form)	Personality profiles
N-Ach	Need-achievement measures
Watson-Glaser (Ym)	Critical thinking ability
Creativity audit	Individual and organization profile
Past-present-future	Identify key constructs
Kolb learning style	Individual learning style
Left brain/right brain	Divergent vs. convergent style
Intellectual test	Ability to solve problems
"Creativity" test series	Divergent thinking ability
Mutual creativity measure	Identify mutual creativity ratings
Self-creativity measure	Self-rating of creativity

The program as outlined may be used to teach creative problem solving techniques. In these techniques it is arranged that conditions are optimized to give the likelihood of creative solutions. Creative problem solving with model problems may be used to heighten the usage of creativity in the group but once expertise has been obtained these same techniques may be used on real problems. An advantage in using group problem solving is that, apart from the fact that more novel solutions may be obtained, there is the benefit of greater commitment to the implementation of the solution. This is especially important if the implementation of the solution involves greater effort and more action than the group is normally inclined to take.

In summary, a core creativity program involves the testing of participants for fluency and other factors, a set of activities that are designed to enhance creativity usage and a second testing at the conclusion of the formal program (it is intended that activities will be ongoing). The results are of interest in seeing whether divergency of thinking can be enhanced by appropriate training. However, the other benefits of the program are of greater interest as far as the management of the R & D resource is concerned. The advantages of the program may be listed as follows:

(i) testing of the participants not only for fluency but also for other factors gives information on the group makeup, which can be used to select the best management style. Knowledge of the skills available in the group allows the best allocation of projects and can be used in the formulation of training programs to round out areas where skills are lacking.

(ii) results from the testing phase may be fed back to individuals and are of help in identifying special skills, problems and so on that the individual can build on or work to correct as necessary. There is, therefore, the opportunity for individual growth.

(iii) the sharing of results among group members can heighten awareness of group differences which can lead to the contribution of different individuals to the solution of a problem rather than having a problem the possession of a single individual (although a single individual should still have overall responsibility for the problem).

(iv) techniques of creative problem solving may be learned and these activities may be used to provide conditions for creativity usage. However, the techniques are useful in their own right and may be applied to real problems once some expertise has been attained. The use of group techniques to solve problems likely gives a greater commitment to the implementation of the solution.

There are many program activities that can be used between the test and re-test phases. The activities undertaken depend on the time available, the philosophy of the program and the resources available. The use of one specialized trainer, for example, may lead only to an emphasis on one particular technique such as Rolfing or Synectics. It may be desirable to concentrate on one technique but only after the R & D group has been exposed to a variety of techniques; the group preference and the type of problem on hand can then guide the selection of a particular technique on which to concentrate. Examples of activities that have been found useful in this regard are listed below:

(i) Description of the personalities of creative individuals including the manner in which discovery was attained.

(ii) Distribution of literature on creativity. Lateral thinking techniques provide a good beginning since these consist of a wide variety of methods (de Bono, 1969; de Bono, 1977).

(iii) Training in techniques such as brainstorming, synectics and circular response.

(iv) Feedback of the results from tests such as those on divergency, personality, I.Q., motivation and climate.

(v) Miscellaneous activities such as movement, color, painting/Gestalt Art, sculpture making, awareness and relaxation. The purpose of these techniques is to make the individual more aware of stimuli and hence more open to originality.

Setting up a creativity program requires the acceptance of the aims of the program by the R & D group. Some attention to strategies is, therefore, required in light of the types of resistance that can be expected (Zaltman and Duncan, 1977). There will certainly be resistance from individuals who might feel threatened, especially those individuals who have a poor self-image in the area of creativity. There will be special resistance to the sharing of results in a group setting. There might also be a genuine belief that creativity enhancement was not necessary.

A multiple strategy should be employed. This may be demonstrated by way of example for an actual case involving an R & D group. A facilitative strategy was first employed along with a re-educative strategy. Some two months before the start of the program the subject of creativity was introduced at group meetings. Literature on creativity was distributed and discussed. The need for creative approaches in R & D were emphasized and the successful use of problem solving techniques demonstrated. Discussions were held about creative individuals. Information was given to show that the usage of creativity could be enhanced by training. A different approach was taken with each member of the group. One member was interested in the use of creativity techniques for children and literature on this subject was emphasized showing how the proposed program materials could be used at home. Another member was interested in creativity from an intellectual point of view and those aspects of the proposed program were emphasized in discussions. A third member had previously worked through a book on lateral thinking and the connections between this approach and the proposed program were emphasized. The fourth member of the group was afraid that he was low on creativity and here the discussions centered on blocks to creativity and it was pointed out that his creativity would increase once those blocks were removed. In the above conversations the Director of R & D acted as the change agent while the R & D group was the target system. A re-education strategy was used to show that there was a problem with creativity in that creativity usage could be improved. The facilitative strategy was used to show that there were activities that could be employed to increase creativity usage. However, segmenting the group and using biased arguments, as well as the fact that only successes in the creativity area were cited, meant that a persuasive strategy was also employed (Zaltmann and Duncan, 1977).

Obtaining group acceptance for the program was not sufficient because there was still resistance to the testing phase. The head of R & D therefore decided to become part of the target system. To do this it was necessary to use a third party who would become the change agent and a management consultant was hired. The consultant had scientific as well as management credentials and was therefore acceptable to the R & D group. The hiring of a consultant had other benefits as outlined below. It may be noted that the inclusion of the Director of R & D in the group for the program may be viewed as an implicit power strategy since if he was willing to go through the testing phase and the program, it was unlikely that any other group member could refuse. Although this eliminated overt resistance it is possible and perhaps likely that covert resistance was present.

Various aspects of arranging group procedures have been detailed elsewhere (Stein, 1975). Where testing is involved the participants are anxious to find out how the tests were constructed. In addition there is usually a desire that the data be anonymous and that there will be feedback of results. There is also the question of whether the results will be circulated in the organization, for example to the personnel department. The use of an outside consultant was useful in answering these concerns. Thus the consultant was able to discuss the background and basis of each test. Each participant was given an identifying code letter known only to the participant and the consultant. Where individual results were presented to the group there was no identification other than the code employed. Each individual was given full feedback on his test results along with an interpretation by the consultant. The Director of R & D obtained all the test results but only identified by code. The results were not passed to anyone else in the organization.

It is recommended that criteria for the evaluation of a program be selected before the program is begun (Stein, 1975). One criterion used was the performance of participants on the tests of fluency and flexibility since this gave a numerical rating of change. A second criterion was the achievement of solutions to the R & D problems discussed earlier. Thirdly the participants were asked to give their opinions on the program and whether or not it should continue.

As part of the agreement to hold the program it was necessary to decide on the length of the program. A long program involves much commitment whilst a short program is unlikely to achieve worthwhile results. It was decided that initially a six-month program would be used and that if the results were worthwhile an extension of six months would be made. From discussions with those involved in creativity programs elsewhere it appeared that six months was the shortest period over which changes would be obtained.

3.12 SUMMARY

Creativity is highly prized. A working definition of creativity is that it is a process which produces a novel product which is found to be useful by a significant group. The need for industrial innovation has placed a great emphasis on the area of creativity especially in new product development and in the R &

D area. Creativity fits into the spectrum of activities from Dreamer to Mandarin discussed earlier. There are ample references in the scientific literature and in the management literature as to the need for creativity.

It must be admitted that the term creativity has been abused. Originally creativity was taken to refer to works of genius but recent usage has led to a dilution of the term. A scale of creativity may be imagined. However industry only demands relatively better novelty than the competition. In fact, many industries could not cope with a continuous stream of highly creative products.

Creativity has been examined in many ways. Biographical and introspective examinations have been especially used with the eminent. Studies have also been made on the eminent and the not so eminent using personality and cognitive viewpoints. There is considerable disagreement over the relative importance of personality and cognitive factors, but it is clear that both are required for an individual to be creative. Mere creative potential is not enough and studies have shown that a high degree of motivation and persistance is required for an individual to perform creative tasks. Characteristic personality profiles have been found for eminent scientists for example. Environmental factors also play an important role and may facilitate or inhibit creativity. However the environment can do no more than encourage the creative potential which exists. Theories of creativity indicate at least four stages to the process; preparation, incubation, insight and verification. However it must be emphasized that there is a later communication stage which is very important in ensuring that the creative product is brought to the attention of the significant group which must make the determination as to its usefulness.

The connection between intelligence and creativity has been extensively examined. It appears that there is a minimum level of intelligence necessary for a person to be potentially creative in any given field. However above a given level of intelligence specific to that field there is probably no benefit as far as creative behavior is concerned. Above this level of intelligence personality factors such as persistance and motivation play an overwhelmingly important role.

Consideration of intelligence and creativity as two independent dimensions has led to a four-fold classification of individuals; this approach has been used in several studies. The results indicate that individuals can be characterized as belonging to one of the four classifications and that there are differences for individuals in the different classes as to personality factors, preferred occupations, levels of creativity and behavioral characteristics. Other studies have led to the classification of individuals as convergers or divergers. Convergers prefer scientific endeavors and it has been argued that creative scientists come from the divergent part of the convergent population. Similarly, divergent individuals prefer the arts and it may be argued that the more successful divergers are those having some of the convergent characteristics such as persistence.

Creative problem-solving is part of the broader area of creativity. Studies of the problem-solving process have indicated that there are several stages and that each stage requires individuals having different ability. Alternatively an individual must learn to switch between different types of behavior for the different stages of the problem-solving process. Different aids or techniques have been identified as being especially useful at different stages of the process. As distinct from individual creativity problem-solving often employs group techniques in which it is hoped that added expertise may be available. Industry is most interested in creative problem-solving due to the need for novel solutions to many problems.

Testing to identify the creative individual has taken place at many levels. Four categories of tests appear to be necessary to cover various aspects associated with creativity. These aspects are intelligence, personality, creativity and other aspects such as the examination of organizational climate. Strictly speaking there is no such thing as a creativity test and tests designed to measure "creativity" are open-ended or divergent tests which measure flexibility and fluence. It has been emphasized in the literature that divergent ability is not the same as creative ability. Divergent tests do appear to measure characteristics not usually tapped by the normal intelligence tests.

The question as to whether the creative ability of individuals can be influenced is of great importance. It would appear that individuals have a certain creative potential but in many cases this potential is not utilized due to various types of blocks. It appears reasonable to assume that the level of creativity usage can be increased. Many of the characterists of the creative personality are known and so by encouraging behavior similar to that of the creative person creativity usage may improve. In addition the various stages of the creative process are well established and so structured aids may be employed. For example, attribute listing may be used in problem formulation, brain-storming may be employed in divergent stages of problem-solving and decision aids used in convergent problem-solving stages. These structured aids provide a setting in which creativity usage may be maximized. It has to be stated, however, that there are very few documented reports of improvement on an individual basis or on a group basis via the structured techniques. Improvements have been reported but most of the evidence is of an anecdotal nature. The area of creativity is still in a state of flux and no one method can be suggested which will optimize the usage of creativity by individuals and groups. However any method which does enhance creativity usage and leads to the formulation of more novel solutions and products is to be encouraged.

REFERENCES

Abell, F.D.	1980	"Defining the Business", Prentice-Hall, Englewood Cliffs, N.J. Chapters 3 and 9.
Anastasi, A.	1976a	"Psychological Testing", Collier MacMillan, New York, 4th edition.
Anastasi, A.	1976b	Ibid, Chapter 9.
Anastasi, A.	1976c	Ibid, Chapter 10.

Anastasi, A.	1976d	Ibid, Chapter 13.
Anastasi, A.	1976e	Ibid, Chapters 17-20.
Barron, F.	1969a	"Creative Person and Creative Process", Holt, Rinehart and Winston Inc., New York, Chapter 4.
Barron, F.	1969b	Ibid, Chapter 9.
Barron, F.	1969c	Ibid, Chapter 12.
Beckhard, R. and Harris, R.T.	1977	"Organizational Transitions: Managing Complex Change", Addison-Wesley, Reading, Mass. p. 4
Brown, M.	1979	"Management Right and Left", Management Today, September, 67-69.
Davis, G.A.	1973	"Psychology of Problem Solving", Basic Books, Inc., New York.
Day, G.S.	1975	"A Strategic Perspective on Product Planning", J. Contemporary Business, Spring 1-34.
de Bono, E.	1969	"The Five Day Course in Lateral Thinking", Penguin Books, Ltd., Harmondsworth, Middlesex, England.
de Bono, E.	1971	"Lateral Thinking for Management", McGraw-Hill Ltd., London.
de Bono, E.	1977	"Lateral Thinking", Penguin Books, Harmondsworth, Middlesex, England.
de Bono, E.	1980	"Opportunities: A Handbook of Business Opportunity Search", Penguin Books, Ltd., Harmondsworth, Middlesex, England.
Dessler, G.	1976	"Organization and Management", Prentice-Hall Inc., Englewood Cliffs, N.J., 328.
Eysenck, H.J. and Kamin, L.	1981	"Intelligence: The Battle for the Mind", Pan Books, London.
Fixx, J.	1978	"Solve It - A Perplexing Profusion of Puzzles", Frederick Muller Ltd., London.

Freedman, R.D.	1980	"Learning Style Theory: Less Than Meets the Eye", Academy of Management Review, 5, #3, 445-447.
Freedman, R.D. and Stumpf, S.A.	1981	"The Learning Style Inventory: Still Less Than Meets the Eye", Academy of Management Review, 6, #2, 297-299.
Furst, C.	1979	"Origins of the Mind: Mind-Brain Connections", Prentice Hall, Englewood Cliffs, N.J.
Gordon, W.J.J.	1961	"Synectics: The Development of Creative Capacity", Harper and Row, New York.
Guilford, J.P.	1964	"Progress in the Discovery of Intellectual Factors" in "Widening Horizons in Creativity", C.W. Tayler (ed.), John Wiley, New York.
Haeffle, J.W.	1962	"Creativity and Innovation", Reinhold Publishing Corp., New York, Chapter 2.
Hudson, L.	1967	"Contrary Imaginations", Penguin Books, Ltd., Harmondsworth, Middlesex, England, Chapter 6.
Kijewski, V.	1978	"Market-Share Strategy: Belief vs Actions", Pimsletter, Cambridge, Mass.
Kingston, W.	1977a	"Innovation", John Calder, London, p.16.
Kingston, W.	1977b	Ibid, 30.
Kolb, D.A.	1976	"Management and the Learning Process", California Management Review, XVIII, #3, 21-30.
Kolb, D.A.	1981	"Experiential Learning Theory and the Learning Style Inventory: A Reply to Freedman and Stumpf", Academy of Management Review, 6, #2, 289-296.
Litwin, G.H. and Stringer, R.A.	1968	"Motivation and Organizational Climate", Division of Research, Harvard Business School, Cambridge, Mass.
Mintzberg, H.	1976	"Planning on the Left Side and Managing on the Right Side", HBR, 54, July-August, 49-58.

Nicholls, J.G.	1972	"Creativity in the Person Who Will Never Produce Anything Original and Useful", American Psychologist, 27, August, 717-727.
Nyfield, G., Saville, P., Field, J. and Hodgkiss, J.	-	"British Supplement to the Watson-Glaser Critical Thinking Appraisal (Form Ym)", NFER Publishing Co., England.
Nystrom, H.	1979a	"Creativity and Innovation", John Wiley and Sons, Chichester, England, Chapter 6.
Nystrom, H.	1979b	Ibid, 38.
Opolot, J.A.	1977	"Reliability and Validity of Smith's Quick Measure of Achievement Motivation Scale", British Journal of Social and Clinical Psychology, 16, 395-396.
Osbaldeston, M.D., Cox, J.S.G. and Loveday, D.E.E.	1978	"Creativity and Organization in Pharmaceuticals R and D", R and D Management, 8, #3, 165-175.
Osborn, A.	1957	"Applied Imagination", Charles Scribners Sons, New York.
Parnes, S.J., Noller, R.B. and Biondi, A.M.	1977	"Guide to Creative Action", Charles Scribners Sons, New York.
Perkins, D.N.	1981	"The Minds Best Work", Harvard University Press, Cambridge, Mass.
Ravenscraft, D. and Scherer, F.M.	1981	"The Lagged Impact of R and D on Profits", Revised Draft Pimsletter, Cambridge, Mass.
Rickards, T.	1974	"Problem Solving Through Creative Analysis", Gower Press, Epping, Essex, England.
Rickards, T.	1979	"A Reappraisal of Creativity Techniques in Industrial Training", J.E.I.T., 3, #1, 3-8.
Rickards, T. and Bessant	1980	"The Creativity Audit: Introduction of a New Research Measure During Programs for Facilitating Organizational Change", R and D Management, 10, #2, 67-75.

Rickards, T.	1980	"Designing for Creativity: A State of the Art Review", Design, 1, #5, 262-272.
Schlicksupp, H.	1977	"Idea Generation for Industrial Firms - Report of an International Investigation", R and D Management, 7, #2, 61-69.
Smith, J.M.	1973	"A Quick Measure of Achievement Motivation", British Journal of Social and Clinical Psychology, 12, 137-143.
Steele, L.W.	1975	"Innovation in Big Business", Elsevier, New York, p. 188.
Stein, M.I.	1974a	"Stimulating Creativity", Academic Press, New York, Vol. 1, p. xi.
Stein, M.I.	1974b	Ibid, Chapter 2.
Stein, M.I.	1974c	Ibid, pp. 6 and 16.
Stein, M.I.	1974d	Ibid, Vol. 1, "Individual Procedures".
Stein, M.I.	1975	"Stimulating Creativity", Academic Press, New York, Vol. 2, "Group Procedures".
Stein, M.I.	1982	Forms and instructions from Science Research Associates, Inc., 259 East Erie Street, Chicago, Illinois.
Steiner, G.A. (ed.)	1965	"The Creative Organization", The University of Chicago Press, Chicago, Illinois.
Taylor, C.W. and Barron, F. (eds.)	1963	"Scientific Creativity: Its Recognition and Development", John Wiley and Sons Inc., New York.
Taylor, C.W. (ed.)	1972	"Climate for Creativity", Pergamon Press, New York.
Twiss, B.C.	1980	"Managing Technological Innovation", Longman, London, Chapter 3.
Vernon, P.E.	1970	"Creativity", Penguin Books Ltd., Harmondsworth, Middlesex, England.
Watson, G. and Glaser, E.M.	1964	"Watson-Glaser Critical Thinking Appraisal", Harcourt, Brace and World, New York.

Welsh, G.S.	1975a	"Creativity and Intelligence: A Personality Approach", Institute for Research in Social Science, Chapel Hill, North Carolina, p. 5.
Welsh, G.S.	1975b	Ibid, Chapter 2.
Welsh, G.S.	1975c	Ibid, Chapter 4 and Afterword.
Welsh, G.S.	1975d	Ibid, Chapters 2 and 3.
Whitfield, P.R.	1975a	"Creativity in Industry", Penguin Books, Ltd., Harmondsworth, Middlesex, England, Chapter 1.
Whitfield, P.R.	1975b	Ibid, Chapter 2.
Whitfield, P.R.	1975c	Ibid, Chapters 3 and 4.
Wilson, R.C., Guilford, J.P., Christenson, P.R. and Lewis, D.J.	1954	A Factor-Analytic Study of Creativity Thinking Abilities", Psychometrika, 19, 297-311.
Zaltmann, G. and Duncan, R.	1977	"Strategies for Planned Change", John Wiley and Sons, New York.

PART 4

THE R & D PROJECT PROCESS

4.1 INTRODUCTION

The previous sections have dealt with various aspects of R & D. The nature of R & D activities has been examined and it was concluded that there were many activities only some of which could be classed as true R & D; in this latter category is the R & D project. The project is the means by which new knowledge, products or processes are generated and it is at the core of the R & D operation. In the classical picture the structure of the organization is determined by the strategies adopted and the positioning of R & D in the company is determined by such factors. Especially for R & D, innovation is routinely demanded. Thus the nature of creativity is of extreme importance. The purpose of the present discussion is to focus on the project process. It is assumed that the company has a well defined organization and a coherent strategy and that R & D is regarded in a rational fashion by the company as a legitimate activity.

The process by which a project is accepted for action by the R & D group is outlined in Figure 17. The R & D strategy is determined by general strategies and long-range plans, by factors such as the nature of the environment and short-range objectives such as profits, sales growth and so on. The project selection process normally takes place each year at budget time unless there are urgent reasons for modification of existing projects or the introduction of new projects. As has been mentioned earlier it may be decided that a particular goal may be more readily achievable via product license or acquisition rather than by internal development using R & D.

It is assumed that there are more projects than can be funded and that some form of project selection is mandated. This may be an unwarranted assumption, however, since in a survey most companies questioned believed that it was more of a problem to find good projects than to select from a project pool (Gee, 1971). There is, therefore, an emphasis in two areas. First the ways in which new projects may be generated and second the manner in which projects are selected. It would seem that most attention has been directed towards project selection and selection methods range from the intuitive to tightly structured mechanical procedures. Many of these evaluation methods depend upon financial assessment and so it will be necessary to outline the financial area. Other methods depend on ranking or decision-tree analysis and a brief treatment will be given. Techniques for generating projects will be discussed in detail. Technological forecasting may be employed to give an overall view of technological developments and may be used to give "alternative futures". Creativity techniques may also be covered. Structured methods for new project (or product) development have also been recently introduced.

Selection of a project depends, in part, on cost and time estimates. The project definition is much more than a title but involves a plan. This is necessary for resource allocation such as personnel, supplies, services and equipment and also to integrate the activity into the entire product development process. For example, the design group must know when to expect a new material, the production department has to anticipate the provision of facilities to produce the design and marketing/sales must have advance notice of the time of product introduction. The provision of a plan also provides a basis for project control. The progress of the project can be judged against milestones achieved and resources employed.

The successful project implies, at some stage, the delivery of a product. There must be continual contact between R & D and other company units such as marketing and manufacturing. These interface interactions will be dealt with in the next chapter as will the management of the R & D group.

4.2 THE GENERATION OF PROJECTS

4.2.1 Technological Forecasting

Within every company some estimate of the future is implicitly made. Obviously it is preferable that the forecast be deliberately sought, carefully formulated and specifically stated. There are many different environmental factors that must be monitored and forecast but with the high rate of technological change in many industries technology is the most potent force and so technological progress must be anticipated. Technological forecasting (TF) has emerged to fill this role.

The basic problem in TF is making decisions today with incomplete and uncertain knowledge about the future. Although TF concentrates on likely technological developments it must be remembered that the forecast has to be put in context. Social, political and economic factors have to be considered; these factors can influence the direction of technological developments by aiding or opposing a particular development or by biasing developments towards alternative futures. This has the important implication that changing the present to fit the image of the desired future is preferable to projecting the present along the direction of logical development; the former weights desirability over mere feasibility. The range of TF techniques that has been developed and tested in the industrial setting enables the analysis of such issues as (Wills et al, 1972):

(i) the market-pull for a particular technology,

(ii) the theoretical potentials and relevance of the technology,

(iii) the rate of progress of the technology,

(iv) the impact on the company and on society.

R & D planning and budgeting has traditionally been an area of uncertainty. TF has an obvious place in providing information as to the relevance of the technologies employed or under development.

The use of TF is not a panacea, however, as no technique is without drawbacks. The limitations of TF are as follows:

(i) there may be unpredictable interactions between technologies which give rise to totally unexpected developments,

(ii) unprecedented demands may arise necessitating the development of new technologies that may themselves shape future demand in unknown ways.

(iii) the discovery of new phenomena. Examples include the transistor, laser and genetic engineering technologies.

(iv) inadequate or biased data. Due to high cost there is a limit to the amount of data that can be collected; time constraints also place a limit to the data acquisition process. The data to hand may be biased. Companies often unconsciously bias data by neglecting to collect data that "is not relevant".

There are preconceived notions of "what business are we in" and wishful rather than actual futures.

Figure 26 indicates a sequence for the organization of TF in a business. There are two major areas of techniques; extrapolative methods use the existing knowledge base and project this into the future while normative forecasts are based on perceived future needs. Morphological analysis does not fit readily into either the extrapolative or normative categories; normex reconciliation is employed to provide consistency between extrapolative and normative forecasts. These techniques will now be briefly described. Further details will be found elsewhere (Jantsch, 1967; Jones and Twiss, 1978; Ayres, 1969; Wills et al, 1972).

Extrapolation has been practiced on a wide scale even before the advent of TF as a formal procedure. There are several reasons for this. The technique is deceptively easy to understand and it is based on present knowledge. The projection of the present into the future gives a feeling of confidence because the forecast is based on "fact"; since the transition from the present to the future is smooth there is a sense of continuity. All of these comments emphasize the dangers of using unevaluated extrapolation as the basis for planning the future.

Extrapolation has made use of three types of data for TF:

(i) functional capabilities: this type of extrapolation is carried out in terms of direct performance terms such as speed or in terms of a figure of merit. It is important that the variable chosen is not directly linked to a particular technology otherwise the forecast will be constrained to that technology.

(ii) capability of specific technologies: here the functional capability of a specific technology is explored. It may be noted that extrapolation of several specific technologies may be plotted on the same graph, where the technologies are linked in the sense of serving the same function. The envelope gives the overall functional capability trend.

FIGURE 26

A SEQUENCE FOR THE DEPLOYMENT OF TECHNOLOGICAL FORECASTING IN A BUSINESS

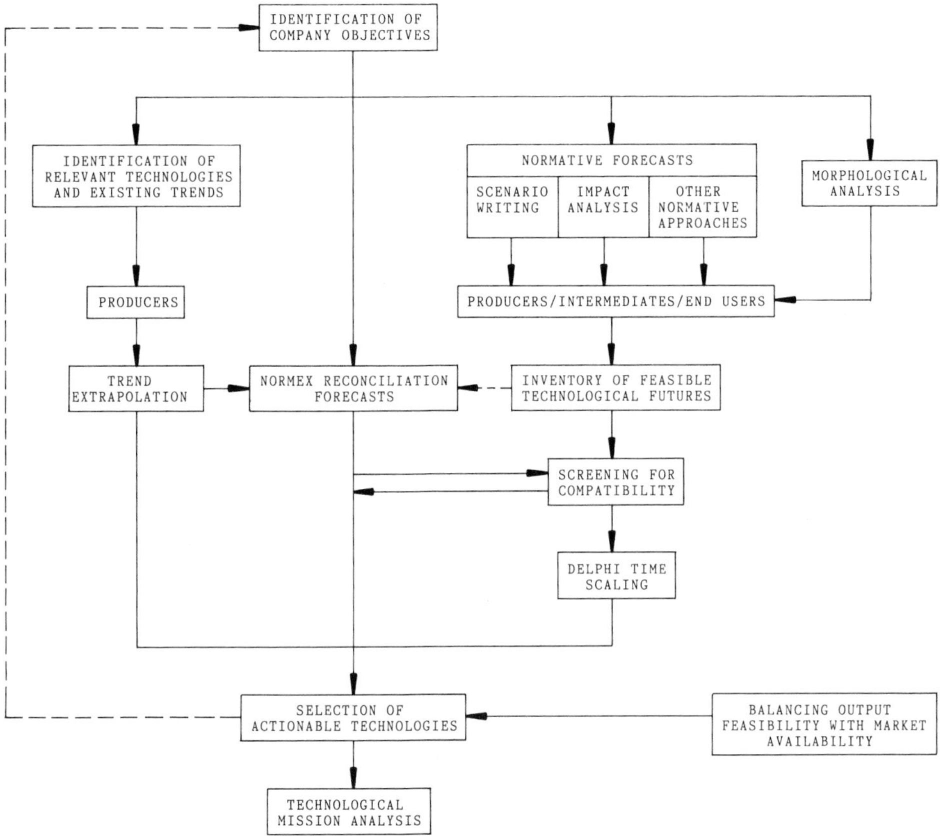

Fig. 26 *Reprinted by permission of Penguin Books Ltd.

(iii) scientific and technical findings as yet unrelated to (i) and (ii).

Figure 27 shows the four main classes of trend curve (Wills et al, 1972). Figure 27(a) shows a linear increase with flattening. Examples are the growth in efficiency of thermal power plants, the effect of mechanization in decreasing labor hours and in the growth of production materials. The most widely encountered curve is that of exponential growth with or without a constraint during the time period under review as depicted in Figures 27(b) and (c). The third main class of curves is the double exponential with subsequent flattening. Examples are found in the early phases of R & D development; early success leads to the allocation of substantial funds and other resources allowing a further acceleration of the development (Figure 27(d)). The fourth class is the slow exponential with sudden increase and eventual flattening (Figure 27(e)). The sudden

FIGURE 27

THE FOUR MAJOR CLASSES OF GROWTH CURVE

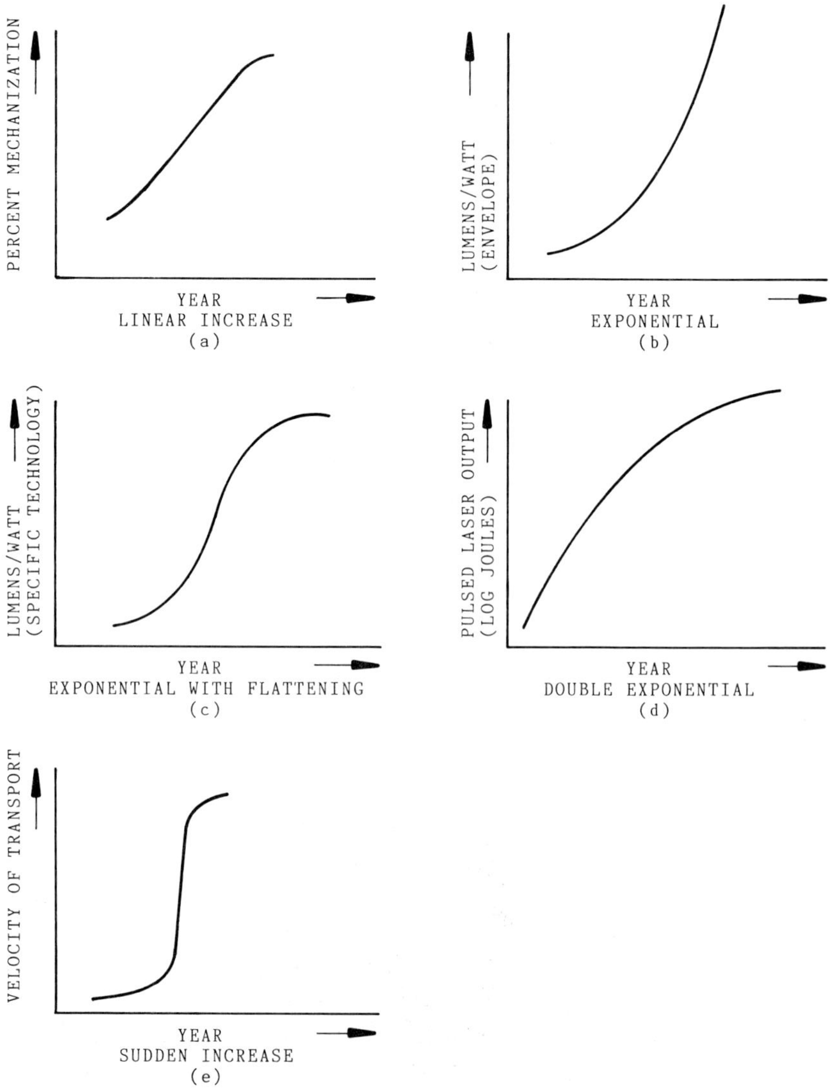

acceleration of performance often occurs when a new technology appears. This appearance may be a scientific breakthrough that occurs "by chance" or may be a consequence of vast expenditures e.g. the trend of explosive power versus time received a sudden increase with the perfecting of nuclear devices.

It should be noted that extrapolation techniques are the only ones that give a time scale for the developments. The time scale is a natural consequence of the extrapolation of historical data. Other techniques may forecast future developments but the time of development must then be determined. The time scale feature of extrapolations may also be used where there is a precursor to a particular technology. At least up to the mid-1960's there was a lag of about ten years between the maximum speed of combat aircraft and the maximum speed of civil aircraft. Thus it would have been possible to predict trends in transport aircraft speeds from the knowledge of the speed-time curve for military fighters. From the point of view of technology the trend could continue but would ultimately be limited due to problems such as aircraft surface heating; increased speed of itself is not an advantage past a certain point due to difficulties with pilot reaction time and limiting 'g' forces on the pilot and structure. From the civil aircraft viewpoint the transfer of technology from the military to the civil area in terms of aircraft speed virtually ceased due to market pressures for cost effectiveness rather than speed; higher speed civil aircraft development has been delayed due to environmental factors.

<u>Morphological</u> analysis is a technique used for identifying and counting all possible means to an end at any level of abstraction or aggregation. Zwicky was responsible for the popularization of the technique and gave the following rules:

(i) the problem to be solved or the functional capability to be achieved must be stated with great precision,

(ii) the full characteristic parameters must be identified,

(iii) each characteristic parameter must be subdivided into distinguishable cases or states; alternatively ranges or regimes may be used,

(iv) there must be an applicable method for classifying the output and for choosing combinations which give the required performance; the feasibility of combinations must also be evaluated.

Figure 28 shows the method as applied to jet engines operating in a pure medium containing three simple elements only and being activated by chemical energy (Wills et al, 1972). Eleven characteristic parameters were identified giving (2 x 2 x 3 x 2 x 2 x 4 x 4 x 4 x 3 x 2 x 2) or 36,864 distinguishable combinations. Elimination of three cases which were self-contradictory left 25,344 combinations. Other analyses indicated that at least one important parameter, having three states, had been omitted; this would multiply the number of combinations by a factor of three. The difficulties of evaluating this number of combinations is readily apparent.

It is interesting to point out the connection between morphological analysis and the area of creativity. One way in which creativity has been approached was as the combination of known ideas which gave a novel result. Morphological analysis is a means for mechanistically combining known elements. The resulting combinations can then be screened for novelty. However, as demonstrated above the number of combinations is exceedingly large. This emphasizes the true gift of the creative person; there must be a directed search involving only those combinations likely to give a novel result.

Despite the difficulties of evaluation, morphological analysis has been used for a wide range of studies ranging from textile products, fertilizers, ocean transport, cigarette lighters to military applications.

<u>Normative</u> approaches to TF begin with a statement of need and then seek to identify the way to satisfy that need. The major normative techniques include scenario iteration, impact analysis and relevance-tree analysis/contextural mapping.

Scenarios attempt to develop a logical sequence of events to indicate how a given goal may be achieved. The approach requires considerable imagination since the influence of many different factors must be taken into account. By using the technique, interactions between different forces - technology, market,

FIGURE 28

MORPHOLOGICAL ANALYSIS OF A JET ENGINE

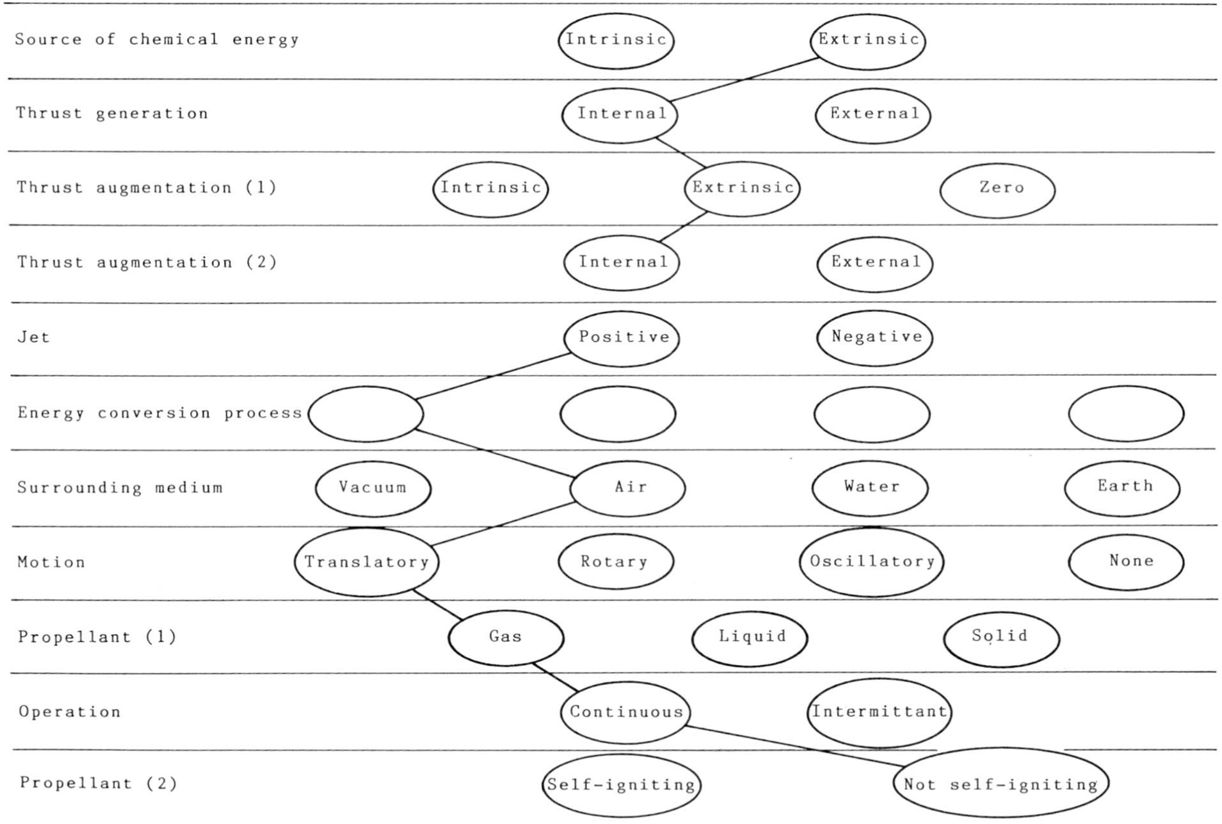

social and so on - are recognized. Gaming involving several teams can be employed to modify the scenarios by iteration. Traditionally, scenarios have been used in war games and to develop strategies for political situations. Scenarios have been employed to a lesser extent in the technology area.

Impact analysis looks at the influence of a technological advance on society. This may lead to the modification of the goal sought or to a series of developmental steps which are adjusted to alter the impact of the new technology as desired. For example, the impact of holography could lead to the development of a radar system giving the air traffic controller a three-dimensional view of all the aircraft in the area. Microscopes could be developed that would allow the three-dimensional display of protein molecules; this would greatly aid in the development of new drugs.

Relevance trees are a checklist approach to demonstrate the relationship of elements in the development of an aggregate level of performance. Figure 29 gives a simplified form of a relevance tree. Starting with a goal such as the development of pollution-free road transport the problem is broken down into a series of alternatives at different levels of hierarchy. For example the goal may be achieved by using an electric car which is powered by a fuel cell of type A, B, C or D. The relevance tree shows all possible paths to the objective and may be used as the basis for forecasts of costs, time scale and probability of success. Contextural mapping is a by-product of the relevance analysis since it enables the business to identify its position and that of current technology in relation to a particular technological objective.

Normex reconciliation is a technique that allows the integration of normative scenarios with the extrapolation of technical capabilities. The technique facilitates the establishment of R & D objectives that will meet market requirements at some specified future time. A measure of both the mean and the variance of technological parameter forecasts is obtainable thus allowing the derivation of the risk associated with the predictions.

FIGURE 29

RELEVANCE TREE

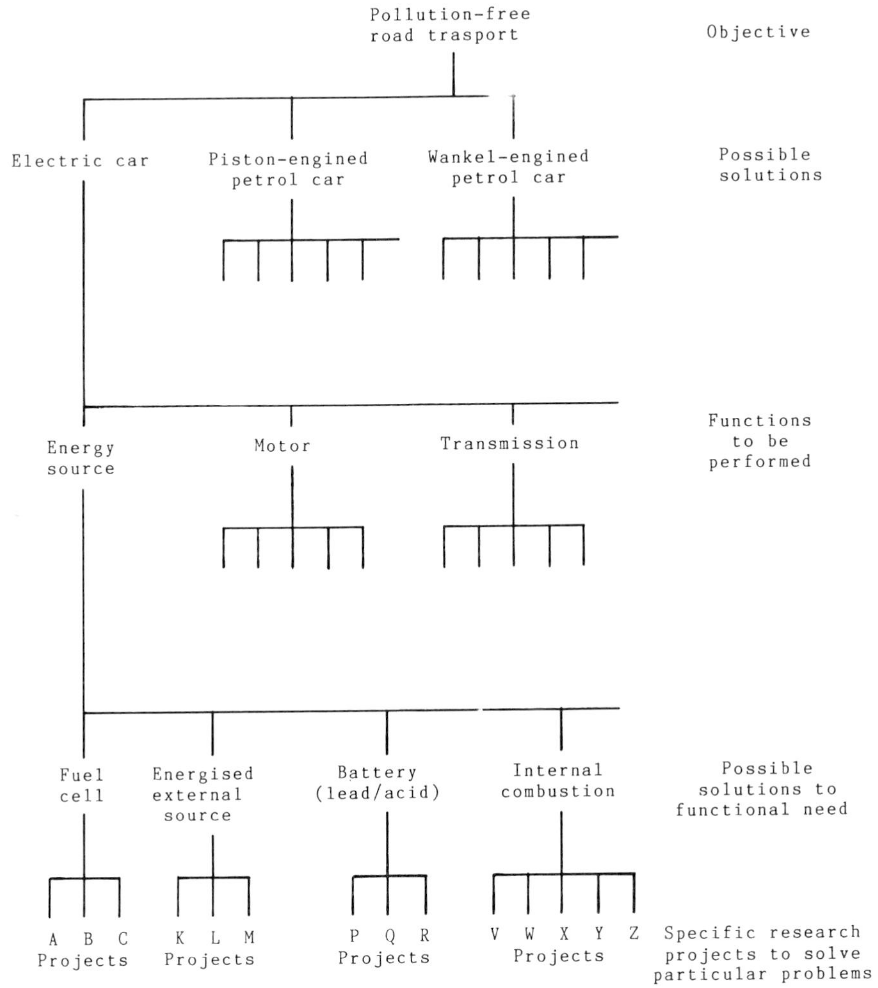

*Reprinted by permission from "Managing Technological Innovation", 2nd ed., by Brian Twiss, Longman Group Limited, publisher.

The Delphi technique is a method which uses expert opinion to forecast future events and the timing of these events. The method begins with a scenario which is distributed to a panel of experts who may or may not be at the same location. Three rounds of forecasting are typically undertaken:

(i) Panel members are asked to place the 50/50 probability of the realization of the scenario within specific time periods such as in ten year intervals in the next fifty years. The answer "never" is permitted or "greater than fifty years".The answers are plotted and divided into quartiles.

(ii) In round two the members are informed of the results of the first round and dissenters are asked to explain their position and/or to modify their forecast. Dissenting scenarios may be circulated.

(iii) panel members are informed of the outcome of round two and asked to proceed as previously. It will be found that reasonable consistency will be achieved. If not further rounds may be undertaken.

The Delphi method has attractions and drawbacks. It seems to be logical to ask experts in a particular area to forecast the future but great advances often occur by developments outside a particular field. The expert has a tendency to take what is known and to extrapolate it into the future. The forecasts can, therefore, be seriously in error. There are also the questions of weighting of the opinions by the more influential members of the panel, especially in face-to-face situations. The way in which the scenarios are phrased is also critical to the outcome.

4.2.2 Creativity and Structured Techniques

Techniques developed to aid in creativity can be used for the generation of new projects. As far as budgeting and approval is concerned, it is simply not enough to define a project in terms of the objective; sufficient detail must be provided to allow resource requirements to be determined and likely timescales to be specified. The techniques to be described allow potential solutions to be identified and this allows a preliminary plan to be laid out. In addition, the techniques can be employed for idea generation. In this case, starting with a general problem such a market need, ideas can be generated for potential solutions; this is then used as the basis for project definition. In the literature the techniques are presented either as creativity techniques or as idea-generation methods. There is obviously overlap and techniques such as morphological analysis can also fall into the TF area.

Table 21 gives a classification of methods of idea generation developed at Battelle in West Germany (Schlicksupp, 1977). Also shown are a variety of modifications for each technique. It may

TABLE 21

Classification of methods of idea-generation

A. Brainstorming and its variations	D. Methods of creative confrontation
A1 Classical brainstorming	D1 Classical synectics
A2 Anonymous brainstorming	D2 Synectics-conference
A3 Didactic brainstorming	D3 Visual synectics*
A4 Destructive-constructive brainstorming	D4 Analysis by stimulating terms*
A5 'And-also'-method	D5 BBB-method* (= B8)
A6 Creative collaboration technique	D6 Force-fit-game
	D7 TILMAG-method*
A7 Buzz-session (discussion 66)	D8 Neighbor-field-integration*
	D9 Semantic intuition
A8 Imaginary brainstorming	D10 Forced-relationship
A9 SIL-method*	D11 Catalogue-technique
B. Brainwriting and its variations	**E. Solution by systematical structuring**
B1 Method 635	E1 Morphological box
B2 Brainwriting-pool*	E2 Functional analysis
B3 Idea-Delphi*	E3 Attribute listing
B4 Idea-card-collection	E4 Morphological tableau
B5 Idea-engineering	E5 Sequential morphology*
B6 Collective-notebook-method	E6 Problem solving-tree
	E7 Process analysis
B7 Trigger-technique	
B8 BBB-method* (= D5)	
C. Methods of creative orientation	**F. Methods of systematic problem-specification**
C1 Heuristic principles	F1 Progressive abstraction
C2 Loosening the problem field	F2 Epistemological analysis
	F3 K-J-method
C3 Bionic	F4 N-M-method
	F5 Hypothesis-matrix*
	F6 Relevance-tree

* These methods have been developed from Battelle research work.

TABLE 22

Grouping of methods of idea-generation according to main features of methodical approaches

Name of group	Main features of the methods of the group
A. Brainstorming and its variations	Unrestricted discussion with deferred judgment; freewheeling phantasy and spontaneous associations are welcome
B. Brainwriting and its variations	Spontaneous and associative writing down of ideas on slips or special forms; mutual exchange of the forms
C. Methods of creative orientation	Search is guided by various different heuristic principles; also in order to look at the problem from a new point of view
D. Methods of creative confrontation	Stimulation of ideas by transmitting structures, terms, phenomena etc. to the problem, which at first do not seem to have any relationship or relevance
E. Solution by systematic structuring	Decomposition of the problem into sub-problems; solution of sub-problems and re-integration to an overall solution; systematic display of all possible ways to solve a problem
F. Methods of systematic problem specification	Revealing the crucial relations within a problem or problem field by systematic and hierarchical structuring approaches

TABLE 23

Definition of the elementary problems constituting complex problem-solving processes

Type of problem	Description
Analysis problem	Perceiving of structures, causes, relations regularities, patterns, etc.
Search problem	Finding out material or immaterial structures, with equivalent or similar characteristics
Constellation problem	Bringing together structures or parts of structures which result in a new, valuable constellation (Gestalt)
Selection problem	Determining which element out of a plurality shows in respect to a set of criteria the highest degree of satisfaction
Consequence problem	Problem solving according to the inherent laws and principles of the problem situation

be noted that Japan developed its own problem-solving techniques, the N-M and K-J methods, using analogies to stimulate ideas in a manner similar to synectics. Table 22 shows how the methods of idea-generation are grouped according to the main features of the approaches. Table 23 indicates how problems may be classified; the methods of idea-generation are best used for particular types of problems as shown in Table 24. As part of the study of methods of idea-generation, extensions to the known techniques were made. This led to the development of techniques such as idea-Delphi, visual synectics, semantic intuition and many others (Schlicksupp, 1977).

A review of creativity and problem-solving techniques from the perspective of the United States listed many of the methods given above (Souder and Ziegler, 1977). However, other techniques are also listed. For example, checklists can be used in which the problem is analyzed against a prepared list of challenges until an idea is sparked. The checklist can consist of the following: How can we magnify, modify, substitute, reverse or combine it? How can we make it look like something else, animate it, take it literally, make it a parody or an imitation? How can we give it convenience of form, time, place, quantity, readiness, combination, automation or selection? What am I trying to accomplish? Have I done this before? How? Could I do this another way? What if I do the opposite? What if I do nothing? What about shape, size? What if reversed, inside out, upside down? What else can it do? What can be left out? What if carried to extremes? Checklists work best when applied to existing products.

TABLE 24

Allocation of methods of idea-generation to different types of problem

	Methods recommended	
Problem type	Primary choice	Secondary choice
Search problem	Methods of groups A and B	Methods of group D
Analysis	Methods of group F	Methods of group E
Constellation problem	Methods of group D	Methods of groups A, B and E

TABLE 25

The whats and whys of the four major techniques

Techniques	Operational mechanisms ('what')	Precepts (links with theory - 'why')
Brainstorming	1 Generate many ideas 2 Avoid evaluation while generating ideas. 3 Seek new combinations ('hitchhike, freewheel')	Fundamental precept - postponement of judgment overcomes habitual responses. Wide range of ideas increases efficiency of idea search. When used in a group may have positive effect on interpersonal relationships and climate.
Synectics: Gordon (1960s-70s)	1 Seek ways of making the familiar strange and the strange familiar 2 Use metaphors and analogies to assist the process	Metaphoric state has been identified as important during the creative process. A metaphoric excursion may compress the 'incubation' period prior to insight.
Synectics: Prince/Gordon (1970s)	1 Identify a range of problem definitions 2 Separate process tasks (group leader) and content decisions (client) 3 Encourage positivity to ideas via 'itemized response' -client notes positive aspects of an idea -client and group try to overcome negative points	Importance of motivating the client has been recognized. Also - value in permitting client to concentrate on seeking new insights in a positive, supporting environment.
Morphological chart	1 List possible dimensions that together describe a system being studied 2 List alternatives in each dimension 3 Examine as many combinations or subcombinations as possible 4 List any promising and unusual new ideas suggested	To widen the area of search to avoid overlooking obvious possibilities. To postpone selection (judgment) until a wide set of alternatives have been listed.

Table 25 (continued)

The whats and whys of the four major techniques

Techniques	Operational mechanisms ('what')	Precepts (links with theory - 'why')
Lateral thinking (a) Random stimulus	1 Sample any rich set of random stimuli ('walk through Woolworths, Dictionary, Science Museum') 2 Seek relationships with your problem needs	Restructures perceptions away from preferred patterns. Enriches content of solution set.
Lateral thinking (b) Concept challenge	1 Consider in depth any important statement usually taken for granted 2 Challenge (signal 'PO' to denote non-evaluation) in all ways possible	Assists suspension of judgment. Helps escape habitual thinking.
Lateral thinking (c) Intermed-impossible	1 Move from a realistic idea to an imaginative impossible one 2 Treat as stepping stone to new realistic idea	Encourages the imagination. Weakens censorship mechanisms of repressed ideas.

Table 25 lists the four major techniques from a British perspective (Rickards, 1980). Also given are the operational mechanisms or the manner in which the technique attempts to stimulate the original idea. Synectics as expounded by Gordon seeks to make the familiar strange and the strange familiar; the use of metaphors and of analogies, especially from the biological field, is an especially rich source of ideas. Lateral thinking uses a range of techniques from sampling a set of random stimuli (paging through the dictionary) to the challenging of assumptions and the moving from a realistic idea to an imaginative impossible one. There is an extensive literature available regarding lateral thinking techniques. Most recently lateral thinking has been applied to business opportunities (de

Bono, 1980). This has been done in terms of an opportunity space which includes all the changes, decisions and choices that can be made. Connected with the opportunity space are idea-sensitive areas which are opportunity-rich areas; in an idea-sensitive area it is worthwhile to look for new ideas since a good idea can make a considerable difference. In other words ideas have considerable leverage in an idea-sensitive area.

Further information on creativity techniques and structured aids to innovative thinking is also available in an industrial context (Whitfield, 1975). A technique not covered so far is the Fundamental Design Method conceived by Matchett as "a highly disciplined form of thinking which a person acquires by a carefully organized and painstaking study of the controls which he already imposes on his own thought processes". The components of the technique include thinking with outline strategies, thinking in parallel planes, thinking from several viewpoints, thinking with concepts and thinking with basic elements. The exercises and methods have been assembled by others into systematic approaches to problems e.g. Problem Analysis by Logical Approach System (PABLA) developed by the United Kingdom Atomic Energy Authority. Most of the techniques have been applied to classical design problems but many of the elements are applicable to more general problems.

Of special note is the technique SCIMITAR (Systematic Creativity and Integrative Modelling of Industrial Technology and Research). SCIMITAR is a structured product development tool that has been used for about ten years (Carson and Rickards, 1980). The technique covers all stages of new product development in the following stages:

(i) Opportunity Search - this consists of three-dimensional modelling, establishment of corporate entrepreneurial groups, industrial search processes, locating new-product ideas and administration of the new-product search stage.

(ii) New-Product Development - initial experimentation, initial commercialization, product development and optimization and administration and interface issues.

(iii) Commercialization - pre-launch activities, new-product launch, commercial development and new-product transfer.

Of these stages, it is the Opportunity Search stage that concentrates on the generation of new product ideas via three dimensional modelling and creativity techniques. Figure 30 illustrates the model: the dimensions are market (M), raw materials (RM) and processes (P). Modification to the dimensions for different types of industries may be made. The axes are sub-divided as convenient. The result is a three-dimensional array of cells: each cell represents an existing product or business unit. Opportunity searches using the model include the following:

(i) Inside Search - the search is carried out within the confines of the modeled space with the purpose of identifying opportunities within the current capacity of the organization to perform. Raw materials, processes and markets are all well known. The output will be extensions to the current range of products.

(ii) Perimeter Search - the search is carried out to explore extensions to the business. The further out from the perimeter the higher the risk. Search may be made to extend one dimension or in more than one dimension; changes in more than one dimension represent a more drastic change in company activities.

It will be noted that the three-dimensional modelling technique is a simplified form of morphological analysis in which

FIGURE 30

THE THREE DIMENSIONS FOR

SCIMITAR MODELLING OF AN ORGANISATION'S RESOURCES

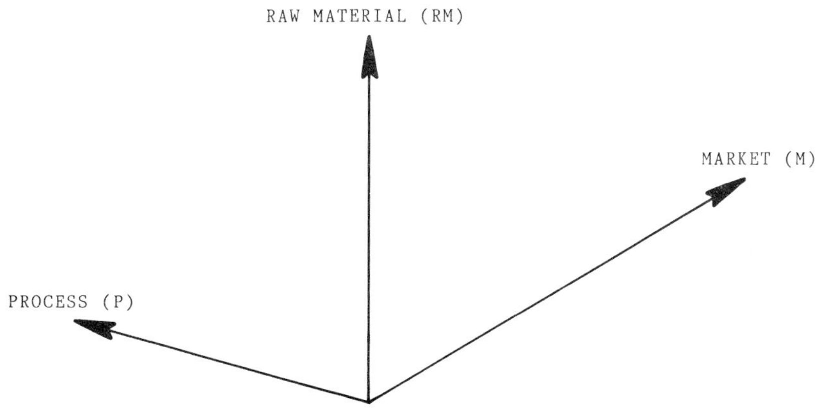

the market is considered as well as the technical characteristics of the products. Restriction of the number of dimensions reduces the number of new product ideas generated but the number may still be in the hundreds. An interesting feature of the SCIMITAR technique is the associated use of attribute listing, technical forecasting and creativity techniques such as brainstorming or synectics. Carson and Rickards give many examples of the actual use of the SCIMITAR method. For example a five-man new ventures team spent twelve months surveying the industrial fields of fertilizers, pigments, building materials, industrial cleaning, animal feeds, effluent treatment and fire retardants. In the twelve-month period 97 preliminary investigations were made. These eventually led to 36 new products submitted for customer evaluation, 26 products were approved, 12 commercialized and 4 were still selling two years later.

4.2.3 Other Methods

There are several other ways in which new projects can be generated. The R & D organization itself may be responsible for the generation of new projects either because of a technological breakthrough or by a scanning of the scientific and patent literature or by personal contacts at scientific meetings. Projects arrived at in the above fashion would tend to be of the technological push variety and a market need, either existing or latent, would have to be demonstrated. Some organizations allow R & D to spend a certain fraction of effort, say 10 percent, on research that is not specifically directed in the hope that projects leading to specific products will arise from this work.

Projects may also arise from top management directives although it is more likely that such directives would emphasize areas of research rather than specific projects; the latter would be the case in the situation of competitive breakthrough in a particular product or product line. Specific programs would then be endorsed by top management. Top management also endorse specific programs that are appealing either in terms of profits generated or in terms of guarantees of long-term survival of the firm. It is unlikely, however, that the project originated at the level of top management.

Other company units may also be approached for suggestions as to new projects. Specifically, production and marketing units should have ideas. However, the type of idea from production will center around improvements to production in terms of higher quality or lower cost; suggestions may be presented in terms of problems to be solved. There will be a different weight placed on suggestions from the production unit depending on whether the company is engaged in a product or process type of business. Requests from production usually involve near-term goals.

The marketing department should be the source of good ideas for new projects. Conventional wisdom holds that customers articulate needs that manufacturers satisfy by the development of new products. Since marketing is in close contact with customers, it should be able to indicate unsatisfied needs. Marketing should understand the strengths and weaknesses of the current products and the relative position of the competitive products. The combination of R & D input as regards technology and marketing know-how should be a powerful combination. Successful innovation requires not only a solid base of technical information but also assessments of user and market needs to spark creative ideas (Holt, 1975).

Recent research on the history of many new products has indicated that in some industries most commercially successful products are developed by product users and not by the product manufacturers (von Hippel, 1978). The fields in which this type of development occurs include scientific instruments and process machinery. The reason for this development by users is that the manufacturer provides a product which does not entirely fill the user's need and so there must be improvements and modifications made. At some stage the manufacturer may decide that the modifications provide sufficient improvement that the product should be formally produced. It should be noted that user-dominated innovation demands a different type of interface which is equipped with personnel trained to understand the user-produced innovation. The interface has been formalized, in those industries dominated by user-innovation, by the setting up of user groups, applications laboratories or custom products groups.

4.3 PROJECT SELECTION

4.3.1 General Remarks

The usual manufacturing company has a business cycle that is less than one year. Cash is used to purchase raw materials that are in turn used to produce goods that are sold. The cash generated is used to refinance the cycle. Two points may be made. The value added in the manufacturing process, higher value added in general giving higher profits, is greater for high technology products. Such products are often the result of R & D activity. The manufacturing cycle is not closed. A company producing the same products year after year will find that, at the very least, the price that the market will bear will decrease as time goes by. This is due to competitive pressures and probably because superior products have appeared on the market. This reinforces the idea that R & D, or alternative mechanisms, should provide a stream of new products on a timely basis. In high technology areas the rate of change is such that there must be high rates of introduction of new products. It may be remarked that new products can act to expand the business rather than simply to maintain the manufacturing cycle; maintenance represents merely survival while expansion represents growth. R & D may be regarded as acting on the cycle at the cash input by injecting the resource of new products.

R & D itself has a cycle. Cash is employed to fund projects that produce new product ideas. These new products are then inserted into the manufacturing cycle are sold and the cash is available to fund further projects. The difference between the R & D cycle and the manufacturing cycle is that the R & D cycle invariably takes more than one year. In fact, the cycle will depend on the success of the particular project. Different projects have different times to come to fruition (or failure) but for the average R & D organization there should be a project mix that ensures a regular flow of new products. It cannot be emphasized enough that the R & D cycle runs on cash. This cash must be generated either by the sale of an asset, for example goods sold, or by the assumption of a liability, for example by way of a short-term or long-term loan. It may be noted that the R & D budget is often tied into the success of the company because of the above factors; a company with decreasing sales or a company that is burdened by debt, and often the two go together, is unwilling to generate the cash needed to invest in R & D. This is often a mistake since the generation of new products is often the only way to reverse the situation. Short-term thinking would convince management to concentrate on reduction in expenses. This cost-cutting can lead to the elimination of R & D which is essentially a long-term mechanism for growth. Even if R & D is not eliminated entirely, cut-backs may be so severe that the mechanism for growth is emasculated and decline of the company becomes a self-fulfilling prophecy.

The manner in which the expenses of R & D in a division are charged has been suggested as follows (Gee and Tyler, 1976):

(i) R & D expense should be charged to "cost of sales".

(ii) Support of established business should be assigned directly to established products.

(iii) Exploratory research applicable to a product line should be assigned to products within that line proportional to investment.

(iv) New venture development expense of a division should be shown against the total earnings of the division rather than against individual products. Investment utilized should be excluded in computing return on investment for the various business segments and for total established business, but should be included in computing total division return.

(v) Where there is corporate R & D, the divisions usually must support the research effort. Expenses for direct product research should be shown against that product line. Exploratory research may not be specifically directed and is usually charged as a general allocation to the division.

The cost of performing research activities includes the following direct expenses (charged as spent for the current year): compensation of research personnel and fringe benefits, cost of materials, outside costs for services. Direct expenses are charged directly to projects. In addition there are indirect expenses: compensation and fringe benefits of supervisory personnel and technical management, expense of operating and maintaining the research facilities, general materials and supplies not allocated to projects. Equipment purchase represents a capital rather than an expense item since the cost of the equipment is expensed over a number of years via depreciation. Further details on accounting methods are given elsewhere (Batty, 1976).

The project selection process is influenced by the funds available for R & D, in other words the R & D budget. It is desirable that R & D funding be relatively stable but in practice fluctuations occur due to the short-term needs of other sections of the company. R & D has to take its turn in the resource allocation process and long-term needs are often sacrificed to short-term pragmatism. There are, however, a number of ways in which the R & D budget may be set:

(i) Competitive comparison: within an industry or industrial segment it is found that companies spend roughly the same proportion of funds (e.g. fraction of sales or profits) on R & D. For example, an industry such as pharmaceuticals spends a relatively high amount on R & D. It must be emphasized that the comparison must be between similar firms. In the pharmaceutical industry, a company engaged in research to produce new drugs would be expected to spend a much higher amount than a company that is a producer of generic drugs; the former company has high initial costs and expects to recoup the expenditures through premium pricing while the second company competes in a cost competitive environment and must emphasize reduction in production costs. Some information on R & D expenditures is given in Part 1. The PIMS data base currently contains the experiences with over 1700 product and service businesses operated by over 200 companies. The data include R & D expenditures and return on investment information (PIMS, 1982).

(ii) Fixed relationship to sales: this method is very often used. The advantage is that R & D spending will usually be constant, or relatively so, from year to year. The disadvantage is that present sales refer to R & D activity in the past. There is no reason to suppose that the next generation of products will cost the same to develop as the present generation.

(iii) Fixed relationship to profit: employing this type of relationship implies that R & D is an expense that is elective. It is true that if the actual survival of the company is at stake then harsh measures must be taken but cuts of R & D personnel are often irreversible. It is difficult to build up a good R & D team and severe cut-backs have a carry-over effect which can greatly hamper later efforts to start up the program again. A reputation as a "hire-and-fire" employer is damaging for many years after the situation has been corrected. It may be pointed out that decreases in profits can be due to many other causes than lack of good R & D, for example currency fluctuations. What is not often mentioned, however, is that a company having a very profitable run may tend to overspend on R & D; this may indeed be the time to spend heavily on R & D but only if the projects are there. Simply spending more on existing projects can lead to complacency and a decrease in efficiency in the R & D organization.

(iv) Reference to previous levels of spending: this method may be used alone, but is often used in conjunction with one of the other methods. For example, a budget based on a proportion of sales may have limits imposed by the previous year's spending taking into account inflation. It should be noted that actual spending rather than budget is taken as the measure. An R & D organization that underspends may be in some difficulty in persuading management that the same thing will not happen with the proposed budget. Since the budgeting process is carried out during the third quarter of the preceding year, an organization that has low spending early on but heavy spending later in the year i.e. the organization will be at budget at year end, will be at a disadvantage in this regard.

(v) With reference to projects: in this case the budget is set with reference to an agreed list of projects. However, there is likely to be negotiation as the amount estimated per project may be contested. This method of funding can lead to large swings from year to year as the number of agreed projects changes. If it is decided that new projects take the place of completed or abandoned projects then the method takes on more of a historical flavor.

In practice the level of funding is determined using several of the techniques. There must be an agreed list of projects which allows the budget desired to be specified. In addition money must be added for general activities such as programs too small to justify separate project consideration, technical service and so on. The total cost will be viewed with reference to funding in previous years. Any large deviations must be explained and the project list may be reexamined. Reduction in sales and/or profitability will put pressure on the budget. On the other hand, acceleration of R & D by competitors will give an impetus to larger R & D expenditures. Project success plays a large role in determining the credibility of the R & D organization. Past success tends to make the budget process easier and leads to fewer budget cuts. It must be emphasized that cuts in the budget can occur even after a budget has been agreed. This generally happens during the consolidation for the whole division or company where it may be necessary to reduce expenses to achieve a budgeted profitability. Cutting on a regular basis from year to year often leads to inflation of the required budget by R & D and the budget process becomes a "second guessing" procedure. The other point to be made is that the budget and the projects are often viewed as independent by management; a budget cut of say 10 percent may be made leaving the projects "as is" with respect to level of effort and attainment schedule.

There is also the question of who should evaluate the projects. From an R & D viewpoint it is necessary to obtain backing for the projects from the different company units such as marketing, manufacturing and engineering even before a formal budget presentation is made. In fact, approval for a project often is a byproduct of either a project idea from a company unit or a project arrived at jointly between R & D and a company unit. It is difficult to obtain approval for a project which does not have this grass-roots backing. Ultimately top management set the R & D projects but the concentration is on the new product timing and costs rather than on the science and technology. An R & D group that has done its homework, so to speak, and has a project list that answers company needs through a judicious mixture of technology-push and market-pull is in a good position to obtain management approval provided R & D has had good success in the past.

4.3.2 Project Funding and Rewards

The common language of business is money. It is, therefore, necessary to express major projects in terms of the funding needed and the rewards that may be expected. Figure 31 shows the typical form of the project funding-reward curve. The figure is a diagrammatic representation and the time axis can be expanded or contracted depending on the length of time needed to bring a product to market and on the time to bring the product to

FIGURE 31

PROJECT NET CASH FLOW DIAGRAM

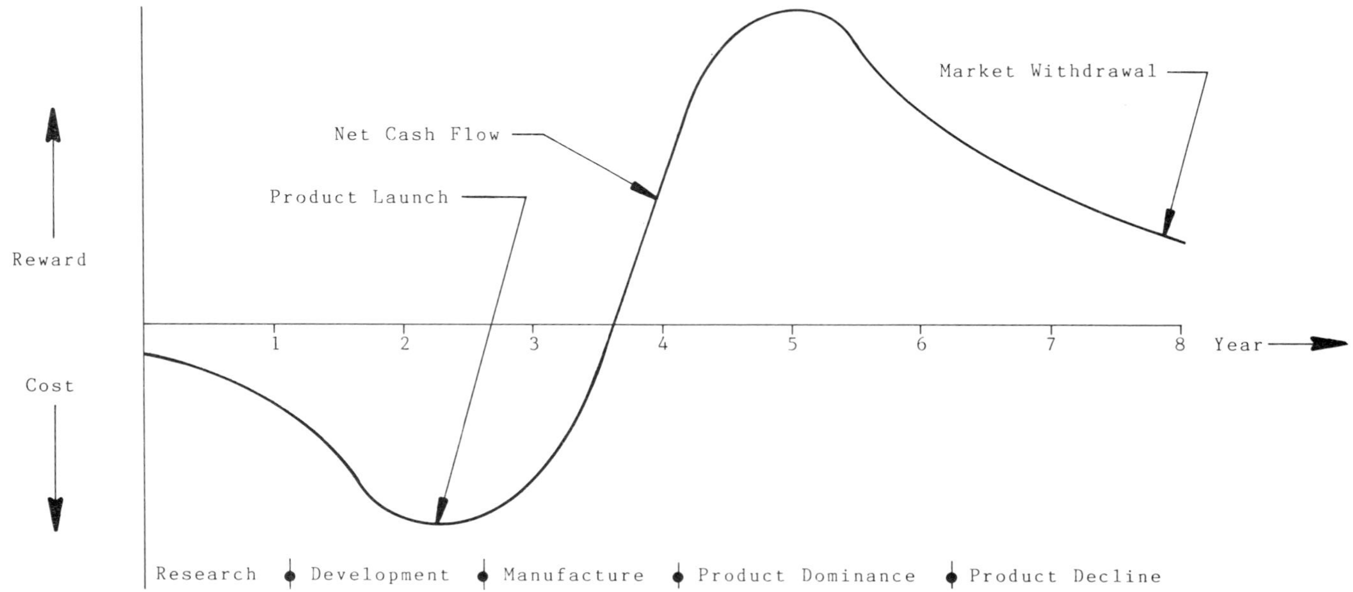

obsolescence. The relative heights of the cost and reward peaks depends on the amount of work needed during development and the success of the product in the marketplace. Early on the cost of the project is low since the studies are mainly concerned with definition. Costs increase as the project moves into the laboratory and increase further with the development of prototypes and then with development of tooling for production. Marketing, packaging and other expenses may be incurred as the product comes to be sold. Sales generate cash but there is a time lag before the cash flow becomes positive due to the continuing stream of expenditures. As the product becomes a force in the marketplace the rewards increase. At some stage the product will be overtaken by a superior product, sales saturate and then decline. The peak in the net cash flow curve need not coincide with the start of product decline since the peak represents profits. Due to the experience gained in manufacture of the product cost savings are often made (the cost-experience curve). Thus profits may increase even though sales volume is fixed. However, another factor is the selling price of the product. There will be a tendency to reduce the selling price as the competition increases and as newer products appear on the market. Reduction in selling price will offset gains from cost reduction. The benefits of product improvement may be noted. An incremental improvement is relatively inexpensive compared to development of a new product, can serve to stabilize the selling price and can extend the product life cycle.

In financial terms, the attractiveness of a project depends on the excess of cash inflows over cash outflows. However, the cash flows take place at different times. In the early years cash is invested with the purpose of a return on the investment at a later time. There are other vehicles for investment, however, that would bring a return. At the very least the money available could be put on deposit and earn interest. Thus the stream of cash outflows for the project must be compared to the stream of cash inflows taking into account the timing and the result compared to alternative investment opportunities. The first step in the process of evaluating investment alternatives is a consideration of the time value of money.

In business it is expected that money invested today will increase in amount as time passes through the earning of profit. It follows that an amount of money available today for investment is more valuable than the same amount available some time in the future. The present value of an amount that is expected to be received at a specified time in the future is the amount that, if invested today at a given rate of return, would cumulate to the specified amount. Thus the present value of $100 to be received one year from now at a rate of return of 10 percent is $90.91. In other words, $90.91 invested now at 10 percent will yield $100 one year from now. The formula for calculating the present value of a payment of $1 to be received N years hence at an interest rate i is:

$$\frac{1}{(1 + i)^N}$$

Table 26 gives the data for various values of i and for different years. It will be seen that the present value decreases as the number of years at which the payment is received increases and the present value decreases as the rate of return increases.

As an extension of the basic equation for the present value of a sum of money, the present value, P, of a stream of N

TABLE 26

Present value of $1

Years Hence	10%	20%	30%
1	0.909	0.833	0.769
2	0.826	0.694	0.592
3	0.751	0.579	0.455
4	0.683	0.482	0.350
5	0.621	0.402	0.269
6	0.564	0.335	0.207
7	0.513	0.279	0.159
8	0.467	0.233	0.123
9	0.424	0.194	0.094
10	0.386	0.162	0.073
15	0.239	0.065	0.020
20	0.149	0.026	0.005

periods of equal payments in amount R received at the end of each period is given by:

$$P = R \frac{[(1 + i)^N - 1]}{[i(1 + i)^N]}$$

The case in which there are different payments R_j in years j and benefits B_j is given by:

$$P = \sum_{j=i}^{N} \frac{B_j}{(1 + i)^j} - \sum_{j=1}^{N} \frac{R_j}{(1 + i)^j}$$

In this case P is called the Net Present Worth (NPW) of the project and the outflows R_j and inflows B_j of money are compared at time t = 0 for the assumed rate of return i. As pointed out previously the stream of costs R_j usually occurs early in the project while the stream of benefits begins at product launch and lasts for the lifetime of the product.

There are two ways of deciding on the rate of return that is appropriate: trial and error or the cost of capital. The higher the rate of return the lower the present value of the cash that is generated. Thus there will be fewer projects that will give cash inflow that meet or exceed the given value of i. If the value of i chosen leads to the rejection of many proposals that management feel are worthwhile then the value of i has been set too high. If, on the other hand, the value of i chosen results in a flood of proposals that meet the criterion then a higher value of i should be chosen. Companies often choose a rate of return based on past experience and based on comparison with other companies in the industry. The cost of capital approach assumes that the company must raise the cash to finance the project. This is done in two ways, either through borrowing or through the sale of stock (equity). The borrowings must be repaid with interest and the shareholders will expect a return on their investment. Unfortunately, it is difficult to separate the cost of capital from a host of complicating factors such as the general condition of the stock market, the investors' estimate of future earnings and dividend policy. For this reason the cost of capital is little used. In fact most companies use a judgmental approach to setting the rate of return. Few companies would use a rate less than 10 percent; higher rates such as 15 or 20 percent may be used in those industries where the profit opportunities are good. The rate of return selected will also depend on the degree of risk in the project (and uncertainty). When a bank loans money it has a high expectation of receiving back that sum with interest. The return from most business projects especially in R & D is much less certain due to the probability that the project will not be a success technically or that the product will be a failure in the marketplace. Conceptually, high risk can be allowed for by increasing the rate of return. In practice, most companies either deliberately shorten the estimate of economic life, lower the estimate of cash inflows or take the risk factor into account as a judgmental matter when the final project decision is made.

Besides the net present worth other measures are used for project worth. The total benefit-cost ratio is given by:

$$P_B/P_C$$

where P_B is the present value of benefits and P_C is the present value of costs.

The payback period is also used as a criterion of project worth and is defined as that length of time required for the stream of cash inflows received by the venture to equal the balance of the original cash outflows. There is no discounting involved. The faster the payback the more desirable the project.

When the NPW method is employed, the required rate of return must be selected in advance. The discounted cash flow (DCF) method avoids this problem by computing that rate of return which equates the present value of the cash inflows with the value of the cash outflows. Thus the rate making the net present value (NPV) zero is computed. This is called the internal rate of return (IRR). The same considerations apply to choosing an acceptable IRR as go towards choosing a required rate of return.

The return on investment (ROI) is also used. Normally there is no discounting and the net income expected to be earned on the project each year is employed. The unadjusted return on investment is given by the ratio of the annual net income to the amount of the investment.

The different economic performance criteria are shown in Table 27 with the advantages and disadvantages of each criterion. It is generally agreed that the most reliable criterion is NPW.

The difference between ROI and benefit/cost ratio should be noted. ROI is employed in industry and so the income is developed on the basis of sales predictions and consequently the profits to be generated. Benefit to cost ratios are employed for public projects. In this case the returns are not straightforward to calculate since the benefits are not easily expressed in monetary terms. Benefit/cost ratios are only used in industry for projects such as pollution reduction or an increase in safety conditions.

An example may be used to illustrate the different criteria.

TABLE 27

Economic performance criteria for projects

Criteria	Definition	Comment
NPW	Present value of benefits less present value of costs	Most reliable criterion
Benefit/cost ratio	Ratio of benefits to costs	Obscures total level of investment. Benefits difficult to calculate.
IRR	Discount rate at which benefits equal costs	May be multi-valued.
Payback period	Time for benefits to equal initial investment	No discounting. Minimum period desirable.
ROI	Ratio of return to investment	No discounting. Time value of return observed.

TABLE 28

Effect of discounting on project selection

Year	Discount Factor	Cost	Cost Present Value	Benefit	Benefit Present Value
Project A					
1	0.909	$20,000	$18,180		
2	0.826	$40,000	$33,040		
3	0.751	$40,000	$30,040		
4	0.683	$40,000	$27,320		
5	0.621	$20,000	$12,420	$40,000	$24,840
6	0.564			$50,000	$28,200
7	0.513			$50,000	$25,650
8	0.467			$50,000	$23,350
9	0.424			$40,000	$16,960
10	0.386			$30,000	$11,580
		$160,000	$121,000	$260,000	$130,580
Project B					
1	0.909	$40,000	$36,360		
2	0.826	$40,000	$33,040		
3	0.751	$40,000	$30,040		
4	0.683	$20,000	$13,660		
5	0.621	$20,000	$12,420	$20,000	$12,420
6	0.564			$40,000	$22,560
7	0.513			$40,000	$20,520
8	0.467			$40,000	$18,680
9	0.424			$70,000	$29,680
10	0.386			$70,000	$27,020
		$160,000	$125,520	$280,000	$130,880

Table 28 gives the cash flows for two projects A and B. A rate of return of 10 percent is chosen for the evaluation. The NPW of project A is $9,580 while that of project B is $5,360. Thus project A would be chosen over project B. The same decision would be made based on the discounted ROI or on the payback period. The use of the undiscounted ROI as a criterion would lead to the choice of B over A. The effect of the rate of return on the NPW is shown in Table 30 and the data are plotted on Figure 32. At low rates of return project B is preferred over project A but there is a cross-over at about 7 percent and above this rate A is the preferred project. Note that for rates of return much above 10 percent the NPW is negative. This is because the benefits are only achieved at longer times i.e. the development time is long. The IRR can be read from the graph and is 11.9 percent for project A and 10.8 percent for project B.

TABLE 29

Criteria comparison of projects A and B

	NPW	R O I		Payback (years)
		Not discounted	Discounted	
Project A	9,580	1.62	1.09	7.4
Project B	5,360	1.75	1.04	8.3

TABLE 30

Effect of return rate on project choice

	0	5%	10%	15%
Project A	100,000	44,310	9,580	- 11,380
Project B	120,000	48,740	5,360	- 20,230

It may be mentioned that for new ventures the risk is usually quite high and this is taken into account by including a probability of success into the criterion. Thus the venture ratio is defined as:

$$\frac{\text{(Probability of success)(Discounted benefits over n years)}}{\text{(Discounted cost of R \& D)}} > T$$

where T is a threshold that has to be exceeded. Typically the value of T is taken as 2.

The subject of non-quantifiable benefits has been discussed already. This leads to the idea of a social discount rate to replace the rate of return for those programs with benefits not easily expressed as monetary returns such as social programs.

There are arguments for and against low or high discount rates in this context (Gibson, 1981a). It would appear that the choice of social discount rate is judgmental.

The techniques employing discounting were developed for capital investment decisions. The type of decision here typically involves the purchase of a new machine which will turn out components faster. The cost involves the initial purchase of the

FIGURE 32

EFFECT OF RATE OF RETURN ON NPW

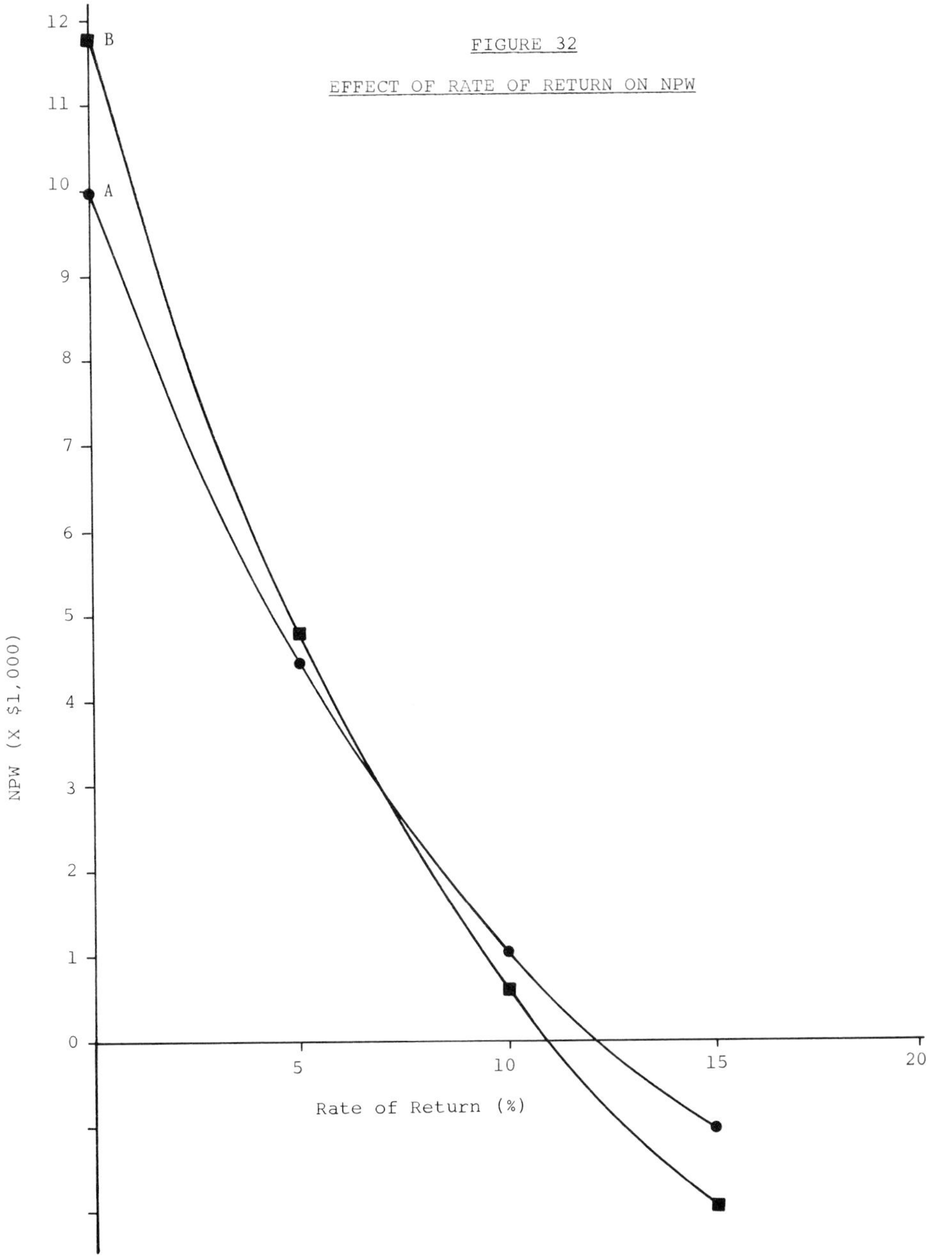

machine. The benefits are in the form of lower manufacturing cost. It will be seen that the costs and benefits are not difficult to quantify and that there is a high probability that the figures will be close to reality. The same situation does not pertain to the R & D project. There is a problem in specifying the costs, the benefits and the timing of the cash flow. Unless the project is merely an incremental improvement type of development, there is a real possibility that there will be failure either technically or in development. A successful development may be a failure in the marketplace. There are many examples of difficulties in projects some of which have been described (Twiss, 1980a). For example, the cost of the Concorde supersonic airliner had a linear escalation in R & D costs from 150 million pounds in November 1962 to 1,065 million pounds in June 1973. An examination of new projects leads to the conclusions that forecasts are almost always optimistic and that the greater the technological advance involved the larger the extent of underestimation. According to Twiss some 475 projects at the British Central Electricity Generating Board, at a large consumer products company with research in the biological and chemical field, and at an engineering process company, showed project durations 1.39 to 3.04 times the estimation and errors in cost estimates from 0.98 to 1.51. In the United States, two drug companies showed cost escalations of 1.78 and 2.11 times and time escalations of 1.61 and 2.95 respectively. It may be mentioned that studies have shown a low ability to predict technical success and that predictions of sales are particularly poor. Estimates made at the beginning of the project are apt to be highly inaccurate. Estimates improve as the project progresses, as the technical problems become more apparent, as the product becomes better defined so that manufacturing data as to costs and sales performance become clearer and because economic factors can be better assessed. The underestimation of project duration and cost is probably a subconscious action on the part of the project champion to ensure project selection and is often based on an ideal set of circumstances that do not obtain such as a smooth flow through research and through development with all deadlines met and no change in economic or market conditions. One way of improving the accuracy of forecasts is to look at the track record for projects in the past in the R & D organization. Due to the distribution of costs and durations about the predicted values for each project, the accuracy per project may not be improved but the accuracy of the total R & D costs and overall durations should be improved. For estimating project costs and duration the following are recommended (Twiss, 1980b):

(i) Devote resources to estimation procedures.

(ii) Study past performance.

(iii) Analyze the causes of error.

(iv) Do not yield to pressure to reduce cost or time estimates.

(v) If possible, obtain independent forecasts to minimize personal bias.

Even when great care is taken there will still be errors in the forecast and it should be realized that:

(i) The absolute value of the figures is questionable but the figures may be used to compare projects in a relative fashion.

(ii) The figures should be regarded as provisional. Updating of the figures should be carried out periodically and the underlying assumptions should be tested.

The above comments apply to the R & D segment of the project. There must be a corresponding degree of attention to manufacturing and the marketplace. This mandates close links between R & D and the other company units.

Although it is difficult to obtain reliable figures for costs, benefits and time scales, it is worthwhile to carry out the exercise especially for major projects. It may be that financial availability is a critical factor and so it is necessary to have the timing of cash flow. At the very least the cost-benefit curve will indicate whether the project is within the capability of the company to finance. If the project is financed the cash flow schedule will allow critical points, at which the project should be reviewed, to be identified.

This section has focussed on the financial aspects of project evaluation. However, there are many other factors to take into account (Anthony and Reese, 1975). According to these authors, "Non-monetary considerations are also important in making actual decisions; they are often as important as the monetary considerations and in some cases so important that no economic analysis is worthwhile." The foregoing statement was made with respect to capital investment decisions but applies equally well to projects.

The return from R & D projects has been examined in the light of decreased R & D funding over the last 15 years. In terms of the ratio of R & D expenditure to sales there has been a consistent decline over the years 1970 to 1979, the last year considered, for product and process research (PIMS, 1982). The technique used was distributed lag analysis in which today's profits are assigned to the R & D investment in previous years. The lag between cash outflows and cash inflows was approximately 4 years but could be as long as 6 years. The pretax return on R & D investment was as follows: 1975 negative; 1976 14-32 percent; 1977 29-44 percent; 1978 27-45 percent; 1979 29-42 percent. For example, a dollar of R & D expenditures spread over the mid-1970's led to an estimated $3.08 in added profit contribution in 1978. For marketing expenditures the added profit contribution was $1.52 but this was mainly attributed to 1977 expenditures. Although the contribution is smaller for the expenditures on marketing the shorter time between investment and income gives a higher internal rate of return. The data from the PIMS base also allow a comparison of product and process returns. The undiscounted return on process R & D was found to be $1.75 per dollar invested while for product R & D it was $1.67 per dollar invested. Process R & D not only paid off at a slightly higher rate but the payoff was faster. The average lag between outflows and inflows was 1.2 years for process R & D while for the product R & D the average was 3.8 years. The pre-tax return on process R & D was consequently 63 percent as against 15 percent for product R & D. Paradoxically the variability on the return on process R & D was almost twice as high as for product R & D, implying that the higher average return on process R & D was subject to higher risk.

It may be noted that for the period covered although the return on industrial R & D was in the range 10 to 40 percent, the returns for the economy as a whole were in the range 60 to 100 percent. Thus there is a social return to industrial R & D, which benefits the purchasers of new and improved products. The implication is that economic growth would be more rapid if more were invested in R & D or if the R & D process could be made more effective and efficient.

4.3.3 R & D Project Selection Methodologies

In 1971 Gee surveyed 27 companies in the United States involved in R & D with the objective of finding the most popular methods of project selection. It was found that practices ranged from intuitive approaches to sophisticated quantitative techniques (Gee, 1971). Since that time the situation has not changed substantially but more techniques have been introduced. There is a dichotomy between academic research and industrial practice. However, the increasing pressure to cut expenses and the competition for scarce resources in the corporation has led to an increase in interest in evaluation methods. In order to examine the various techniques some form of classification is helpful. A five-category classification with the categories becoming increasingly more quantitative and analytic has been proposed and will be employed here (Gibson, 1981b). These categories are as follows: ranking, scoring or rating methods, economic rating methods, formal optimization methods and risk analysis/decision analysis methods. Each of these categories will be described and will be illustrated by examples. It must be remembered that there are a large number of techniques for each category but that many of the methods are substantially the same. It is, therefore, only necessary to give a small number of examples to illustrate the principles.

(i) Ranking: a ranking procedure can simply consist of an ordering procedure in which the manager places the projects in order of preference. More elaborate schemes may be used such as "Q-sort," "dollar metric" and "standard gamble" (Pessemier and Baker, 1971). According to these authors the methods are "systematic procedures for obtaining and integrating subjective and objective benefit data." Each of the methods requires the manager to compare one alternative project to another alternative or to a subset of alternative projects. This is repeated for all pairs and a set of project values is logically computed using the standard preferences. The values have meaning only relative to the set of alternative projects evaluated i.e. if another alternative is introduced the preferences must be recomputed for the new set of alternatives. Pessemier and Baker, in a comparison of the various methods, found that the dollar metric showed the most promise. In this technique, the manager carries out a pair-wise comparison of projects and indicates how much the cost of executing the preferred project would have to increase before the preference would be reversed.

Fundamental difficulties in ranking methods have been discussed by Gibson (Gibson, 1981b). It must be emphasized that a rank ordering is dependent on the set of alternatives. A ranking that is achieved may result in the lowest projects being dropped. However, the ranking process must be redone even though it was the lowest ranked projects that were dropped. This is because ranking methods depend upon the assumption of transitivity which in general does not hold true.

Methods of ranking are opposed by theorists because the bases on which choice is made need not be revealed. Thus the manager may take one project over another for reasons other than probability of success, or return on investment. For the manager this is the strength of the method since ranking allows many factors to be taken into account intuitively. There may be excellent reasons why one project has to be selected over another. The manager has to balance needs against the resources available. It may be quite permissible to choose a project on the basis of available resources even though the return on the project would not be as high.

(ii) Scoring or Rating Methods: these methods are an extension of the ranking methods in which an explicit set of criteria are employed. Each project is rated against the set of criteria and the score totalled. A different weight may be given to each criterion. The advantage is that the judgment criteria are made explicit. To the manager this may be a disadvantage. The set of criteria may not include all the factors and the weighting may be incorrect; the choices are subjective so that the subjective choice of project rank has been replaced by a score or rating based on a subjective set of criteria. However, it is helpful to consider the criteria important for project selection. At the very least the criteria judged important by R & D personnel may be compared with criteria judged important by upper management so that consensus may be achieved. It is recommended that the method be kept as simple as possible due to the difficulties in justifying a set of criteria. Due to the perceived simplicity of scoring or rating methods many different approaches have been described. Some of these may be described.

Ansoff has proposed a figure of merit (profit) and a figure of merit (risk) (Ansoff, 1964).

$$\text{Figure of merit (profit)} = \frac{(M_t + M_b)\ E\ P_s\ P_p\ S}{(C_d + J)}$$

where

M_t = technological merit

M_b = business merit

E = estimate of total earnings over product lifetime

P_s = probability of project success

P_p = probability of successful market penetration

S = strategic fit of proposed project with other projects, products and markets

C_d = total cost of development including capital and facilities

J = savings factor resulting from the use of existing facilities and capabilities.

$$\text{Figure of merit (risk)} = \frac{C_{ar}}{RM_p}$$

where

C_{ar} = total cost of applied research

F = total cost of resources (facilities, staff, etc.)

M_p = figure of merit (profit)

The formulae illustrate the point made earlier that a false sense of security may result from the application of a numerical method. There is no a priori reason to justify the choice of the factors, the equal weighting, to the exclusion of all other factors.

An interesting approach recently described uses a simple multiplicative figure of merit expressed in probabilistic terms (Merrifield, 1978).

Merrifield index = (probability of commercial success)
X (probability of technical success)

The probability of technical success is not considered further but the probability of commercial success is broken down into the sum of 12 factors, 6 under the heading of business attractiveness and 6 under the heading of company strengths. Business attractiveness factors address the question, "Is this a good business for anyone to be in?" While company strength factors address the question "Is this a good business for us to be in?"

The business attractiveness factors are as follows:

Sales/Profit Potential. The maximum score is 10 and to achieve this score the project should generate within 5 years of commercialization an additional 10 percent in sales and a 40 percent ROI before taxes on a discounted cash flow basis.

Growth Rate. To achieve a score of 10 the growth rate should be at least 10 percent per year in unit volume or in sales adjusted for inflation.

Competitor Analysis. Key considerations are (a) what measures can or will the competition take when their market share is threatened by the new development (b) how strong is the patent that will bar the competition at least in the early stages of commercialization and (c) what is the rate of technological change in this area (an indication of the product life cycle). Scores are (a) 4, 3 for (b) and (c) each, giving a maximum of 10.

Risk Distribution. Are all the eggs in one basket or does the development have at least four or five potential applications of a significant size either from a marketing or from a manufacturing viewpoint. Displacement of the product by a competitive product in one area would not be catastrophic if there are multiple applications. The score is 10 if there are at least 4 or 5 significant segments for the development.

Industry Restructure Opportunity. The ideal situation is one in which a significant technical development with strong patent protection can be made in an otherwise technically stagnant industry that is fragmented between many small competitors but with an attractive aggregated potential. A clearly superior product would have fast penetration and would allow significant restructuring to be achieved in the industry. A score of 10 is given for a major restructuring opportunity.

Special Factors. This category covers other special circumstances which could include political, social and economic developments. The score is 5 if there are neither positive nor negative factors, less if there are negative factors and more if there are positive factors e.g. subsidies for R & D.

The company strength factors are as follows:

Capital Needs vs Availability. A project with low capital needs has a lower risk than one with high capital needs especially for a company with limited cash flow. If cash is readily available then the company is in a strong position and the capital intensity of a project or development might well be a barrier to others. A score of 10 is given for a strong cash flow in a capital intensive situation. Fewer points are assigned if less capital is needed. When a company has a low cash flow, 5 points are given for a situation of low capital intensity and fewer points as the capital intensity increases.

Marketing Capability. If the new development fits directly into a strong in-house marketing, distribution and technical service capability, the time required for commercialization will be significantly shortened and the probability of success increased. A score of 10 is assigned to a strong in-house marketing capability and fewer points for a weaker situation. Note that marketing strength is measured relative to marketing strength of the competition.

Manufacturing Capability. Strong in-house capability can greatly reduce the time needed to reach commercialization and will reduce expenditures until the probability of commercial success is well established; manufacture of prototypes outside adds to the cost and time of development. A score of 10 points is assigned for existing full-scale manufacturing capability with 5 points for an "interim" manufacturing capability.

Technology Base. A means of assessing the technology base is given in Table 31. There are five functions. A score of 2 is given if the percentage of effort allocated approximates the percentage of effort needed. If there is a significant gap between what is needed and what is available then score 0 or 1. The maximum score is, therefore, 10. Note that the first three functions fall under a product maintenance category while the last two functions refer to new product development. Not only the total score is important but also the distribution of scores.

Raw Materials Availability. An assured supply of basic components or raw materials is an essential element that determines whether a new product can be profitably commercialized. Assured availability is given a maximum score of 10. However, availability is not enough since there may be availability but at a constantly escalating price.

Management and Other Skills. In-house availability of critical management skills is needed for effective commercialization. A project champion is of great benefit (see Part 1). For complete satisfaction with this area, a score of 10 is given.

TABLE 31

Strength of the technology base

Function	Effort allocated (%)	Effort needed (%)	Score
Market support			
Manufacturing support			
Product maintenance systems			
New component development			
Science support of applied R & D and engineering objectives above			
Totals	100%	100%	10

The factors to be assessed are summarized in Table 32. In the layout three projects are being evaluated for attractiveness. Figure 33 shows the Merrifield diamond on which the scores are plotted. Axis A gives business attractiveness and axis B company strengths. Based on experience with application of the model Merrifield states that a project with a score less than 70 has a low chance of commercial success while a score greater than 80 has a high probability of success. It should be remembered that the model only considered commercial success in detail. It is quite possible that the probability of technical success is low and this would reduce the overall attractiveness of the project.

A further illustration of the Merrifield diamond is given by an example. Three alternatives are shown with scores in Table 33 for the development of a product from R & D of company X (our company). The product promises to restructure two or three major industries each of which represents $300 million in annual sales and with a good opportunity for a substantial restructuring. The business attractiveness factors sum to 49 (out of a maximum of

FIGURE 33

MERRIFIELD DIAMOND

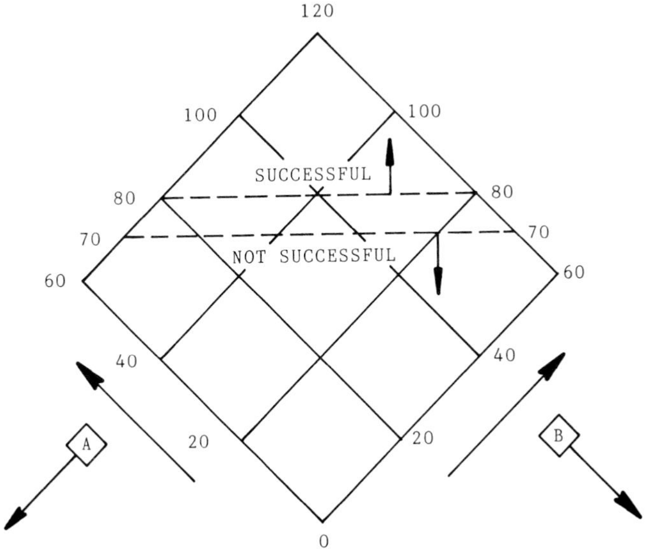

PROJECTS WITH LESS THAN 70 POINTS
HAVE RARELY BEEN SUCCESSFUL

TABLE 32

Summary of business attraction and company strength factors

	Projects*		
	I	II	III
A. Business attractiveness			
1. Sales/profit potential	____	____	____
2. Growth rate, %/year	____	____	____
3. Competitive situation:			
- Competitor reactivity	____	____	____
- Activity index of technology	____	____	____
- Patent position	____	____	____
4. Risk distribution (segments)	____	____	____
5. Opportunity to restructure an entire industry	____	____	____
6. Special political and social factors:			
- Antitrust			
- Ecology			
- Energy			
- Foreign exchange			
- Geography			
- Sovereign rights	____	____	____
Totals	====	====	====
B. Company strengths**			
1. Capital requirements	____	____	____
2. Marketing capabilities	____	____	____
3. Manufacturing capabilities	____	____	____
4. Technology base	____	____	____
5. Raw material availability	____	____	____
6. Skills availability:			
- Champion			
- Technical, legal, financial, etc.	____	____	____
Totals	====	====	====

*Rate each factor on a scale of 1 to 10

**Company strengths (fit factors)

TABLE 33

Analysis of three alternatives A, B and C

	Score		A	B	C
A. Business value factors		B. Company fit or strength factors			
1. Sales/profit potential	10	1. Capital needs	0 *	10	10
2. Growth rate	10	2. Marketing	0	2	8
3. Competitor analysis	8	3. Manufacturing	3	5	7
4. Risk distribution	6	4. Technology base	5	6	6
5. Industry restructure	8	5. Raw materials	10	10	10
6. Ecology, etc.	7	6. Management	4	6	8
Total	49	Totals	22	39	49

*Boxes indicate weaknesses requiring additional strength

60). However, company X has weaknesses in capital availability, marketing and manufacturing capability giving a score of 22. The position of the project on the diamond is shown in Figure 34 and it will be seen that the project is marginal. Alternative B represents a joint venture with company Y that has a strong cash flow and some technical and management strengths. The venture is at position B in Figure 34, and is just above the threshold for acceptability. Alternative C represents a joint venture between X, Y and Z where Z has strong marketing and other skills; Z will be an exclusive sub-licensee in one business sector only. The position of the project is now very strong and the project is very likely to succeed. It should be noted that although the fit of the company has been improved, the profitability is not so high since profits must be shared with the other partners in the venture. This aspect is not taken into account but could be by a revised business attractiveness analysis. Alternative B would then be marginal but alternative C would still be very strong.

The Merrifield approach is attractive but suffers from the objections given earlier. There are six factors for business attractiveness and six for the fit of the company. Each factor has equal weight. In a ranking method the basis of choice is not made explicit. In this rating method the basis of choice is explicit but the basis on which the choice of the criteria was made is not given. Even so the approach does give an indication

FIGURE 34

HOW CONSTRAINT ANALYSIS LED TO LICENSING STRATEGY

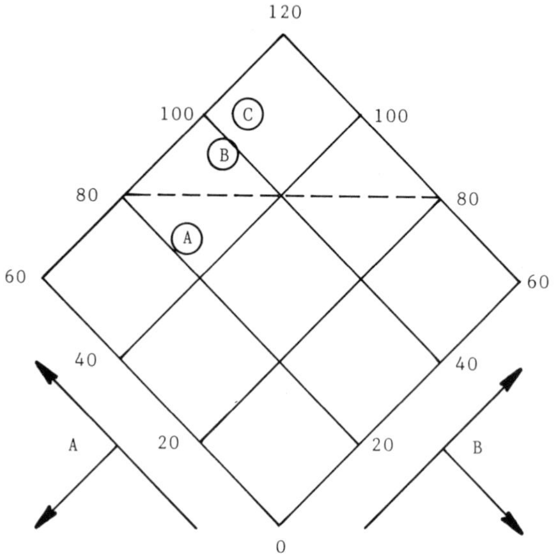

A = COMPANY (X) ALONE

B = COMPANY (X) + COMPANY (Y)

C = COMPANY (X) + COMPANY (Y) + COMPANY (Z)

of project attractiveness and is useful so long as the final choice is not made exclusively on the business attractiveness and company fit scores. Attention must also be given to the probability of technical success. However, there is some evidence to suggest that the risks on the technical side are less than on the commercial side (Mansfield, 1981). Mansfield looked at the R & D programs of three companies (one in chemicals, two in proprietary drugs). He considered that the likelihood of economic success was the product of three separate factors, the probability of technical success, the probability of commercialization given technical success and the probability of economic success given commercialization. Only 12 percent of the R & D projects were economically successful. However, 60 percent of the R & D projects were technically successful, 30 percent were commercialized but only 12 percent earned a profit. Most R & D projects aim for fairly modest advances and are thus fairly assured of success. Development carries a smaller risk still.

(iii) <u>Economic Rating Methods</u>: these methods employ calculations of ROI, NPW and IRR and may be regarded as a subset of the scoring or rating methods described above. However, the reason for considering economic rating methods separately is that the calculations do consider the project worth in terms of money and so personnel outside R & D feel more comfortable with such a calculation. Money and specifically return on investment provide a method of project evaluation that is acceptable throughout the whole company. A full venture analysis may be conducted (Gibson, 1981c) or simple economic criteria may be used:

$$\text{Value of project} = \frac{[\text{Estimated percentage of sales or savings}]\;[\text{Estimated probability of success}]}{\text{Estimated cost of R \& D}}$$

$$\text{Project index} = \frac{[\text{Estimated probability of technical success}]\;[\text{Estimated probability of commercial success}]\;[\text{Expected ROI}]}{\text{Estimated cost of R \& D}}$$

All projects must be judged on the same basis. The probability of success must not be overestimated and it has been recommended that 60 percent of earnings and 50 percent annual return before taxes be used (Gee and Tyler, 1976).

The 7 most commonly used indicators for selecting R & D projects to be funded are as follows (Schwartz and Vertinsky, 1977):

Estimated cost of the project as a proportion of the total R & D budget, I_1

Estimated IRR, I_2

Estimated potential market share, I_3

Probability of technical success, I_4

Probability of commercial success, I_5

Availability of government funding, I_6

Estimated payback period, I_7

These factors may be linked to long-range corporate goals to give a modified matrix scoring model (Gibson, 1981d). For example, the long-range goals may be as follows:

To improve the company's market share by developing new products and services in the company's chosen areas of activity, G_1

To promote product and service innovation in the chosen market sectors, G_2

To improve product and service profitability, G_3

To improve product and service productivity, G_4

To develop new products and services that display a social consciousness, G_5

To improve the management of new products and services, G_6

To select projects P_1, P_2, P_3 .... from the set of projects the first step is to define a matrix, I, whose coefficients define the rating of each project P_i against each factor I_i. Next a matrix G is defined in which each factor I_i is rated against each factor of long-range corporate goals, G_i. Last a weighting W_i is given to each of the long-range goals according to its importance to the company. The ranking of the projects is given as:

$$\begin{bmatrix} A \\ B \\ C \\ D \end{bmatrix} = [I] \times [G] \times [W]$$

Note that this is a scoring model which has been described in this section because of the economic indicator IRR and the payback period employed.

(iv) Formal Optimization Methods; these techniques involve the use of the computer to optimize project selection by means of linear programming, non-linear programming, integer programming, dynamic programming and so on. Essentially the situation may be described as follows (Mansfield, 1981):

List of n possible R & D programs.

Project i has a cost of C_i, probability of success of P_i and, if successful, a profit gross of R & D costs, of R_i.

The objective is to maximize

$$\sum_{i=1}^{n} X_i \left(P_i R_i - C_i\right)$$

where

$$\sum_{i=1}^{n} X_i C_i \leq C$$

and

$$X_i = 0,1$$

Thus X_i equals 1 if the project is accepted and 0 if it is rejected. The goal is to choose all of the X_i so that the profit is maximized, subject to the constraint that the total sum spent on R & D does not exceed C. This is an integer programming problem.

An example of the application of linear programming to R & D project selection has been given for the constraint of resource allocation (Bell and Read, 1970). There is a description of a linear program for a multiple project portfolio that has been used to aid project selection at the Central Electricity Generating Board laboratory in the United Kingdom. A number of constraints are built into the model as well as the division between wages and capital costs. It is stressed, however, that the results are only an aid to managerial judgment. The method lends itself to sensitivity analysis by varying among other things personnel level, type of skills, alternative project versions and equipment constraints.

The main difficulty in using these techniques is that much effort is needed to supply all of the data for the analysis; the data may not be available or may be of questionable validity. The techniques have been little used by the practicing R & D manager.

(v) Risk Analysis and Decision Analysis Methods:

Decision analysis (DA) or risk analysis are easier to apply than complex optimization techniques and are based on the logical thought processes of the manager.

The level of risk has been considered in several of the formulae given earlier. Where the investment is small projects can be compared by modifying the expected benefit:cost ratio to take account of the overall risk:

$$\text{project index} = \frac{B}{C} P_t P_c P_E$$

where B/C is the benefit:cost ratio and the other factors refer to the probabilities of technical, commercial and economic success respectively. The area of benefits and costs can also be examined. Instead of a single benefit or cost number a distribution can be given as a function of probability. This is shown in Figure 35(a) where the probability distribution of developments costs is given. A similar distribution could be given for

FIGURE 35 (a)

RISK ANALYSIS FOR DEVELOPMENT COST

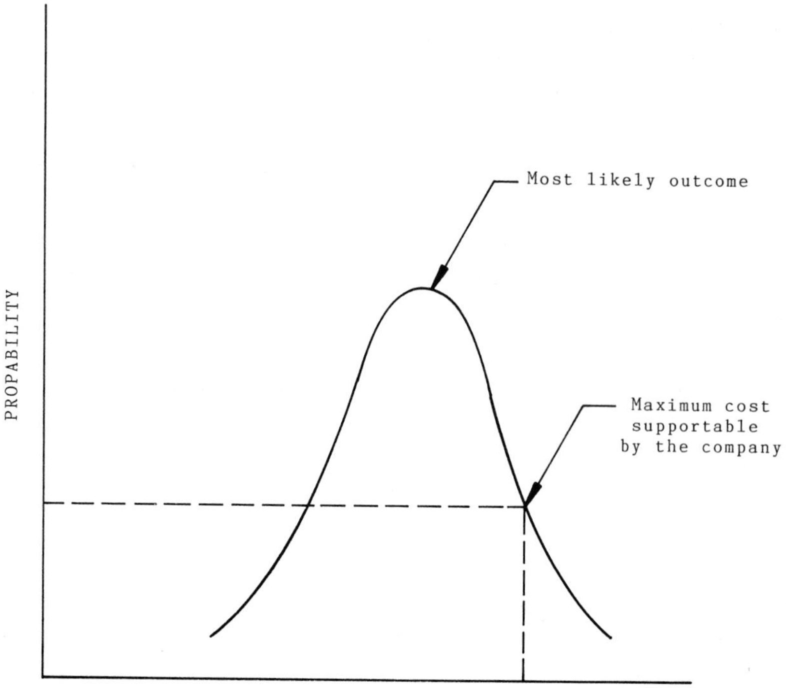

DEVELOPMENT COST

FIGURE 35 (b)

EVALUATING RISK IN PROJECT COMPARISONS

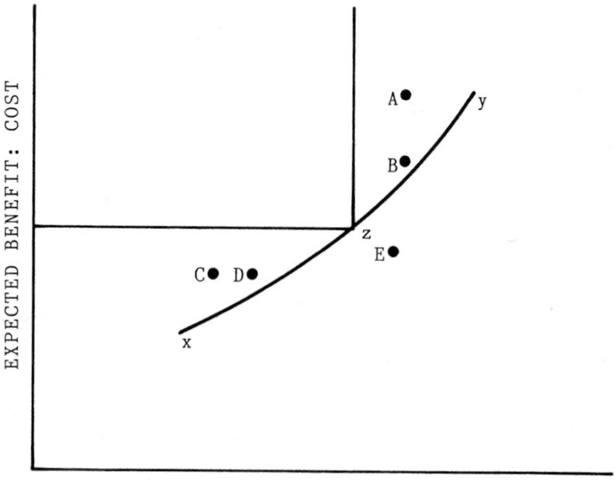

RISK

*Reprinted by permission from "Managing Technological Innovation", 2nd ed., by Brian Twiss, Longman Group Limited, publisher.

benefits. Figure 35(b) indicates the variation of benefit:cost ratio as a function of risk. In this way selection can be made. Other things being equal project A is preferable to project B because of a higher benefit:cost ratio at the same risk. Project C is preferable to project D since it has lower risk at the same benefit:cost ratio. If line XY represents the maximum return for any project, then project E is immediately eliminated. Projects may be placed in three classes by this type of analysis: those that are rejected due to a failure to meet the criteria, those which can be selected if certain non-financial criteria are met, those that are marginally acceptable and should be scrutinized either to reject or accept the project.

DA requires that the possible outcomes of an action along with the probabilities be specified by the decision-maker. This is, in fact, the difficult part of the analysis since DA is particularly useful in complicated series of decisions and so extensive work is needed to ensure that all outcomes have been covered. Multi-attribute DA is the most realistic way to treat decisions (Kenney and Riaffa, 1976) but single-attribute decision analysis is much easier to explain and does illustrate the basic principles (Riaffa, 1968).

The basic diagram used in DA is shown in Figure 36. A decision is represented by a square box and a subsequent chance outcome is represented by a circle. There may be a cost associated with making the decision. Probabilities are placed on each outcome and the sum of the probabilities must be unity. When the problem is set up in the form of Figure 36, the choice of option can be made by calculating the expected monetary value (EMV) of each outcome. The EMV is the product of the pay-off and the associated probability of achieving the outcome.

An actual example is shown in Figure 37. A company is considering whether to develop and market a new product. Development costs are estimated to be $100,000 and there is a probability of 0.7 that the development will be technically successful. If the product is marketed it is estimated that:

(a) there is a probability of 0.4 that the product will be highly successful and that it will produce an income of $400,000 ($300,000 after subtracting the development cost).

FIGURE 36

BASIC DIAGRAM OF DECISION ANALYSIS

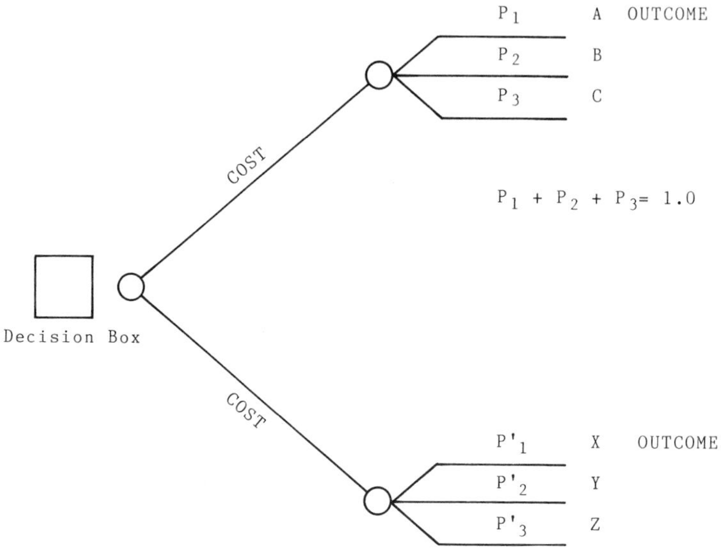

(b) there is a probability of 0.4 that the product will be moderately successful and that the income will be $100,000 (break even).

(c) there is a probability of 0.2 that the product will be a failure and that it will lose $100,000 (total loss of $200,000 including development costs).

The EMV of each outcome is as follows:

(a) for failure in the technical stages, the EMV equals -$100,000 x 0.3 = -$30,000 (a loss of $30,000)

(b) development succeeds but the product is a failure, the loss is $100,000 development costs plus $100,000 marketing costs or a total loss of $200,000. The total probability of this outcome equals 0.7 x 0.2 or 0.14. The EMV of this outcome is 0.14 x -$200,000 = -$28,000.

(c) development succeeds and the product is moderately successful, the probability is 0.7 x 0.4 = 0.28 and the net income is $0 giving an EMV of 0.28 x $0 = $0.

(d) development succeeds and the product is highly successful, the net income is $300,000 and the probability is again 0.7 x 0.4 or 0.28. The EMV is 0.28 x $300,000 = $84,000.

FIGURE 37

SIMPLE DECISION TREE FOR PRODUCT DEVELOPMENT

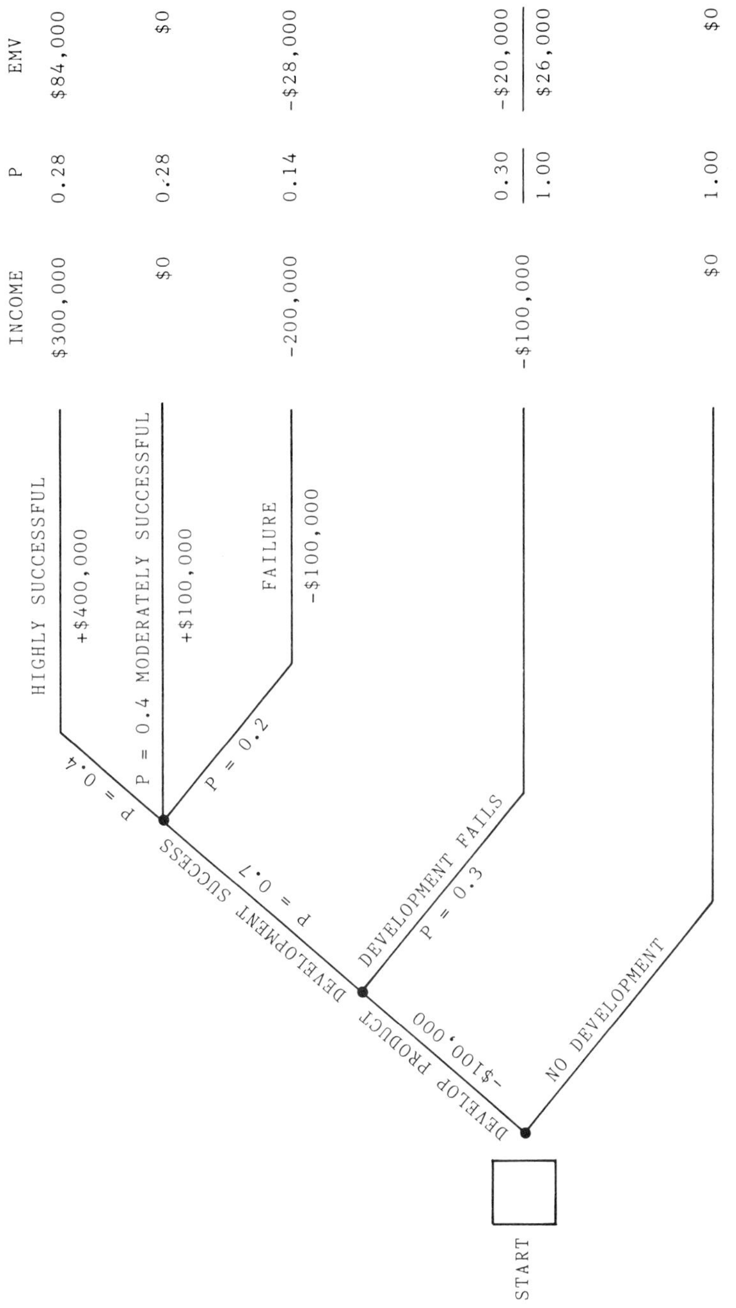

*Reprinted by permission from "Management Accounting", 5th ed., R.N. Anthony and J.S. Reece, copyright © 1975 by Richard D. Irwin, Inc.

The total EMV for the value of the decision "Product Development" is the sum of all the expected outcomes, $26,000. The value is compared to the EMV of the alternative decision "Do Not Develop Product", which has a value of zero. The decision will, therefore, be to proceed with the development of the product. This does not mean that the income of $26,000 is guaranteed but the likelihood is that there will be a profit in the venture. Note, however, that only one out of the four outcomes gives a profit and that the probability of this outcome is 0.28. Also note that the sum of the probabilities (0.3 + 0.14 + 0.28 + 0.28) equals 1.

Although the EMV of the previous example was $26,000, a manager might not want to accept the risk of a loss especially if there were other alternatives. For example, the $100,000 might be put in the bank at 10 percent interest. The probability of a return of $10,000 is 1 and a risk-averse manager might well prefer this option. Aversion, or otherwise, to risk is dealt with in utility theory (Gibson, 1981e). A decision-maker who requires a greater return than the EMV is called risk averse while a manager who is content with a return less than the EMV is called risk prone. A decision-maker's risk curve or utility curve may be developed by the following procedure. First the boundaries within which the decision is made must be set. In the example (Figure 38) the range is set at plus or minus $50 million. A utility of zero is assigned to -$50 million and a utility of 1 to $50 million.

U (-$50 million) = 0; U ($50 million) = 1.

Figure 38 illustrates an equiprobable lottery, such as tossing a coin for heads or tails, with outcomes of $50 million or -$50 million. The question is to determine the certain monetary equivalent (CME) that would make the individual indifferent to entering the lottery. The risk averse manager might pay to be excused from playing but the risk prone manager would ask for money to withdraw from the game. The lottery is an artificial example but many business situations approximate to this case in which there is not only opportunity for gain but also the chance of a substantial loss. In the modern corporation success is rewarded modestly but failure is punished harshly. This tends to make the corporate manager risk averse.

FIGURE 38

AN EQUIPROBABLE LOTTERY USED FOR ESTABLISHING A POINT ON AN INDIVIDUAL'S UTILITY CURVE

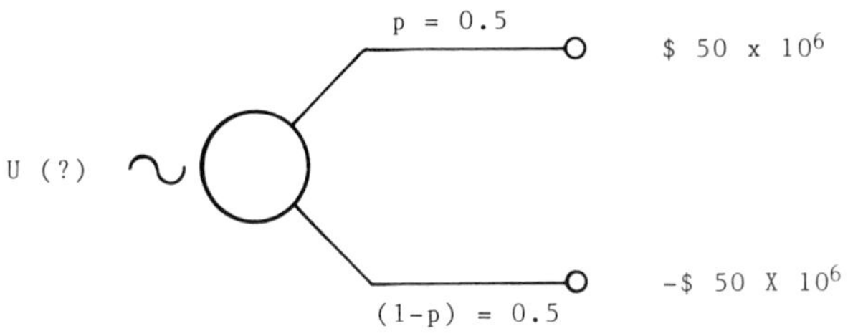

An alternative way of looking at the lottery is in terms of an actual case. Suppose that a product is being developed and that the profit is estimated to be $50 million. However, the market has changed and there is an equal probability of losing $50 million. Should the development be halted? How much would this cost? Suppose that it would cost $15 million to halt the project. Then

U (-$15 million) = pU ($50 million) + (1-p) U (-$50 million)

U (-$15 million) = 0.5 (1) + 0.5 (0) = 0.5

Figure 39 shows the utility curve with the end-points marked. Also shown is the point U (-$15 million) equals 0.5. The actual curve can be built up by repeating the process of the lottery. Suppose that the equiprobable outcomes are $50 million or -$15 million. The EMV of the lottery is $17.5 million. Perhaps the decision-maker would accept a CME of $10 million instead. Then:

U ($10 million) = pU ($50 million) + (1-p) U (-$15 million)
= 0.5 (1) + 0.5 (0.5)
= 0.75

and another point on the curve is established. Note that the decision-maker is risk-averse. The straight line represents the neutral situation and the risk-prone individual would have a curve that was concave downwards.

FIGURE 39

THE UTILITY CURVE OF A RISK-AVERSE MANAGER

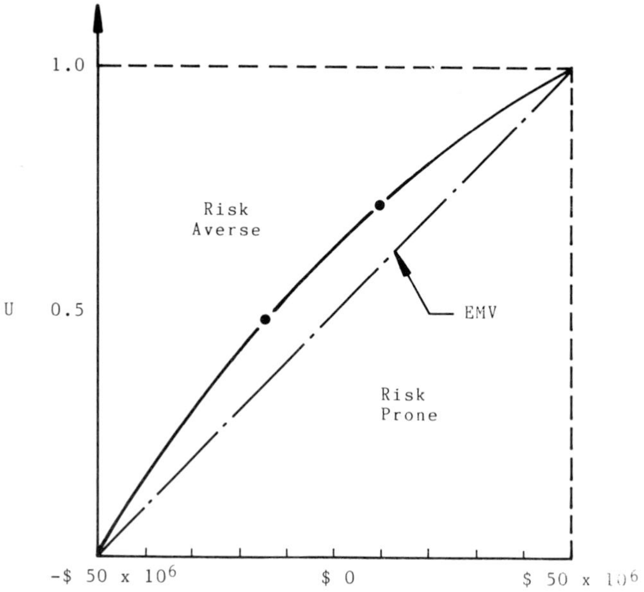

The utility curve developed cannot be applied to the case considered in Figure 37 because of the limits chosen. For that case a utility curve with limits of say $300,000 and -$300,000 would be developed. Note that the net income is used. For each outcome the net income is used to determine the utility value from the curve. This U is multiplied by the probability p of the outcome to give the expectation. The expectation values are summed for all outcomes of a particular decision (development success outcomes in Figure 37 for example). The decision option having the highest expectation is the desirable choice.

4.3.4 Other Comments on Project Selection

The development of a project and its evaluation has been presented as a single action; the project is developed as part of a portfolio of projects and there is an evaluation process that either approves or rejects the project. It is true that the project must at some stage, preferably early on, be prepared with a description, time scale and estimated costs. However, before this is done it is usually necessary to carry out some feasibility evaluation which will involve literature searches and possibly a limited amount of laboratory work. This implies that there should be money available for research that will result in projects - sometimes this is called "seed money". The amount can be small since the definition part of a project is normally inexpensive. The definition stage may be quite long depending on the state of the technology.

Similarly the approval of a project is not a single action. It is likely that the project, if it is approved, will only get a qualified go-ahead. The project will be subject to periodic review especially at critical technical milestones or at transition stages e.g. from research to development or from development to manufacturing. There will be special interest of management as the commitment of funds increases. As the amount of cash required increases, especially if large capital purchases are required, the decision process will move higher and higher in the corporation hierarchy.

The dilemma with R & D projects is illustrated in Figure 40.

FIGURE 40

THE VARIATION OF RISK AND COST WITH PROJECT TIME

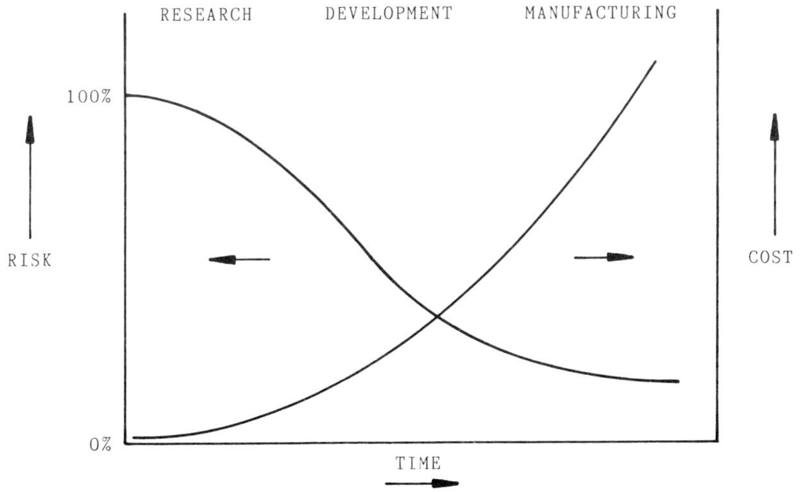

Early in the program the risk is high but the amounts expended are low. As more is spent the risk decreases. Hopefully, the risk is low by the manufacturing stage. Unfortunately, the risk is usually quite high at the point at which large expenditures of funds are required. This is one of the reasons for the great interest in project evaluation methods. As the project moves towards development and then on to manufacturing the methods of evaluation become increasingly financial with an emphasis on ROI, IRR, payback period and so on. What must be remembered is that the evaluation of a project which has just been formulated is also exceedingly important. Poorly defined programs and sloppy selection methods can result in projects being undertaken when the project should be rejected. Top management may not be overly interested since the expenditures are relatively small but this is a mistake since unnecessary projects divert scarce R & D resources and above all waste time. It may take several years to reach a decision point at which a commitment of a large quantity of funds is required. At this stage there may be a careful appraisal of the project but it is then too late. The competition will have had several years in which to develop new products and the company may never be able to catch up. About 70 percent of money spent in R & D is spent on salaries and fringe benefits. Thus substantial sums are wasted if R & D personnel engage on frivolous programs.

The message is that project evaluation is a continuing process. Projects should be carefully screened at inception and should be reviewed against goals on a regular basis. A project that is not achieving its goals or which is later found not to have the right fit should be cancelled.

4.4 PROJECT PLANNING AND CONTROL

4.4.1 <u>General Comments</u>

Many different project selection methods have been discussed but what all these have in common is a need for data on which to make the selection judgment. A project description is needed. For exploratory areas the description may be minimal provided great resource commitments are not required. For product development there is an emphasis on financial as well as on technical aspects. The project activities extend over a period of time. Along with the project description a project plan is required, which indicates the timing of activities and the dates at which significant activities will be achieved (milestones). The timing of resource commitments may also be indicated. The project plan is used to assist in project control by allowing the comparison of progress to the plan. Significant deviation from the plan calls for action. Projects can and should be cancelled if the technical goals are not being achieved and do not look as if they can be achieved. If there is one criticism of R & D that is justified it is the proclivity of the group to continue projects that should have been cancelled and to hang on to successful projects that should have been transferred to the next stage in the development process i.e. to product development or to manufacturing. The project plan provides a method for assisting in the decisions regarding a program.

A large number of methods have been proposed for the planning and control of R & D projects. Only a few of these methods will be described here. It is important to remember that the method chosen should reflect the philosophy of the company and the type of R & D. It would be a fruitless exercise to use a complex computer-generated project plan for an exploratory program or to use such a project plan in an organization geared to informal methods. The overall project plan is suggested as follows:

(i) Project Abstract - a one-page document that gives a project description with manning level and expenses.

(ii) Project Rationale/Protocol - a multi-page document that provides much more detail and is used for major projects. In a simpler form the document may be used for other programs or for internal R & D use.

(iii) Project Plan - depending on the level of resources and the project complexity the plan may be highly detailed or given in outline form only.

A more extensive description of the planning and control methods will now be given.

4.4.2 Project Abstract and Protocol

A project abstract should contain the following:

(i) Project Title or Name.

(ii) Indication whether the project is new or is continuing.

(iii) Objective - for industrial R & D the objective is a new product or process and a description as focussed as possible should be given.

(iv) Business Rationale - an indication of why there is a market need for the new product or how the new process will benefit the company. There should also be a brief description of other competing products/processes and a description of competitive activity.

(v) Technical Rationale - details of the approach giving the project status (especially if a continuing project) and indicating why the approach is likely to succeed.

(vi) Milestones - description and dates of significant milestones should be given.

(vii) Resource Needs - expenses and manning levels for the life of the project should be given year by year. If possible the cost of the manning should be given. Capital equipment needs should be indicated.

The Project Abstracts should be one-page documents and are intended to be part of the R & D operating plan for the year.

Abstracts will be reviewed with top management and provide a ready means of reviewing the direction of the R & D effort.

For in-house R & D review it is necessary to have more detail than is given on the Project Abstract. One form of protocol is the following:

(i) Title of Project.

(ii) Objectives - a more detailed statement than is given on the Project Abstract.

(iii) Rationale - the need for the project both from a technical and a market viewpoint.

(iv) Status - detailed description of the present status of the project.

(v) Approach - detailed technical approach.

(vi) Resource Data - details on expenses, manning, capital equipment.

The Protocol is accompanied by an Activity List that gives details of the activities needed for the project to completion. Major tasks are broken down into sub-tasks each of which is identified by name. Planned starting date and ending date are given; actual dates are compared to plan. If necessary resource information is given such as individual responsible, number of man-days for each task and cost of each task; actual resource use may be compared to plan.

The Activity List provides a detailed view of project activities. It is often useful to present this list graphically to see at a glance the project status. Three convenient ways of doing this will be described: the Flow Diagram, the Gantt Chart and the PERT Diagram.

Details on the Flow Diagram for project activity description are given elsewhere (Gibson, 1981f; Davies, 1970). Figure 41 shows the notation for the nodes. The first node is the AND gate. The node must be reached along all the incoming paths before further progress can be made. Thus the AND node serves as a "hold". The other input node shown by the convergence of arrows enables progress to be made provided the node is reached along any of the arrows. There are two types of output node. In the first there is a simple divergence of paths with all outgoing paths followed. The second output node is the decisive diamond. The answer to a question written inside the diamond determines which of the output paths will be followed. All information for the decision must be available by the time the decision diamond is reached and all analysis must be completed. Under these circumstances the time taken to reach a decision may be neglected. Activities are represented by rectangular boxes. Activities are specified in the boxes in the form of commands. The activity box may represent a sub-program that is detailed elsewhere. Replicated processes are indicated at the side of the box. Parallel processes carried out by changing process or other variables are indicated by placing the sets of parameters in parentheses at the side of the box. Figure 42 gives an example of a flow diagram for the process of seeking a suitable glaze as a high temperature protective coating. Progress along the central core activities leads to success. Loops allow activities to be repeated such as choice of new compositions or minor change in composition. The use of a flow diagram may be extended by putting probability values on some of the activities. Times for different activities may also be placed on the diagram but there are difficulties if activities have to be repeated. There is also no guarantee that the final goal will be achieved since failure can occur at multiple points along the way. The Flow Diagram approach may be used even for small or moderate programs. On the other hand PERT techniques to be discussed later are considerably more complicated and are not worth applying to projects of less than 5 man years' effort (Hardy, 1965).

FIGURE 41

RESEARCH PLANNING DIAGRAM NODES

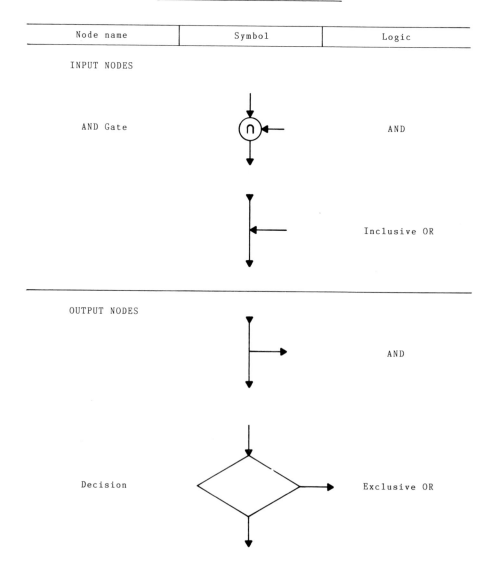

Node name	Symbol	Logic
INPUT NODES		
AND Gate	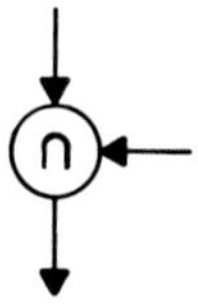	AND
	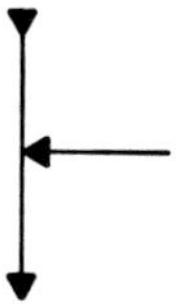	Inclusive OR
OUTPUT NODES		
	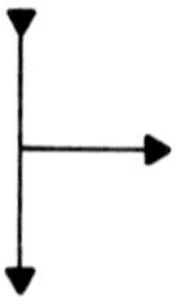	AND
Decision	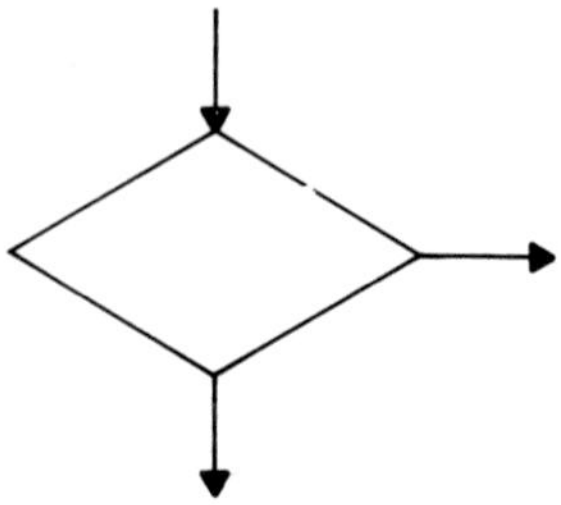	Exclusive OR

FIGURE 42

FLOW DIAGRAM FOR GLAZE PROGRAM

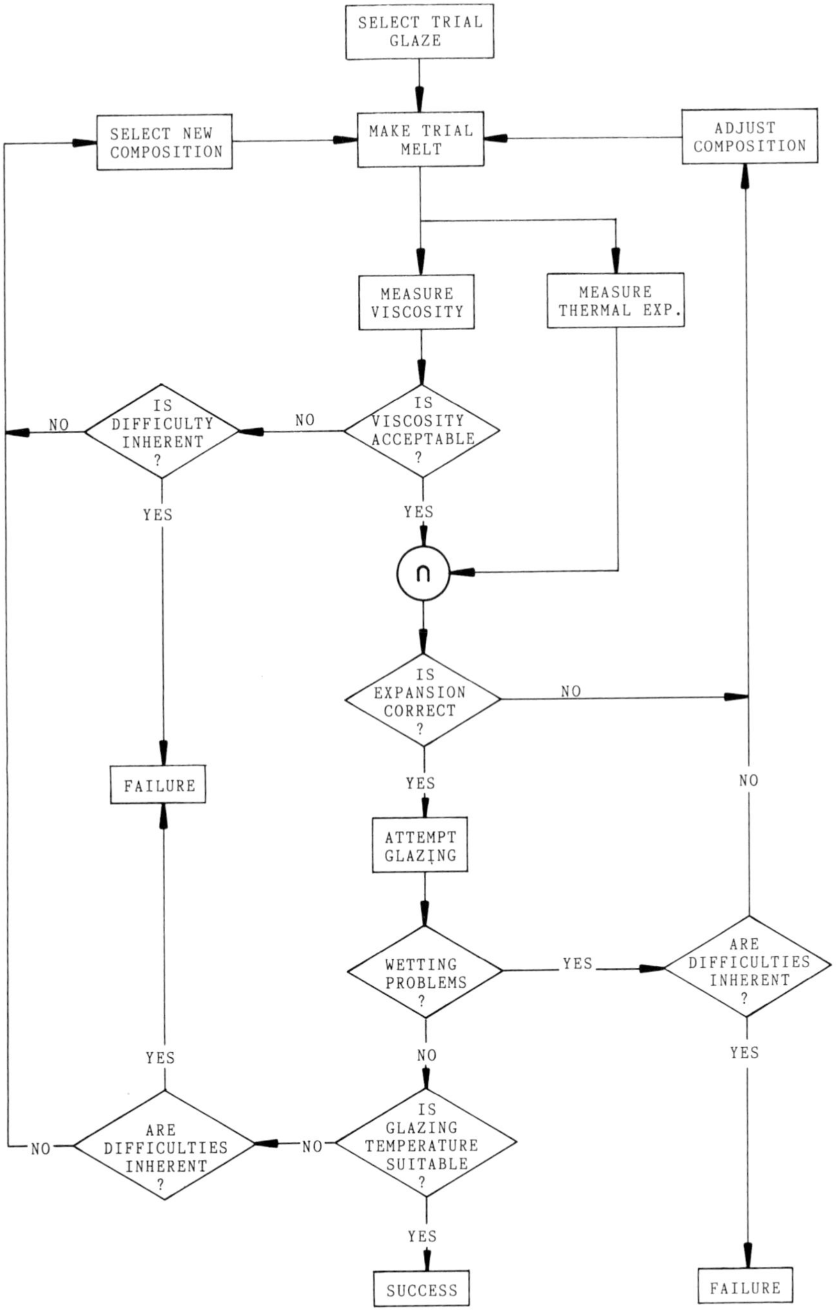

The Gantt chart also provides a simple method of representing research programs (Gibson, 1981g). The horizontal axis is time, usually years broken down into months while the vertical axis carries the description of the activities. The activities are laid out in sequence with the first activity of the program at the top. Figure 43 gives a Gantt chart for a typical R & D program. The chart has many advantages and allows the following to be shown: overall display of activity linkage, critical dependencies (for example, if a task cannot be started until other specified tasks are complete), milestones can be labelled, the critical path can be identified (the path that actually determines the life of project). The chart is self-explanatory and allows continuous monitoring of the project. The chart can be used for management to indicate progress against plan. Boxes may be shaded as activities are completed and a "NOW" cursor can be moved across the chart to indicate on a daily basis whether the project is on time. The layout of the activities shows what the effect of increase in effort would have on the project; increased effort on one of a set of activities in series will speed up the overall effort but effort would have to be increased on all activities that are in parallel. Potential problem analysis may be undertaken once the Gantt chart is prepared. Various scenarios may be tested ahead of problems developing so that action can be taken to avoid difficulty.

It may be remarked that the preparation of activity lists and project plans provides a good basis for the evaluation of performance by R & D personnel provided the activities, milestones and time scales have been mutually agreed. Consistent achievement of goals should be rewarded. Failure to meet goals on a consistent basis should be investigated and appropriate action taken. The clear statement of goals and the objective measurement of achievement is essential both for project performance and personnel success.

PERT (Program Evaluation and Review Technique) and CPM (Critical Path Method) were first developed for large complex programs in the defense area but have been used since for much smaller programs (Davis, 1976). The model is of the linear programming type and the basic concept is of a network of interrelated activities necessary to achieve prescribed events.

FIGURE 43

GANNT CHART FOR PREPARATION OF A PROJECT PROPOSAL

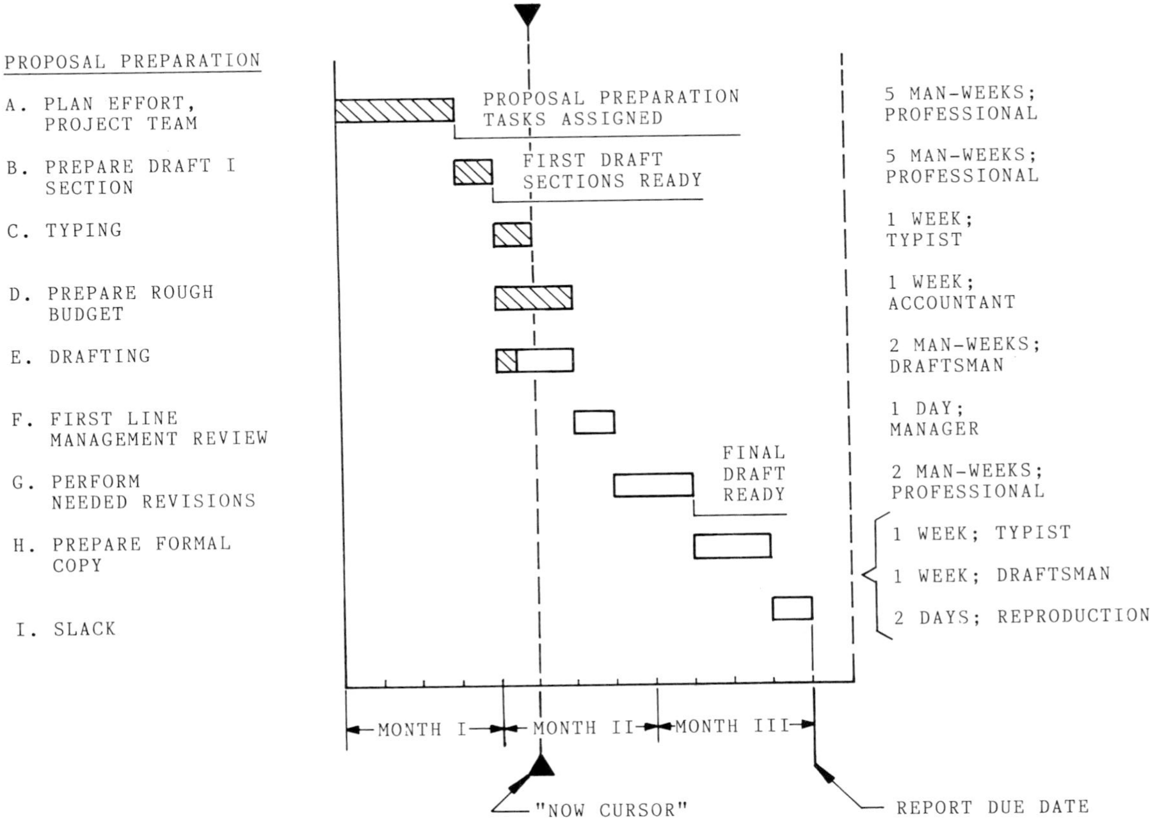

The original convention for setting up the network was activity-on-arrow (AOA) where the activity was represented by the arrow joining events; events are represented by circles and signify the completion of all activities that lead into that event. Figure 44 shows a simple network of the AOA type. The activities are represented by a-j and the events by 1-7. The time for each activity is associated with the arrow. In any particular case the network for a project is set up as follows:

(i) divide the project into mutually exclusive activities

(ii) define the ordering of the activities

(iii) give a time estimate for each activity

On Figure 44 the critical path is indicated by heavy arrows. The critical path is that set of activities critical to project completion on schedule; if any activity on the critical path is delayed the whole project completion will be delayed. In the example given the critical path consists of events 1, 3, 4, 5, 7.

FIGURE 44

PERT CHART WITH ACTIVITY-ON-ARROW (AOA)

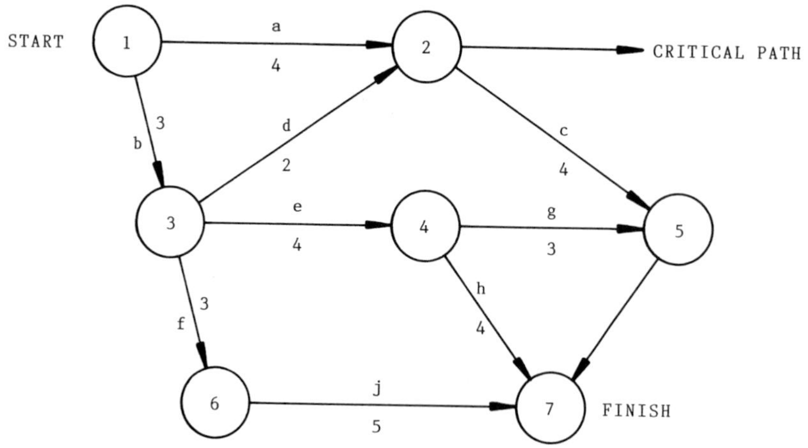

The second convention for setting up the network is the activity-on-node (AON also called Precedence Diagramming). There, the circles or nodes are activities i.e. the users of resources. The arrows only indicate the precedence. It is recommended that the AON system be used since activities are scheduled and not events. For the AOA system it is necessary to

FIGURE 45

AON REPRESENTATION ILLUSTRATING SLACK

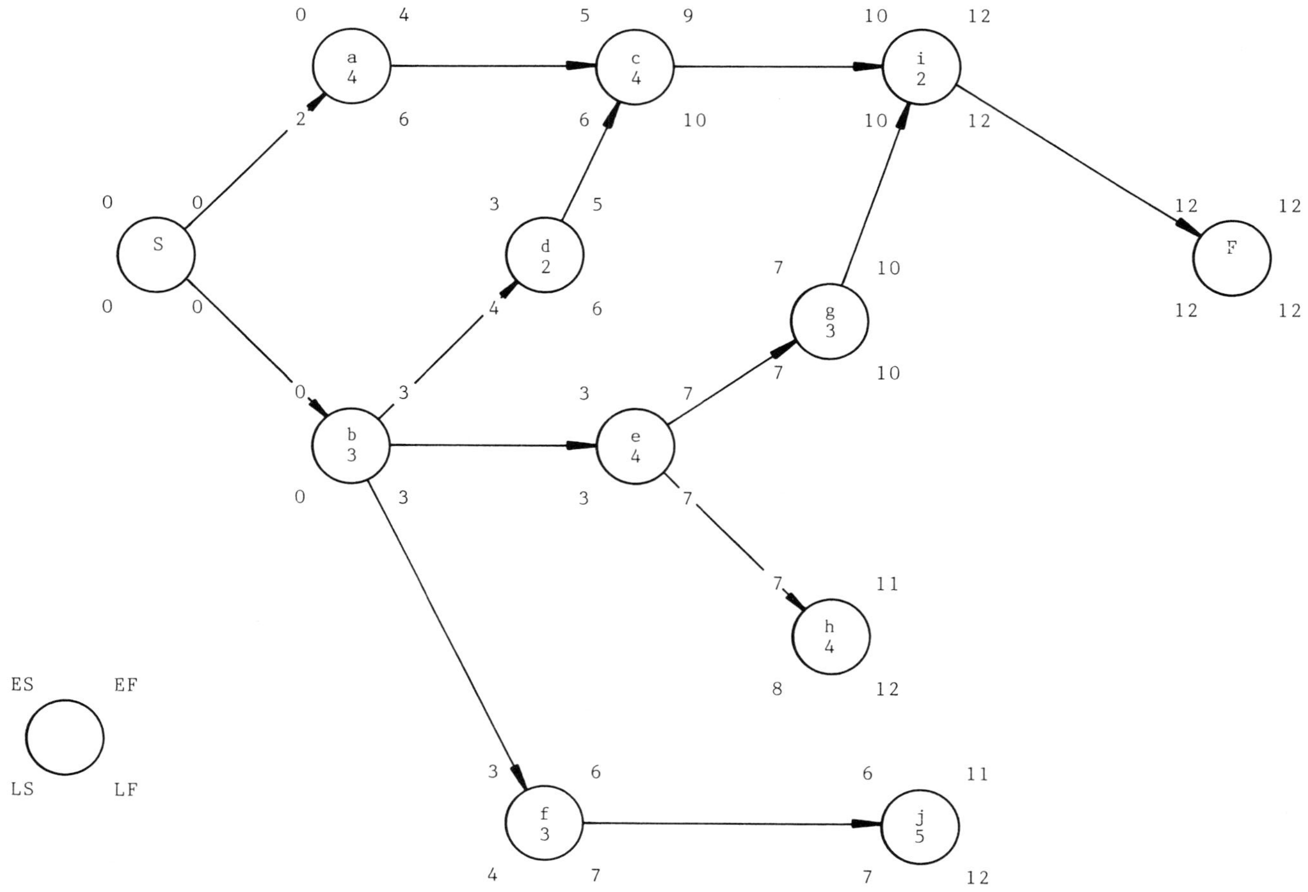

use dummy activities to indicate relationships. Table 34 shows the conversion from AOA to AON for the network shown in Figure 44. The AON network is shown in Figure 45. Here the concept of "slack" is illustrated. The activities are a-j with a Start (S) and Finish (F). Reading clockwise around the circle the numbers refer to early start, early finish, late finish and late start (note that start on day T means at the end of day T). Slack in an activity is given by LS-ES or LF-EF. The number in the circle refers to the length of time for the activity to be completed.

TABLE 34

Activity-on-arrow (AOA) to activity-on-network (AON) conversion

Activity	Precedence	Time to completion
Start (S)	-	0
a	S	4
b	S	3
c	a, d	4
d	b	2
e	b	4
f	b	3
g	e	3
h	e	4
i	c. g	2
j	f	5
Finish	i, h, j	0

PERT is often applied to R & D activities where precise time estimating is impossible. The method provides for the determination and use of probability distributions for projected activity duration. These usually take the form of three time estimates for each activity designated optimistic, most likely and pessimistic (a, m and b respectively). If it is further assumed that the Beta probability distribution applies then the expected activity duration is given by:

$$T_e = \frac{a + 4m + b}{6}$$

and the variance by

$$\sigma = \frac{b - a}{6}$$

where σ is the standard deviation of the expected time distribution. It is then possible to compute such quantities as:

(i) the earliest time a particular event can occur and the probability of it occurring at any particular time,

(ii) the latest time a particular event (milestone) can occur if the scheduled completion date is to be met,

(iii) the probability that the entire project can be completed by a specified date. This is done by summing the expected activity completion times and variances along the critical path and using the Normal distribution assumption to get the expected value and variance for the completion date.

(iv) allowable schedule slippages for any activities or events not on the critical path,

(v) analysis of the network for potential problem areas.

Figure 46 gives the representation of a network that will be used for the purpose of illustrating "crashing" i.e. project completion speedup by increasing resource commitment. Table 35 shows the costs and times for this network. The crash time is the shortest time needed to complete the activities and associated with this is the crash cost. Note that the crash time is the shortest time possible and cannot be improved on no matter how much is spent. The cost slope refers to the cost per unit

FIGURE 46

AON REPRESENTATION OF PROJECT FOR ILLUSTRATION OF "CRASHING"

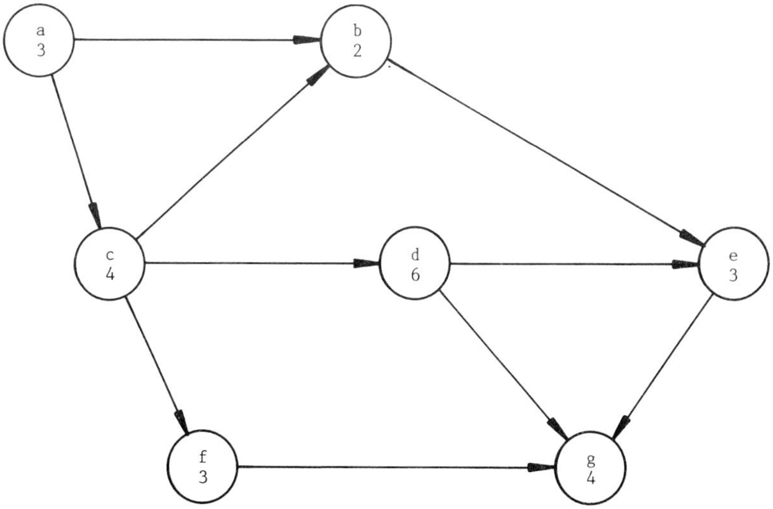

time for reduction in the time to completion of the activity. A linear relationship is usually assumed for simplicity. The project shown has a completion time of 20 weeks at a cost of $84,000. To reduce the time the least cost sensitive activity on the critical path is chosen. For example, the total time can be reduced to 19 weeks by acting on activity d. This will increase the cost to $86,000. To go 18 weeks another week can be taken from d and the cost increases to $88,000. Reductions can be made to further speed up the project but at each step the network must be examined to see whether the critical path has changed. It may be necessary to manipulate several activities (even increasing some) to get the required completion date.

TABLE 35
Effect of time on project cost (time in weeks. Cost slope is cost per week)

Activity	Normal Time	Crash Time	Normal cut	Crash cost	Cost slope
a	3	1	10,000	20,000	5,000
b	2	1	5,000	25,000	20,000
c	4	2	20,000	40,000	10,000
d	6	4	30,000	34,000	2,000
e	3	1	8,000	48,000	10,000
f	3	2	1,000	2,000	1,000
g	4	1	10,000	40,000	10,000

In some cases there is more than one critical path. Table 36 gives the illustration of building a house. If the network is laid out it will be found that there are four critical paths:

a b c i j k r s t w
a b c i j n r s t w
a b e h j k r s t w
a b e h j n r s t w

In this case the path with maximum variance is used. The mean time for completion is 36.8 days and the standard deviation is 2.43 days. To find the probability of completion in 30 days find (36.8 - 30/2.43), i.e. 2.8, in the Normal Tables. The probability is 0.0026 and there is virtually no chance of completion in 30 days at the given level of resources.

TABLE 36

Pert scheduling - house building illustration

Job	Description	Predecessors	a	m	b
a	Start	None			
b	Lay foundation	a	4	6	10
c	Wooden frame and rough roof	b	3	4	6
d	Brick work	b	2	4	6
e	Basement drains & plumbing	b	1	2	4
f	Pour basement floor	e	1	2	3
g	Rough plumbing	c,f	2	3	5
h	Rough wiring	c,f	1	2	3
i	Heating & ventilating	c,f	2	4	5
j	Plaster board & plastering	i,h,g	7	10	17
k	Install finish flooring	j	2	3	6
l	Install kitchen fixtures	j	1	2	4
m	Install finish plumbing	j	1	2	3
n	Finish carpentry	j	2	3	6
o	Roofing and flashing	d,c,f	1	2	4
p	Gutters and downspouts	o	1	2	3
q	Storm drains	p	1	2	5
r	Sand and varnish flooring	k,l,m,n	2	3	6
s	Paint	r	1	3	5
t	Finish electrical work	s	1	2	4
u	Finish grading	q	1	2	3
v	Walks and landscaping	u	3	5	8
w	Finish	v,t			

The changing of resource levels in the form of manning is shown with reference to the project outlined in Table 37. Here the objective is to keep the manning fixed over the life of the project. Table 38 shows that by appropriate grouping level manning requirements can be achieved. This is done by looking at starting times and the amount of slack in each activity. It is likely that the requirement of level manning will lead to sub-optimization of the completion schedule in other projects than the one considered.

The examples given have been chosen to illustrate the principles of PERT and CPM. These techniques are really only worthwhile for large programs and much of the complexity of setting up the network, calculating times and probabilities and optimizing resource allocation can be done by computer. However, the activities still have to be listed and needless to say must be

TABLE 37

Manpower requirements for project example

Activity	Duration	Manpower/week	Precedence
1	4	9	S
2	2	3	S
3	2	6	S
4	2	4	S
5	3	8	2
6	2	7	3
7	3	2	4,6
8	4	1	5,7
F	-	-	1,8

TABLE 38

Leveling of manpower resources by activity grouping

Week	Activity	Manning	Activity	Manning	Activity	Manning
1	1,2,3,4	22	2,3,4	13	3,4	10
2	1,2,3,4	22	2,3,4	13	3,4	10
3	1,5,6	24	5,6	15	2,6	10
4	1,5,6	24	5,6	15	2,6	10
5	5,7	10	5,7	10	5,7	10
6	7	2	7	2	5,7	10
7	7	2	7	2	5,7	10
8	8	1	1,8	10	1,8	10
9	8	1	1,8	10	1,8	10
10	8	1	1,8	10	1,8	10
11	8	1	1,8	10	1,8	10
F	-	-	-	-	-	-

carried out successfully for the project to succeed whatever the planning method employed.

4.5 SUMMARY

The project is the basic mechanism whereby R & D directs resources towards the development of new processes or products. The basic project cycle involves the project definition which includes a description of the project, the technical and business relevance, the approach to be taken and the benefits that will accrue from successful completion (usually sales and profits from the developed product or process). The objective in industrial R & D is to maximize the value added leverage from high technology products. Cash from existing product sales is employed to develop new products that in turn generate cash to support further R & D.

The generation of new projects is not straightforward. A variety of methods have been used. Technological forecasting may be employed to define the future technological environment either by extrapolating present technology into the future or by exploring scenarios of the future. Social and economic issues must also be discussed. Creativity techniques may be used to develop ideas for projects. Other techniques involve search of the environment by literature and patent review, attendance at scientific meetings and personal contacts. Projects may also be generated by top management decisions, competitors' actions or by technical developments within R & D itself.

Much attention has been directed both in industry and in academia to project selection. Methods proposed fall into four classes: ranking, scoring/rating, economic rating and formal optimization. No one method or class of methods has universal acceptance although most companies attempt some form of project review. It must be pointed out that the generation of projects is the most difficult part of the process. Usually there is not a large project pool from which selection is made.

Project review is not a once-and-for-all procedure. Projects must be reviewed at appropriate intervals determined by time, cost or achievement constraints. For the selection process to work it is necessary to have a project description and plan. The plan is then used as part of the control process. Project descriptions may be presented via Abstracts and Protocols (or in some similar format appropriate to the industry). Activity Lists with milestones indicate expected progress and may be graphically depicted by Gantt charts or Flow Diagrams. For complex programs of large magnitude PERT/CPM techniques have been extensively used. There are now many software packages available for project management with various degrees of complexity.

Project activity is the core activity of R & D. Where R & D differs from units such as manufacturing however is in the openness of the system and in the need for individual contribution. Personal relationships within the R & D group are extremely important as are the interfaces with other company units and the external environment. These subjects will be addressed next.

REFERENCES

Ansoff, I.H.	1964	"Evaluation of Applied Research in a Business Firm", in "Research, Development and Technological Innovation", Bright, J.R. (ed.), Irwin, New York.
Anthony, R.N. and Reese, J.S.	1975	"Management Accounting", Richard D. Irwin, Homewood, Ill. 5th Edition, p. 641.
Ayres, R.U.	1969	"Technological Forecasting and Long-Range Planning", McGraw-Hill, New York.
Batty, J.	1976	"Accounting for Research and Development", Business Books, England.
Bell, D.C. and Read, A.W.	1970	"The Application of a Research Project Selection Method", R and D Management, 1, No. 1, October, 35-42.
Carsons, J.W. and Rickards, T.	1980	"Industrial New Product Development", Gower Publishing Co., Ltd., Farnborough, Hampshire, England.
Davies, D.G.S.	1970	"Research Planning Diagrams", R and D Management, 1, 22-29.
Davis, E.W. (ed.)	1976	"Project Management: Techniques, Applications and Managerial Issues", American Institute of Industrial Engineers, Norcross, Georgia. Publication Number 3.
de Bono, E.	1980	"Opportunities: A Handbook of Business Opportunity Search", Penguin Books, Ltd., Harmondsworth, Middlesex, England.
Gee, E.A. and Tyler, C.	1976	"Managing Innovation", John Wiley and Sons, New York pp. 114, 117.

Gee, R.E.	1971	"A Survey of Current Project Selection Practices", Research Management, 38-45, Sept.
Gibson, J.E.	1981a	"Managing Research and Development", John Wiley and Sons, New York, 204-208.
Gibson, J.E.	1981b	Ibid, Chapter 10.
Gibson, J.E.	1981c	Ibid, Chapter 8.
Gibson, J.E.	1981d	Ibid, 308-314.
Gibson, J.E.	1981e	Ibid, 316-319.
Gibson, J.E.	1981f	Ibid, 265-272.
Gibson, J.E.	1981g	Ibid, 278-282.
Hardy, D.D.	1965	"PERT for Small Projects", Royal Aircraft Establishment Report, No. 65271, Farnborough, England.
Holt, K.	1975	"Information and Needs Analysis in Idea Generation", Research Management, XVIII, No. 3, 24-27.
Kenny, R.L. and Riaffa, H.	1976	"Decisions with Multiple Objectives", John Wiley and Sons, New York.
Jantsch, E.	1967	"Technological Forecasting in Perspective", O.E.C.D., Brussels.
Jones, H. and Twiss, B.	1978	"Forecasting Technology for Planning Decisions", MacMillan, London.
Mansfield, E.	1981	"How Economists See R and D", Harvard Business Review, November-December, 98-106.
Merrifield, D.B.	1978	"How to Select Successful R and D Projects", Management Review, December, 25-28 and 37-39.
Pessemier, E.A. and Baker, N.R.	1971	"Project and Program Decisions in Research and Development", R and D Management 2, No. 1, 3-14.
PIMS Letter 29	1982	"Is R and D Profitable", Strategic Planning Institute, Cambridge, Mass.
Riaffa, H.	1968	"Decision Analysis", Addison-Wesley, Reading, Mass.

Rickards, T.	1980	"Designing for Creativity: A State of the Art Review", Design Studies, <u>1</u>, No. 5, 257-272.
Schlicksupp, H.	1977	"Idea-Generation for Industrial Firms - Report on an International Investigation", R and D Management, <u>7</u>, No. 2, 61-69.
Schwartz, S.L. and Vertinsky, I.	1977	"Multi-Attribute Investment Decisions: A Study of R and D Project Selection", Management Science, <u>24</u>, No. 3, 285-301.
Souder, W.E. and Ziegler, R.W.	1977	"A Review of Creativity and Problem Solving Techniques", Research Management, <u>XX</u>, No. 4, 34-42.
Twiss, B.	1980a	"Management of Technological Innovation", 2nd edition, Longman Group, Ltd., London, 122-130.
Twiss, B.C.	1980b	Ibid, 130-131.
von Hippel, E.A.	1978	"Users as Innovators", Technology Review, <u>80</u>, No. 3, 31-39.
Whitfield, P.R.	1975	"Creativity in Industry", Penguin Books, Ltd., Harmondsworth, Middlesex, England, Chapter 4.
Wills, G. with Wilson, R., Manning, N. and Hildebrandt, R.	1972	"Technological Forecasting", Penguin Books, Ltd., Harmondsworth, Middlesex, England.

PART 5

THE R & D PROCESS: PERSONAL FACTORS

5.1 INTRODUCTION

Many aspects of industrial R & D have been discussed. A definition of R & D has been given. Strategy/structure relationships have been discussed and the role of technology in the firm has been presented. The goal of industrial R & D is to produce products or processes. This is accomplished via the R & D project which has been covered in detail. Novelty is associated with R & D and so the subject of creativity has received attention. However, the areas covered have omitted close consideration of R & D personnel, the personal contacts necessary between R & D personnel and those in other units, conflict, motivation, management styles and so on. There are, therefore, many personnel and process-related factors that impinge on the effectiveness of R & D performance.

It has been mentioned earlier that there is a plurality of social systems in any organization (Burns and Stalker, 1966). Rational considerations would suggest that individuals work in a cooperative fashion so that the ends of the organization are attained. Individuals, however, have private agendas. There is the need for career advancement and personal fulfillment. Any organization has a network of political relationships that can result in actions quite at variance to what would be predicted by a rational observer.

Figure 47 illustrates the view of an organization from the perspective of the individual (Lorsch and Morse, 1974a). The individual is viewed as a system operating within the organizational system that, in turn, is embedded in an environment depending on economic, business, political and social factors. This view emphasizes the value-set of the individual and the way in which the individual fits into the context of the organization. In the R & D organization, the interest is in the competence of the individual, the motivation to produce to potential, the ability to work with others within the R & D unit and the

FIGURE 47

A SCHEMATIC VIEW OF SYSTEMS AND ENVIRONMENTS

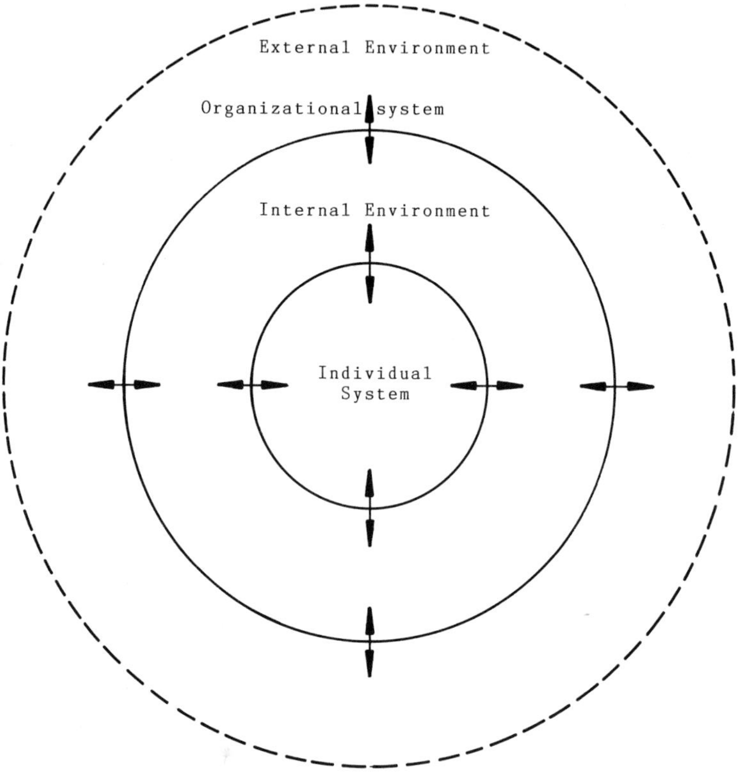

ability to interact positively with individuals in other units and in the external environment; an individual in R & D, of course, does not have to possess strengths in all these areas.

R & D is an open system. In order to survive and prosper R & D must interact with other units of the company and also with the external environment. In the general organization context, differentiation into functional units such as production, R & D, marketing and sales is recognized to produce difficulties unless integration of the various activities is achieved. "Integration is the achievement of unity of effort among the major functional specialists in a business" (Lawrence and Lorsch, 1967). In this study four questions were posed:

(i) How should integrators be oriented and motivated?

(ii) What patterns of conflict resolution and influence should be employed?

(iii) What authority should integrators possess and how do they acquire it?

(iv) Who are the most qualified persons for the position of integrator?

The interface in general for any unit and in particular for R & D is seen as a region of special difficulty. The solution is to employ integration mechanisms but integration involves people and so integration reduces to the selection of persons, having appropriate qualities, who act in the most appropriate manner. The questions posed show that integration involves not only individual qualities but all of the social systems of the organization. In particular, power and influence are involved and conflict is to be expected.

Apart from differentiation, interfaces are created by decentralization. Thus marketing may be decentralized so that each marketing unit faces a relatively homogeneous environment (homogeneous in relation to geographical area, product line or rate of technological change). Integration mechanisms are required. Integration in the form of cooperation between heretofore autonomous divisions may be desirable so that resources such as R & D and computer facilities may be shared. The cost of failure in some product areas is now so high that it makes sense to encourage contact across divisional and even company lines to provide the most complete information before a decision is made (Duncan, 1979).

Relationships other than those generally accepted in strategy-structure theory have also been considered (Hall and Saias, 1980). As these authors contend, strategy does not unambiguously lead to structuring of the organization. Once a structure is in place it constrains the strategic choices that are available. The constraints, in fact, are not simply due to structural hierarchical considerations (the way in which the organization is represented as a network of positions and units) but are also to do with career and political aspirations of the organizational members. Who would believe that a director of R & D would agree to a strategy involving the disbanding of the unit so that members could work directly for manufacturing, engineering, product development etc. where they might be more effective?

Strategic perceptions are conditioned by structure (Hall and Saias, 1980). Structural characteristics act as filters and limit what the organization perceives in the environment. Incoming information is paid greater or less attention by various structural elements depending on whether the information is judged to be interesting and on whether the organization has time to spare to give the information attention. Information is continually transferred not only across the organization interface but also between units (ultimately between individuals). The transfer process is not neutral. Information is often changed by the way it is transmitted. Again factors other than rational factors are seen to be at work. This and many other aspects of the innovation process are described elsewhere (Sayles, 1974).

The R & D manager must be aware of the different social systems in the organization. The success of a project depends on the selection of individuals, the formation of groups or teams with the right mix of talents and the choice of the best management style. The manager must know the best methods to employ in any given situation for interaction with other company units. It is often possible to avoid difficulty by laying the right groundwork. These are not matters involving merely the hierarchy of the organization and often do not involve the processes of communication formally available. It is appropriate to consider some of those aspects that can make or break an R & D manager. General management treatments may be read as background (Galbraith, 1977; Handy, 1967a; Dessler, 1976).

5.2 INDIVIDUAL AND GROUP

5.2.1 Introduction

There are many aspects of personal relationships that are of relevance to the smooth and successful working of the R & D organization. There is a voluminous literature on many aspects of inter- and intra-personal factors. It is not the aim here to review this literature but rather to describe prevailing views on aspects such as motivation, the working of groups and leadership. In the following sections the approach of Handy will be followed (Handy, 1976a). After this general treatment studies specific to R & D will be described.

5.2.2 Motivation

Why do people act the way they do? What motivates them? Specifically, from the point of view of the organization the question is posed as: "Why do people work?" The interest in motivation theory stems from the fact that a knowledge of the ways that individuals are motivated can lead to a direction and influencing of their actions. Individuals can be motivated to act in a manner that benefits the organization. Whether this is manipulation or management will be left to the reader to decide.

Early theories of motivation may be grouped into three categories: satisfaction theories, incentive theories and intrinsic theories (Handy, 1976b). The assumption for satisfaction theories is that a satisfied worker is a productive worker. There is little evidence that a satisfied worker actually works harder but there is evidence that a satisfied worker tends to stay in the same organization. Incentive theories are based on the assumption that individuals work harder when given specific rewards for good performance. Incentive theories can work if the increased reward is worth the effort, if the performance can be measured and attributed to the individual, if the individual wants the reward and if the increased performance does not become the new minimum standard. Intrinsic theories depend upon the satisfaction of human needs. Maslow categorized human needs as follows: physiological needs, safety needs, belonging and love needs, esteem needs and self-actualization needs. Needs are only motivators when unsatisfied. The needs are structured in the order given. Thus the highest need, self-actualization, cannot be addressed until all the four lower needs are satisfied. The idea is to create conditions where effective performance is a goal in itself. The worker obtains satisfaction from a job well done and this is motivation for effective performance. Intrinsic theories should work best where people are intelligent and independent and where the problems are challenging - these are the conditions which should be present in the R & D organization.

The theories all stem from underlying assumptions about man:

(i) Rational-economic man. Man is a passive creature primarily motivated by economic needs. Man is essentially irrational and must be controlled and organized (obviously there are men who are not in the same mold and who know how to control and organize the others).

(ii) Social man. Man is a social animal and gains his sense of identity from relationships with others. Meaning in work is obtained not from the work itself but from the social relationships developed with others. This approach will place great emphasis on management style.

(iii) Self-actualizing man. Man is primarily self-motivated and self-controlled. Given a chance man will integrate his own goals with those of the organization.

(iv) Complex man. Man is variable. He has many motives. The hierarchy of his motives changes from time to time and he does not necessarily need fulfillment in any one situation. He can respond to a variety of managerial strategies but may not unless they are viewed as appropriate to the situation and his needs.

(v) Psychological man. Man is a complex unfolding creature who passes through different stages towards an ego ideal. Man must be allowed to work towards this ideal through the tasks that the organization provides.

Based on the foregoing Handy developed an individual model as shown below:

needs

"E" factors ————> motivation calculus

results

This model states that each individual has a set of needs and a set of desired results. The individual decides how much "E" (effort, energy, expenditure, etc.) to invest by calculation.

Maslow's hierarchy of needs has already been mentioned. The list can be lengthened by adding needs such as justice, independence and so on. The relative importance of needs has been discussed by Herzberg, who grouped needs into two categories: hygiene factors and motivation factors. Hygiene factors such as salary, physical working conditions, supervision and company policy are potential dissatisfiers that must be dealt with. However, dealing with them does not turn them into satisfiers but makes them neutral factors. The satisfiers (motivators) are achievement, recognition, task, responsibility and advancement. McClelland has grouped needs as follows: need for achievement, need for power and need for affiliation. Each individual will have his own set of needs that will differ from individual to individual and will change with time.

The motivational calculus has three separate elements:

(i) The saliency of the need.

(ii) The expectancy that the "E" will lead to a particular result.

(iii) The instrumentality of the result in satisfying the need.

The motivational calculus operates within the confines of a contract, a psychological contract, that the individual has with the group (organization). Specifically, the organization has a set of expectations for the individual which may be specific and open as in a written job description but more often is only communicated in a general fashion. For example, R & D is expected to develop new products. The individual also has a set of expectations from the organization. There will be difficulties if the psychological contract is not viewed identically by both parties.

Organizations may be categorized according to the type of psychological contract that predominates. These categories of contract are: coercive, calculative and cooperative. Coercive contracts prevail in prisons, custodial mental hospitals and other institutions where the inmates are not free to leave. Calculative contracts involve an exchange of goods or money and the contract is voluntary. Most industrial organizations have this type of contract. However, more "E" requires more money. The result will be viewed as coercive if the organization tries to get more "E" for the same pay or tries to reduce pay or fringe benefits (give-backs). Cooperative contracts hold where the individual can identify with the goals of the organization.

However, what is seen to be important by management may not be important for the workers. It is expected that the type of contract will differ across the organization and a contract in R & D will not be of the same type as for the shop floor.

The individual decision model gives guidelines over how to motivate the individual in general but specifically the method of motivation will depend on the individual and the specific situation. The model also gives information on the outcome of efforts by the individual. An individual who reaches a result after considerable effort but who does not obtain the desired reward will experience dissonance which leads to stress. The individual can increase efforts to obtain still higher results or reduce efforts deciding that the achievement of the result will not satisfy his need.

Specifically for the R & D environment, the general literature has been discussed and a model similar to that above has been proposed (Farris, 1973):

Technical Performance = f (A x M x T x Eg x Ig)

where
- A - abilities
- M - motives
- T - technical information
- Eg - expectation that performance leads to goals
- Ig - incentive value of these goals

According to Farris the R & D manager should view his job as "conducting a continuing series of negotiations with his technical people in which work goals are established, technical collaboration in an informal organization is encouraged and career opportunities with the organization are defined as clearly as possible"

5.2.3 Leadership

It is undoubtedly true that the quality of supervision plays an important role in the success of an organization. The term "leader", however, implies something more than supervision. While the art of supervision may be learned there is some disagreement as to whether leaders are born or made. The search for the definitive solution of the leadership problem has provided no clear answer. Just as for motivation, there has to be an element of contingency. A good leader in one situation is perhaps an indifferent leader in another situation.

Approaches to the definition of what makes a good leader have fallen into three categories (Handy, 1976c):

(i) Trait theories - the assumption is that the leader is more important than the situation. The traits of an individual decide whether or not the person is a good leader. Studies have defined a long list of attributes that cannot all be held by a single individual. However, it appears that a leader has above average intelligence, is good at solving complex problems, has initiative and has self-assurance. These appear to be necessary but not sufficient conditions for leadership.

(ii) Style theories - these theories assume that employees will work better for a manager who uses a particular style of leadership. Styles may be compared on a scale one end of which is the authoritarian style and the opposite end the democratic style. For the authoritarian power resides with the leader whereas for the democratic leader power is shared along with responsibility. These styles, and the intermediate mixed styles, imply the different assumptions on motivation covered in the previous section. There have been many classifications of style. Generally the extremes of the spectrum of styles are now referred to as structuring and supportive. There is evidence that supportive styles of leadership give higher employee satisfaction and this style should work better in an unstructured environment (e.g. R & D). It has to be pointed out that some individuals like to be directed and structured and a structured style of leadership may well work better, at least in the short term, for routine and repetitive tasks.

(iii) Contingency theories - these theories take into account other factors of the leadership situation such as the nature of the task, the work group and the position of the leader within that group.

According to Handy, leadership can best be described by a modification of the contingency approach. In any situation there are four sets of influencing factors: the leader with his traits and preferred style; the subordinates with their preferred leadership style; the type of task; the environment including the organizational setting of the leader, the group and the task. Leadership will be most effective when there is a good fit between the leader, the subordinates and the task. These three factors can be placed on a scale that runs from tight to flexible. For example, if the task is unstructured (flexible) and the subordinates prefer control over their own work then the leader should adopt a flexible approach to supervision.

The preferred style of the leader will depend on many factors including his value system, confidence in subordinates, habitual style, feeling of his own importance to the task, need for certainty, degree of stress and age (older managers tend to be more structuring). Taking all of these factors into consideration will allow the leader's style to be placed along the continuum from tight to flexible.

The subordinates' preference for a particular style of leadership will depend on their estimate of their own confidence, the psychological contract with the leader, their interest in the task, their tolerance for ambiguity, past experience of the group and cultural factors (R & D personnel tend to want more responsibility for their work).

How the task is arranged (structured or flexible) depends on whether it is routine or demands creativity, on the time scale, the complexity, the admissibility of errors in performance and on the task importance. Open-ended tasks tend to be towards the flexible end of the scale, short time horizons lead to structuring, complexity of the task may lead either to flexibility or to a high degree of structuring. Unimportant tasks may be tightly structured but it may not be desirable to leave important tasks to chance - it all depends.

The environment will play an important role in the amount of freedom the leader has. There are six key factors: the power position of the leader in the total organization; the relationship of the leader to the group; the organizational norms; the structure and technology of the organization; the variety of tasks; the variety of subordinates. These aspects of the environment have been considered in pairs:

(a) The power position of the leader in the organization and the relationship of the leader and his group - these two elements allow the leader to adjust his style and the nature of the task. The redefinition of the task requires that the leader have a power base in the organization. Power based on expertise and achievement gives the leader more freedom in his choice of style than purely legal power. If the leader has much credit with the group then he can indulge in deviant behavior as suits the task. So can the leader with no credit because he has nothing to lose. It is the leader in the middle status with the group who has to be careful and conform to the group.

(b) The organizational norms and the structure and technology of the organization - the organization limits the degrees of freedom of the leader because of requirements such as authorization procedures, reporting procedures, hours of work, manner of dress, mode of behavior and so on. Decentralization gives more scope for individual managers to define their own style and task. The type of technology also influences the structuring of the task and the style that is appropriate.

(c) Variety of tasks and variety of subordinates - a wide variety of tasks from short-term to long-term, routine to complex and closed to open-ended implies that no one style is appropriate. This may not be allowed since variability of leadership confuses subordinates. The membership of the groups continually changes as employees are hired and others leave. This may confuse the issue of the subordinates' preference for a particular style of leadership.

The conclusion is that leaders are to some extent born and not made. However, it is possible to learn about the art of leadership. At the very least a manager should be able to analyze the situation as regards his own habitual style, the group, the task and the environment and make an informed choice as to the preferred style of leadership.

5.2.4 Groups

Organizations employ groups for many purposes (Handy, 1976d): for the distribution of work, for the management and control of work, for problem-solving and decision-making, for information processing, for information gathering and idea collection, for testing and ratifying decisions, for coordination and liaison, to increase commitment, for negotiation or conflict resolution and for inquest or inquiry into the past. Examples include task forces for special projects in R & D, new product committees, committees that bridge departments for coordination purposes and so on. There are difficulties when the group is expected to perform a variety of very different tasks.

Individuals use groups for many purposes: for satisfaction of affiliation needs, to establish a self-concept, as a means of help and support and for sharing in common activities such as making a product. For individuals groups have social as well as productive functions and these may be in conflict.

Studies indicate that groups are necessary to provide a psychological home for the individual. Groups provide less ideas in total than the individuals working separately but the ideas are better in the sense of being better evaluated (see Part 3 for group creativity processes). A group can be especially effective in those cases in which it is a lack of information that is hindering problem solution. If properly chosen the group can act as an information resource. Groups take riskier decisions than the individuals would have taken if acting separately (perhaps due to shared blame).

The determinants of group effectiveness depend on the givens (the group, task and environment), the intervening factors (leadership style, processes and procedures, and motivation) and the outcomes (productivity and member satisfaction). While productivity generally leads to member satisfaction, a satisfied group is not necessarily a productive group. These factors will now be discussed:

(i) The group - the size of the group is important. The larger the group the greater the array of knowledge and skills but the lower the chance for individual participation. The threshold for participation varies with the individual. As the group gets larger the pattern of influence shifts towards those who can participate easily. However, it is not necessarily the most competent or knowledgeable individuals who are head. For best participation a group size of 5 to 7 individuals seems to be optimum.

Member characteristics are important including the characteristics of the leader. People with similar attitudes, values and beliefs tend to form stable groups. Homogeneity provides satisfaction and a sense of psychological belonging. Heterogeneous groups exhibit more conflict but are generally more productive than homogeneous groups. Group members should not be dissimilar in all characteristics, however, since this can lead to conflict without productive activity. Variety but compatibility seems to be the key to effectiveness and there can only be one leader. A group with two or more individuals seeking to be leader will accomplish nothing.

The objectives of the individuals making up the group will influence the outcomes of group activity. If all members have similar objectives the group will be more effective but individuals will bring "hidden agendas" to the group such as covering up errors, scoring off an opponent, forming alliances and so on. There has to be a trade-off between individual and group objectives. An individual can play many roles in group interactions such as innovating, supporting, defending/attacking, bringing in or shutting out and many others. The roles depend, in addition, on the position of the group on the development cycle. There are some four stages:

(a) Forming - the group is a set of individuals. The parameters of group activity are being decided (perhaps group title, composition, leadership, and life-span). Each individual wants to make an impression at this stage.

(b) Storming - conflict occurs as leadership roles, work patterns and group purpose are challenged. At this time hidden agendas become visible. If the group successfully transitions through this stage, then there will be a more realistic setting of objectives, procedures and norms.

(c) Norming - norms and practices are established including the procedures for decision-making, level of work, degree of openness and type of behavior.

(d) Performing - if the previous stages have been successfully completed, the group will have reached maturity and will be at its most productive stage. A group will mature rapidly when the issues are of extreme importance. If there is excessive conflict and politics which is driven underground during the storming stage, the group will never mature. There are many techniques that are now employed specifically for group building (T-groups, Coverdale and process consultation).

(ii) The task - the type of task will influence the type of

group. Problem solving will demand a smaller group than may be allowed for information dissemination. The task will determine whether the member is present as an individual, as a representative of his department or as a representative of the organization.

The time scale for a result determines the way in which the group will work. Short time scales lead to more structured working. Criteria must be available to define success so that the group knows when the task is satisfactorily accomplished.

The importance of the task will influence the commitment of the group members. The less ambiguous the task the more structured the leadership can be. Ambiguous, open-ended tasks demand supportive leadership.

(iii) The environment - the organization will impose norms about the way the group should hold meetings and report results.

The position of the leader within the organization and with respect to the group will influence the working of the group. The group will be more effective where it is seen that the leader can effectively represent the group in the organization. If the group is accepted in the organization as a whole and it is accepted that the task is an important one, the group will be more effective. The group will be more effective if group interaction is encouraged. This may be done by physically locating members close to one another or by holding meetings off-site.

(iv) The intervening factors - leadership style has already been discussed. For any group to be effective there is a set of processes or functions that have to be done by some person or persons in the group. It is the responsibility of the leader to see that these are done. Groups should tackle problems systematically rather than by leaping from problem statement to solution. According to Handy the task functions are structured as follows: initiating, information seeking, diagnosing, opinion-seeking, evaluating and decision-managing (see Part 3 for other formulations of the problem solving process which depend on whether the problem is open or closed). The decision-making process is also important. Alternatives are decision by authority, decision by majority, decision by consensus, decision by minority and decision by no response.

Groups also require maintenance and this is also the task of the leader although he does not have to carry out the function himself. Handy lists the maintenance functions as follows: encouraging, compromising, peace-keeping, clarifying, summarizing and standard-setting. Groups that are well maintained are more effective.

The leader can influence the pattern of interaction in the group. If the leader is prominent then he is the center of communication. The solution will be achieved more quickly but it may not be satisfying to the group and may not be the best solution. The leader is still important in guiding the discussion and moving the group towards a solution.

Motivation of the group members is important. An individual will be satisfied in a group if he likes and is accepted by the group, approves of the purpose and work of the group and wishes to be associated with the group in the organization. Motivation is something more than satisfaction and the individual calculus will determine whether the individual will be motivated to work hard towards the group goal. However, the composition of the group may make it impossible for the individual to achieve the desired results despite a high amount of "E". This will lead to demotivation.

Groups play a considerable role in organizational life. In terms of time and effort groups can consume a large part of an organization's resources. More attention should be paid to group composition and dynamics. Groups should only be set up when needed and should be disbanded when the task is complete. A group should never be set up as a temporizing measure as a substitute for decision-making.

5.3 SCIENTISTS IN ORGANIZATIONS

There are few substantive investigations of scientists in organizations. However, two sources of information may be cited. Pelz and Andrews studied the effect of the organization of the laboratory on scientific performance (Pelz and Andrews, 1966). Morse and Lorsch examined individuals, organization and environment from a contingency viewpoint (Lorsch and Morse, 1974b). The findings of these two studies will be discussed before moving on to consider specific topics such as the interface and political relationships.

The study of Pelz and Andrews was based on the premise that R & D organizations provide more than facilities for their members. The environment may either stimulate or inhibit scientific performance. Information was gathered on technical performance, working relationships and motivations for over 1300 scientists and engineers. Conclusions were not based on opinions alone but the data were analyzed to determine what conditions, individual or environmental, actually accompanied high or low levels of performance.

All data were collected from laboratories in the United States. Choice of laboratory was based on the ability to gain entry. 1311 scientists and engineers in 11 different laboratories were studied. There were 5 individual laboratories with 641 professionals covering a wide range of technologies. Also included were 144 faculty members from 7 departments of a mid-western university and 526 scientists and engineers from 5 government laboratories with missions ranging from weapons to agricultural products. Missing from this heterogeneous mix were basic research laboratories in industry and independent not-for-profit institutions.

The first consideration was how the research personnel were to be divided into more homogeneous subgroups. Exploratory studies indicated that three dimensions could be used: orientation of the department towards research or development, possession of a doctorate and domination of the department by individuals with doctorates. From these factors five primary analysis groups were obtained:

(i) Ph.D.'s in development-oriented laboratories (group A),

(ii) Ph.D.'s in research-oriented laboratories (group B),

(iii) non-Ph.D.'s in development-oriented laboratories not dominated by Ph.D.'s (group C).

(iv) Non-Ph.D.'s in Ph.D.-dominated laboratories (group D),

(v) Non-doctoral scientists in research-oriented laboratories not dominated by Ph.D.'s (group E).

For groups A and D the number of scientists/engineers was split half in industry and half in government. All of group E were in government laboratories. Two-thirds of group B were in academia and one-third in government. Three-quarters of group C were in industry and one-quarter in government.

Data on performance were gathered from two sources. In most cases a questionnaire approach was used. Some of the performance data were based on the judgment of a person's work by people in the same laboratory. Judgment was based on contribution to the general technical or scientific knowledge in the field and on over-all usefulness in helping the organization attain its goals. More objective criteria such as number of scientific papers, patents, reports etc. over the last 5 years were also employed.

The study thus examined the impact of personal and environmental factors on performance. Several specific factors were examined and a brief summary of some of the more important findings will be given.

(i) Freedom - a combination of freedom and coordination produced the highest performance. However, the scientists/engineers had to feel that they exerted considerable influence on decisions affecting them. Not only should there be a goal setting between supervisor and subordinate but other individuals should be involved as this allows a diversity of viewpoints to be espoused. "Flat" rather than "tall" organizations are more effective since there has to be a chance of face-to-face interaction between the scientists and other significant people in the R & D organization. There is low performance if the superior sets the goals without reference to the subordinate's input. Over-direction stunts growth and independence.

(ii) Communication - effective scientists both seek and receive more contact with colleagues (finding for Ph.D.'s in research and development laboratories and for engineers not for assistant scientists and non-Ph.D. scientists). Contacts occur in many ways besides talking. The basis for this finding seems to be that contacts expose the individual to new ideas, supply information that may be essential to progress, help an individual to think through a course of action, provide friendly criticism and evaluation and perhaps even provide a respite from the job at hand through conversations on other subjects. The results also showed that frequent contacts with many colleagues was more beneficial than frequent contacts with just a few colleagues. The conclusion seems to be that contacts should be encouraged and facilitated both within the group and with outside individuals. It is important that mutual helpfulness and support prevail.

(iii) Diversity - a scientist or engineer is more effective if there is a diversity of tasks. The important fact is not the number of projects that an individual undertakes but the number of skills or specialties that are required. This suggests that a worker should be encouraged to move outside an area of specialization and should perhaps engage in a mix of research and development activities. The study findings showed that the scientist would benefit from a mild exposure to administrative tasks. For example, the individual might participate in review committees where each staff member periodically outlines future directions. This would also aid in communication. The findings are suggestive of the earlier discussions on creativity where it was found that turning away from the problem often allowed insight that led to a solution.

(iv) Dedication - individuals who were highly involved in their work were found to be high performers. The key to this appears to be the influence that an individual has on the decision-making process. The outstanding scientist is excited about his work but it also helps if other people are also excited. The supervisor who can see meaning and significance in what their subordinates are doing thereby helps to reinforce their own enthusiasm.

(v) Motivation - the findings were that persons who relied on inner sources (their own ideas) for motivation were high effective whereas those who relied heavily on their supervisors for stimulation were below par. Emphasis on self-reliance and independence are mandated yet organizational systems of reward are typically creators of dependence. Persons should be allowed ownership of their programs and results. This can be through named authorship of reports, papers or memos and by presentation of results at meetings where individuals outside of R & D e.g. upper management are in attendance. It is not unreasonable that non-performers be prodded into producing. Pay and status should be commensurate with achievement but only have limited motivation benefits. Only at the bottom of the prestige ladder, for assistant scientists, did strong striving for status benefit performance.

(vi) Satisfaction - effective scientists reported good opportunity for professional growth and higher status but were not necessarily more satisfied. Ambition to rise in status was of dubious benefit for effective performance. The best scientists saw only moderate congruence between their personal interests and those of the organization. Dissatisfaction when manifested as a creative tension can be beneficial. Extrinsic rewards cannot be relied on to motivate achievement but when achievement occurs the extrinsic rewards should be consistent and may stimulate further achievement.

(vii) Similarity - the findings here were not strong but were consistent. Scientists tended to perform better if they named as colleagues individuals from whom they differed in the strategy of tackling technical problems and in the style of approach to the work, abstract versus concrete for example (see the discussion on learning styles in Part 3). Scientists did a little better if they named as colleagues persons with similar orientations towards sources of motivation. Similarity or dissimilarity to the immediate superior did not seem to matter. All of this seems to suggest that colleagues of different orientation in the technical field provide diverse inputs and stimulation for problem solving while colleagues of similar motivational orientation provide a supportive social environment.

(viii) Creativity - it was found that creative ability enhanced performance on new projects with free communication but seemed to impair performance in less flexible situations. This suggests that creative individuals be used as a resource when original ideas are needed. Other individuals who might be creative are perhaps engaged in programs that inhibit the use of creativity.

(ix) Age - the influence of age on performance is a highly emotive issue. Earlier findings suggest a peak in performance with a fall-off as the individual aged further. Pelz and Andrews found a saddle-shaped relationship and postulated that the earlier peak corresponded to the more innovative individual whereas the later peak represented the work of the more convergent or integrative type of individual. There was evidence that the decline in performance was not due to a decrease in the abilities of the individual but rather due to a decrease in motivation. This would be due to several reasons. Scientists may have achieved considerably and may be satisfied. There is a tendency to become more conservative with increasing age and there would be less of a willingness to try new ideas. There may be technical obsolescence unless the scientist keeps up to date. Of course technical fields have life cycles. It may be that the great discoveries in a particular field have been made.

As far as the technical director is concerned a mid-career review for a scientist is of great importance. If a sag has set in, steps should be taken before morale is damaged. Transfer to more applied work might be productive. If the person is willing to make the commitment then there could be transfer to another technical field. An alternative is to have the person act as a consultant. Another idea is to have the individual work with young researchers acting as a mentor and in turn being exposed to new ideas and approaches. The data suggest that the decline in performance is not inexorable, there is no irreversible process of decay, steps can be taken to maintain the performance level of an individual over a much broader time period.

(x) Age and Climate - there is overlap here with the previous topic. The period just before 40 (34 to 39) seems to be critical. Corrective action maintains the level of performance. However, it should be emphasized that young scientists should be encouraged to stand on their own feet as soon as possible as it has been found that later performance is improved if the time needed to achieve effectiveness is not protracted. It is noted that corrective action should be taken early rather than late. The young scientist not guided and encouraged may never mature. The older worker who has not developed a desire by age 40 for self-direction or has not developed self-confidence to venture into risky areas may never be a high performer after this age.

(xi) Coordination - a high level of autonomy was found to be useful in the middle-range of situations, those neither tightly coordinated nor loose. For the manager the message is that different techniques should be employed when dealing with different segments of the organization. Specific prescriptions are not available from the study.

(xii) Groups - the data were limited due to the small number of groups studied. For young groups the supervisor can supply technical ideas but must encourage members to suggest their own ideas so that the group can grow. For older groups the effective supervisor is not the technical superior but rather a neutral sounding board drawing out ideas. Older groups decline in performance over several years but less so if members become cohesive and intellectually competitive.

Other studies have followed that of Pelz and Andrews. Smith studied the relationship between team composition and performance (Smith, 1971). Group composition was measured along several dimensions: members' orientations towards professional or institutional achievement, members' preferences among problem-solving approaches and members' contributions to group problem-solving. Heterogeneity was measured by both the variances across all group members and by the differences between the highest and lowest individual scores in the group and the average scores of the rest of the group. The measure based on variances was called the group heterogeneity and that based on the differences between the group leader and his subordinates was called the leader-member heterogeneity. It was found that if the group was homogeneous the supervisor should stress innovation, reject established approaches, offset pressures towards conformity and provide new ways of solving problems and original ideas. With older groups, group problem-solving aids, adoption of abstract approaches and de-emphasis of practical approaches improves performance. For young groups supportive leadership provides a modulation of the high degree of stimulation and encourages more scholarly pursuits. Further elaboration of this theme has been given by Margerison (Margerison, 1978).

A review of creativity, groups and the management of creativity has recently been presented (Stein, 1982). Stein gives the characteristics of the forceful administrator and the resourceful (creative) researcher but of more interest is the list of requirements for the creative researcher: the researcher should be assertive but not hostile or aggressive; the researcher should be aware of superiors, colleagues and subordinates as persons but should not become too personally involved with them (psychological distance); the researcher may be a lone wolf but should not be isolated, withdrawn or uncommunicative (the higher the creativity the greater the freedom in this regard); on the job the researcher should be congenial but not sociable, while off the job sociable but not intimate; the researcher should "know his place" with supervisors without being timid or submissive but should be able to "speak his mind" without being aggressive; the creative researcher may be subtle but not cunning and in all relationships should be sincere, honest and purposeful without being inflexible or Machiavellian; in the intellectual area the creative researcher is to be broad without spreading himself too thin, deep but not pedantic and sharp without being over-critical. This is quite a list.

The area of researcher age and performance has been recently readdressed (Katz and Allen, 1980). Some 50 R & D project groups in one large isolated R & D facility of a large U.S. corporation were studied in terms of group performance, tenure and communication patterns. It was found that group performance increased with group age up to about 2 years, stayed relatively constant from 2-4 years and declined after 4 years. It was proposed that two processes were at work: an initial team-building component that led to increased group performance and a "not invented here" component that decreased performance. The product of these two components was responsible for the maximum in the performance-time curve. A project was found to be best carried out when the team was not completely stable but when there was a degree of personnel turnover; if the project member tenures were too widely dispersed, performance was reduced. The correlations found were to group age and not to member age. The age of members or the length of tenure with the organization did not influence the group performance. One reason for the decreased performance with group age greater than 4 years was due to the change in communication patterns. Surprisingly it was found that intragroup communication was greatly reduced as group age increased. A possible explanation is that as members continue to work in the same group, they tend to become more specialized in their special areas and in their project assignments, resulting in greater role differentiation and less interaction among team members. The group becomes a stable complacent entity and action should be taken to provide stimulation (Farris, 1973).

The approach of Lorsch and Morse was based on a contingency

view of the organization. The contingency view focuses on understanding why some organizations are more effective than others and has led to the conclusion that there is no single best way to organize for effective results. The key to success is the matching of internal organizational characteristics to the tasks that must be performed for the organization to reach its goals. Up to the study of Lorsch and Morse, the thrust of contingency theory was to examine the fit between organizational characteristics and the nature of the task. Little attention had been paid to the way in which the personal characteristics of the members of the organization related to these organizational characteristics and to the nature of the organization's tasks. It was the task of Lorsch and Morse to examine the characteristics of organization members for research and manufacturing units in industrial concerns. In the context of the study it will be noted that the external environment is defined as that environment external to the unit. Part of this environment will be external to the firm; the remainder is the internal environment of the firm that is external to the unit under consideration. A schematic view is given in Figure 47.

The organizational units chosen were at two ends of a spectrum of characteristics. Production units should have relatively certain environments with short-term performance feedback and a relatively high requirement for coordination. Research units are characterized by high uncertainty, long-term performance feedback and relatively low demands for intra-unit coordination. Units were also chosen on the basis of successful performance or not such successful performance. In the following the findings for the research laboratories will be emphasized but results for the production units will be used for comparison.

Three pairs of successful/not so successful laboratories were examined. All laboratories performed industrial research. The research areas were drugs, communications technology and medical technology. Interviews and evaluations by questionnaire gave an indication as to environment and as to success of the laboratory. Short-term feedback on personal effectiveness was obtained via published papers, presentations, books and patent disclosures. Long-term feedback was by the success of products developed in the laboratories. An overall score on uncertainty was obtained from three indices: time-span of performance feedback, clarity of task information, and programmability of task. The certainty-uncertainty scale ran from 0-20. The overall research uncertainty score was 15.4 while the overall manufacturing uncertainty score was 6.3. For the research laboratories the total score was: medical technology 14.0; proprietary drugs, 15.1; communications technology, 17.2. Thus between the laboratories themselves there was a spread in uncertainty of the environment but the manufacturing units (6.0 and 6.6 score) had a much more certain environment. It should be noted that the rating as to success was on a more subjective basis. As mentioned, papers, books, patents, etc. could be employed as criteria but the major criterion revolved around the kind of impact that the laboratory was making on the operating divisions and on the company goals of growth and profit. In summary, data for the study were collected as follows:

(i) External environment - interview with the top executive in each unit and with their superiors. The latter provided information on unit performance.

(ii) Members' individual attributes - collected by tests and questionnaires given to a cross-section of managers and professionals as well as by interviews with about half the managers and professionals.

(iii) Internal environment - top managers provided information on the organizational structure. Formal documents and manuals provided other information on the structure. Members' perceptions on the internal environment were gathered by questionnaires and interviews.

The first topic studied was the relation between the characteristics of members in particular units and the unit's external environment (Lorsch and Morse, 1974c). It was found that members in successful laboratories had a higher feeling of competence than members in less successful laboratories (this was also true for manufacturing units). There were three explanations considered for this finding:

(i) Members with an already high sense of competence may have chosen to join the high performing units or have been chosen for such jobs.

(ii) High performance may have provided members with feedback that enhanced their feelings of competence.

(iii) Members in the successful and not so successful units differed in their personality predispositions. This explanation states that a consistency between people's personality predispositions and the nature of the external environment which shapes their work can lead to high feelings of competence.

Table 39 provides a summary of external characteristics of the manufacturing and research environments. The first of the personality dimensions, mentioned in the third explanation, deals with the cognitive structure of members. Individuals vary in the ability to take differentiated pieces of information from the environment and integrate them into a whole (also called integrative complexity). There are two identifiable aspects of this information processing: the number of attributes or dimensions that individuals differentiated and the complexity or degrees of freedom individuals use to combine the differentiated dimensions. If an individual perceives many elements and combines them (integrates) in many intricate ways then the individual is processing information at a high level of integrative complexity. It is expected that an uncertain environment would call for persons having a high level of integrative complexity in order to have a high feeling of confidence.

TABLE 39

Summary of external environmental characteristics and their impact on unit members

Manufacturing environment	Research environment
Provides rapid, clear objective performance feedback	Provides infrequent and often ambiguous performance feedback
Presents well-defined work and procedures	Presents few constraints as to how work should be done
Presents simpler and more stable problems	Presents complex and changing problems
Presents work which requires high coordination of effort	Provides work which can be done relatively independently of other projects in the unit

The second personality dimension is tolerance for ambiguity i.e. the preference for well-defined, stable and relatively unchanging conditions as against an uncertain environment. It is expected that a high tolerance for ambiguity would be required for an individual operating in an uncertain environment. A high tolerance for ambiguity would also be a requirement where feedback from the environment was long-term. A fit between a person's tolerance for ambiguity and the appropriate environmental factors is more likely to lead to a feeling of confidence.

Persons having a feeling of competence might also differ in their preferred ways of relating to authority. Individuals dealing with an uncertain environment and having direct influence over their work would probably desire more autonomy.

Finally, individuals might vary in their attitude toward individualism i.e. does a person prefer to work alone or as a member of a group? In environments in which little interdependence among unit members was required, people who preferred being alone might feel more competent than if forced to interact with other individuals.

The expectation of Lorsch and Morse was that members of high-performing units would have personality dimensions better suited to the external environmental requirements than members of lower-performing units. Such was <u>not</u> found to be the case. High-performing and low-performing units consisted of members having similar personality dimensions. There was, however, a significant difference between members of manufacturing units and members of research units. This is shown in Table 40. The differences between manufacturing and research personnel reflect the differing external environments of the two types of unit. However, the only difference between members in high-performing and low-performing units operating in similar environments (manufacturing or research) was in the sense of competence. The answer to the question of what gives a successful unit does not lie in the fit between the members' personality predispositions and the external environment. There are some implications for management. In a low performing unit one approach might be to hire individuals with a strong sense of competence. However, such individuals might not find the struggling unit an attrac-

TABLE 40

Members' characteristics for manufacturing and research units

Characteristic	Manufacturing	Research
Integrative complexity	Low	High
Tolerance for ambiguity	Low	Hiwh
Attitude to authority[(1)]	Low	High
Attitude to individualism[(2)]	Low	High

(1) A low attitude indicates that strong, controlling situations are accepted. A high attitude indicates a preference for autonomy.

(2) A low attitude indicates a preference to working in groups.

tive place to work and would be frustrated during the turnaround period. In order to improve the performance of the unit, the manager must learn how to manage and organize his existing personnel more effectively.

Lorsch and Morse next examined the fit between members' personality predispositions and the internal environment of the unit. As defined by them, "The internal environment can be thought of as man's implicit and explicit social invention to help members relate to the work of the organization in dealing with the external environment." The internal environmental variables studied were as follows (Lorsch and Morse, 1974d):

(i) Time orientation

(ii) Goal orientation

(iii) Influence and control - measured by the degree of formality of the internal environment, members' perceptions of the influence pattern and perceptions of the supervisory style characteristics of the internal environment.

(iv) Coordination of work activities - measured by the degree of formality of structure and by members' perceptions of the coordination achieved and the mode of conflict resolution used to achieve such coordination.

It was found that the internal environments of the high performing units differed substantially from those of the low performers. The results were as follows:

(i) High performing laboratories held formal review and reported on project milestones at less frequent intervals while in the low performing laboratories reports of progress occurred more frequently and were seen by scientists as constraining and inconsistent with the long-term nature of their work and with their own personality dimensions.

(ii) With regard to goal orientations it was found that the high performing laboratories encouraged stronger scientific goal orientations. Reviews, reports and evaluations emphasized that the ultimate goal of the laboratory was to contribute to the profit and growth of the company by maintaining the laboratory's own identity as a strong scientific group interested in influencing operating divisions.

In the lower performing laboratories there was more emphasis on economic goals than on scientific goals. There was confusion over goal orientation. Laboratories were seen as moving heavily toward satisfying plant and marketing conditions and away from the more technically sophisticated work for which the scientists had been trained.

(iii) The high and low performing laboratories differed in the formal structure employed in the organization. The high performers exhibited more flexible and less constraining formal structures and practices than the lower performing units. Managers in the high performing laboratories relied on researchers' self-discipline and self-control, while managers in the low performing laboratories assumed the need for more pervasive rules and other formal practices to control members' activities (note that this agrees with one of the findings of Pelz and Andrews).

Members perceived less formality of structure in the high performing laboratories so that they could behave flexibly and adaptively consistent with the rapidly changing information in the external environment and with their high levels of integrative complexity, tolerance for ambiguity and preference for autonomy. Members in high performing laboratories felt that they exercised more influence than did members in low performers. Decision-making was broadly spread across many hierarchical levels for the high performers while persons in the less effective laboratories saw decision-making confined to the upper managerial levels.

It may be noted that both high and low performing laboratories gave members a free rein in carrying out the complexities of a research project, subject to time and budget constraints. High and low performers did differ markedly in the amount of say researchers felt they had in the initial selection of the project. For high performing laboratories, researchers had more say in project selection and in the setting of performance parameters and milestones than did researchers in low performers. Thus high performing laboratories more so than low performers had internal environments in which the participatory style of supervision fitted the nature of the information in the external environment and the members' own personality dimensions.

(iv) Members of the high performing laboratories perceived a significantly lower degree of coordination of work effort than did their counterparts in the low performing laboratories (this was true for two out of the three pairs of laboratories; there was no difference for the third pair). However, lower coordination did not mean fewer interactions among colleagues for sharing ideas and knowledge and for testing the feasibility of research approaches.

Members of effective and less effective laboratories behaved differently in resolving conflicts. Of the three modes for resolving conflicts, confrontation, forcing and smoothing, the successful laboratories used confrontation in preference to the other techniques while the less successful laboratories tended to use forcing as the preferred method. The use of confrontation is more effective since the basis of the disagreement is addressed. Forcing implies that the solution to the conflict is imposed at the will of the most powerful party.

The findings on the internal environment of research laboratories and the relationship to members' personal characteristics are summarized in Table 41. The conclusion of the study was that the high performing laboratories had a better fit between the internal environmental characteristics, external environmental requirements and members' predispositions than did the low performing laboratories. It may be noted that the management reaction to a poorly performing unit may be in entirely the

TABLE 41

The findings about members' characteristics and internal organization environments in the research laboratories

	Uncertain external environment
High-performing research laboratories	Members' data High integrative complexity High tolerance for ambiguity High on attitude toward authority (prefers independence and autonomy) High on attitude toward individualism (prefers to be and work alone) High sense of competence Internal environment Long-term time orientation Strong scientific goal orientation Influence and control: Low formality of structure Perceptions of low structure Perceptions of much influence widely diffused throughout the laboratory Perceptions of much "say" and a participative style of supervision Coordination of work activities: Low coordination through low formality of structure Perceptions of low coordination required and achieved Confrontation mode of conflict resolution
Low-performing research laboratories	Members' data High integrative complexity High tolerance for ambiguity High on attitude toward authority (prefers independence and autonomy) High on attitude toward individualism (prefers to be and work alone) Low sense of competence Internal environment Shorter-term time orientation Weak scientific goal orientation Influence and control: High formality of structure Perceptions of high structure Perceptions of influence concentrated at top of hierarchy Perceptions of little "say" and a directive style of supervision Coordination of work activities: High coordination through high formality of structure Perceptions of high coordination of activities More forcing mode of conflict resolution

wrong direction.When a unit is not performing well the commonly held assumption is that tighter control is a good remedy. This will tend to reduce the differentiation between the research laboratory and other company units and will improve integration at the expense of further deterioration of performance of the laboratory.

Although the findings with respect to research laboratories have been emphasized, similar conclusions were drawn regarding the manufacturing units, namely that success was due to a better fit between internal environment, external environmental demands and members' personal characteristics. However, all three factors are different in a production setting to a research setting. The results are summarized in Table 42. It is clear that there could be considerable difficulties in research personnel working together with production personnel i.e. there are likely to be problems at the manufacturing interface due to the differing orientations of research and manufacturing personnel. It may be noted, however, that in the one pair of manufacturing plants surveyed for conflict resolution, the successful plant employed confrontation while the less successful plant employed forcing.

The general implication of the study for research directors is that there is no one best way to organize and operate the laboratory. What is needed for successful performance is a fit between the internal environment, external demands and members' personality predispositions. For the research laboratories serious doubt must be cast on the assumption that man is basically an economic creature and that there must be an emphasis on financial incentives. As Pelz and Andrews showed, rewards must be appropriate to success but are not the overall motivating factor in high performance. There are some specific suggestions that can be made in the situation where a research laboratory is not performing as desired. These are as follows:

(i) One possibility is to redefine the unit's purpose so that the external environment in which it operates is more consistent with internal environmental characteristics and/or individual members' predispositions. However, it is unlikely that top management would allow the redefinition of the laboratory

TABLE 42

Members' chracteristics and internal organization environments in the production plants

	Certain external environment
	Members' data
	Low integrative complexity Low tolerance to ambiguity Low attitude toward authority (not uncomfortable in strong, controlling authority situations) Low attitude toward individualism (prefers to be and work in groups) High sense of competence
	Internal environment
High-performing plants	Short-term time orientation Strong techno-economic goal orientation Influence and control: High formality of structure Perceptions of high structure Perceptions of influence concentrated at top of hierarchy Perceptions of little say, and a more directive supervisory style Coordination of work activities: High coordination through high formality of structure Perceptions of high coordination achieved Confrontation made of conflict resolution in the one plant measured
	Members' data
	Low integrative complexity Low tolerance for ambiguity Low attitude toward authority (not uncomfortable in strong, controlling authority situations) Low attitude toward individualism (prefers to be and work in groups)
	Internal environment
Low-performing plants	Short-term time orientation Weak techno-economic goal orientation Influence and control: Low formality of structure Perceptions of low structure Perceptions of more influence exercised in all levels in hierarchy Perceptions of much say, and a less directive, more participatory supervisory style Coordination of work activities: Low coordination through low formality of structure Perceptions of low coordination achieved Forcing mode of conflict resolution in the one plant measured

mission unless it could be demonstrated that the new mission were congruent with the mission of the company.

(ii) A performance achievement might be achieved by a minor alteration in the external environment of the unit with the hope that a better fit was obtained. This is more of a fine-tuning approach.

(iii) The internal environment seems to be the most promising area for intervention. A better fit between internal environmental, external environmental characteristics and members' predispositions may be achieved by changing the formal organization. This would provide a new set of signals to persons about what is expected in terms of influence patterns, coordination and so on. If the new set of expectations better fits members' predispositions, performance will improve and the changes will be self-reinforcing.

A closely related possibility is that of a change in the style of supervision. This also could lead to a higher level of congruence and improved performance.

The approaches of changing the formal organization and/or the leadership style are the most promising when individual characteristics and the internal environment do not fit but there is a fit with the external environment and members' characteristics. There remains, of course, the possibility of changes in individual characteristics either by education, the infusion of new personnel and the departure of personnel.

Guidelines for preferred internal environments in successful laboratories are available from the study of Lorsch and Morse but it must be remembered that these guidelines refer to laboratories with environments similar to those of the laboratories studied. Laboratories mainly engaged in development, for example, might be expected to have characteristics closer to those of production than laboratories engaged in basic research. Contingency theory only states that for high performance there should be congruence. Before intervention in any less successful laboratory, a manager should attempt to determine the external demands on the unit, the internal environment and to measure members' personality predispositions.

5.4 THE R & D INTERFACE

5.4.1 Introduction

The importance of interfaces has been emphasized by many

writers. The differentiation of organizations into units having different expertise results in interfaces that must be crossed for integration of the differing activities to be achieved. Nowhere is the interface more important than for R & D. Reference to Figure 6 shows the number of interfaces with the R & D organization and other company units.In addition there are interfaces between R & D and organizations external to the company.

In a general sense, integration via lateral relationships has been suggested as a way to bridge the interface (Lawrence and Lorsch, 1967; Duncan, 1979). There are also specific recommendations regarding innovation. Thus, "It may be possible to reduce the lag, not by accelerating R & D activity itself, but by establishing stronger communication links between R & D and the other functions as the product is readied for market. Better coupling of R & D to engineering, marketing and new product development would make the internal rate of return on R & D higher, and it could also improve a business's competitive position." (SPI, 1982). This comment refers to the fact that there is a long payback period for funds expended on R & D. "Innovation is not a single action but a total process of interrelated subprocesses. It is not just the conception of an idea, nor the invention of a new device, nor the development of a new market. The process is all of these things acting in an integrated fashion....." (Marquis and Myers, 1969). This view emphasizes the interfaces crossed in the stepwise development of an idea to a product sold in the marketplace. The importance of the interface depends on the degree of novelty of the innovation. There will be increased patterns of resistance for innovations that call for the establishment of new behavior patterns.

The way in which the interface plays a role and the methods employed to deal with the integration depend on the nature of the organization as well as the innovation. Table 43 illustrates the manner in which routine and innovative problems are dealt with (Harvey and Mills, 1970). In an innovative competitive environment the degree of formalization of internal communications is relatively unformalized. This indicates that there should be a flexibility in dealing with the problems at the interface.

TABLE 43

Internal and contextual variables affecting the use of routine and innovative solutions on both routine and innovative problems

Organizational Variables affecting propensity to impose particular solutions on particular problems	Tendency to impose routine solutions on both types of problems when:	Tendency to impose innovative solutions on both types of problems when:
CONTEXTUAL		
Size of organization relative to competitors	Relatively large	Relatively small
Age of organization relative to competitors	Relatively old	Relatively young
Degree of competition in market situation	Relatively uncompetitive market	Relatively competitive
Rate of technological change	Relatively slow	Relatively rapid
INTERNAL		
Diffuseness of organization's technology (size of product line)	Relatively specific	Relatively diffuse
Degree of formalization of internal communication system	Relatively formalized	Relatively unformalized

There are further indications of the importance of relationships between R & D and other units. According to Sarett, "In the first place, if innovation is not complete until an idea is marketed, the laboratories must be surrounded by, and closely related to, development and testing, regulatory expertise, manufacturing and especially marketing" (Sarett, 1979). The importance of the interface is again emphasized (Kottcamp and Rushton, 1979). "Every interface crossed in the process of technological innovation necessitates a technology transfer, and hence problems of acceptance, ownership and control arise. Each interface can become a potential barrier to innovation unless managed and coordinated with great skill."

According to Roberts two kinds of gaps exist in the overall process of profitable innovation (Roberts, 1979). A lack of key inputs to the innovation process leads to ineffective outcomes in attempts to innovate. Poor followup to effective technical R & D results limits the profitability of the outcome. Three kinds of input are needed:

(i) Effective information must be delivered in a timely manner to the R & D organization. The input must be relevant in content as to technical information, market-oriented information and manufacturing information.

(ii) A significant amount of entrepreneurial energy must be employed to advance the innovation in the organization.

(iii) Appropriate skills (technical and managerial) must be available to take the variety of inputs and meld them together in a coherent fashion.

The development of key input information emphasizes the importance of the interface with the technical environment, marketing and manufacturing. However, the development of effective results may not lead to a product in the marketplace because the transfer can be blocked by other units. For example:

(i) Manufacturing may not accept the results of R & D by emphasizing difficulties in production.

(ii) Marketing and sales must enthusiastically accept the R & D output and push the activities forward.

(iii) The field service organization must promote and service the product in the field otherwise the organization will soon be convinced that the product is inappropriate and there will be premature withdrawal from the market.

The solution to problems at the interface is viewed by Roberts in terms of key personnel. "In perhaps an elitist view of what takes place in an innovative organization, I believe that only 20 to 30 percent of the people in an R & D laboratory provide what uniquely matters in achieving innovation." These people fulfill critical functions and at least five roles are identified each of which is critical in order for innovation to occur. These roles are as follows:

(i) Idea generator. This is the creative individual who frequently comes up with new and bright ideas for developing new products and processes. The entrepreneur who takes an idea, whether his or from somewhere else, and carries the idea forward is also a well-known role. In many situations it is the entrepreneur who is more evident in the firm.

(ii) Gatekeeper roles. There are at least three types of gatekeeper. The technical gatekeeper bridges the inside organization to the outside world of technology and R & D advance. The market gatekeeper understands what the competitors are doing, the regulatory environment and what is happening in the marketplace. The market gatekeeper ensures that the technical activities are directed towards development of a product that will be accepted successfully in the marketplace. The manufacturing gatekeeper understands environment and capabilities of the manufacturing plant and keeps R & D up to date on the realities of what can be accomplished in production of the new product. This activity avoids the development of a product that either cannot be made (for example due to unrealistic tolerances) or that can only be made at high cost thus reducing profitability.

(iii) Program Manager. This role refers to the formal and informal coordination, supervision and integration of the efforts of individual contributions. These skills are often in short supply in the organization.

(iv) Sponsor. This role is similar to the product champion discussed in Part 1.

An alternative view of interfaces has been given (Sayles, 1974) who has stated that, "Decentralization and Integration Doesn't Work." This view emphasizes that organizations committed to innovation must become more flexible and adaptive. Placing all the elements of some technology development activity within a single organization unit will not ensure the development of new products. Sayles states, "We have all learned the painfully difficult process by which basic research, often originating in widely scattered laboratories and scientific minds is gradually and unpredictably transformed into some practical technique, hardware or material. In turn these 'developments' have unpredicted and potentially diverse applications. There is no way of either containing organizationally or predicting by budget or plan the pathways that this process takes of moving from theory - to basic research to developmental programs - to user applications." Managers in diverse units must learn to deal laterally with one another. Sayles advocates an open system that is flexible and periodically adapts to the innovation need. The innovations referred to by Sayles appear to be major innovations such as development of the Xerox process. In the typical R & D organization the innovations are more modest but nevertheless vital to the survival of the firm. What is to be expected is that looser, less defined organizations will be found for situations of high uncertainty. This is as true for R & D units as for other units of the company. However, in any given situation the structure of the R & D unit will be relatively less formalized than that of other units in the firm. What Sayles appears to be saying is that the activity of innovation cannot be delegated to a single unit with the hope that new products will appear. This is not really at variance with other approaches to the problem of innovation. The emphasis on lateral relationships stresses the importance of the interface. In the view of Sayles the interface is perhaps not so well defined in extent and can be fluid over time. Other writers would tend to see structural units as relatively more defined and permanent. It must be remembered that flexibility and fuzziness in organizational responsibilities may be more effective in situations of high uncertainty but that there must be organizational slack available in the form of duplicated and additional resources. In any given organization there will be an effort to reduce what many outside of the R & D process would consider as waste. There is, therefore, a tendency for greater formalization and structuring of the organization to minimize resource duplication.

Whatever the actual organization adopted there will be differentiation into specialized units. The innovation process must cut across organizational lines and involve the participation of many units. It is well recognized conceptually that the interface can be a major barrier to innovation. Case studies have now confirmed this.

5.4.2 The Internal Interface

The internal interface refers to the interface between the R & D unit and other units of the firm. Typically the interfaces of interest are with top management, marketing and manufacturing. Depending on the nature of the R & D there may be interfaces with development, applied research or engineering. In a divisionalized organization there may be corporate R & D as well as divisional R & D units; these units differ in span of activity and time scale of results.

One of the earliest difficulties involving the interface to be recognized was that of transferring research results to operations. Some 20 years ago it was stated that the critical problem was to transfer R & D outputs to operating units (Quinn and Mueller, 1963). It is interesting that this study also adopted a contingency approach by stating that no single solution was best for all companies. Lateral flows were emphasized as well as diagonal flows between technical groups at different levels in separate divisions. According to Quinn and Mueller, it is very important to examine resistances at critical technological transfer points (interfaces). To move the technology effectively across each interface requires information about the technology, enthusiasm for the technology and the authority to use the technology. The following rules are recommended:

(i) Keep the number of transfer points in any given flow to a minimum.

(ii) Maintain as much continuity as possible in the personnel carrying the technology forward.

(iii) Give the people involved in the change enough information to keep irrational reactions to a minimum.

Lack of information about the technology can arise for many reasons: isolation of research, physical decentralization, company size, number of people involved in the transfer, differences in the perspective of people from different units and complexity of the technology. Understanding the technology is not enough, however, Many restraints come from improper motivation due to short-term management incentives, lack of urgency in research, entrenched ideas and vested interests, the NIH (not invented here) complex and overly long lines of formal authority resulting in long approval delays for expenditures. In order to overcome problems in transfer, Quinn and Mueller recommend the creation of specific formal organizations with explicit responsibility for taking a given new technology through to successful commercialization and having the authority and resources to motivate existing line organizations to give the technology the attention it deserves. Examples of specific organizational forms that have been found useful include:

(i) Task-force groups - these groups are usually composed of a small group of personnel from research, development, marketing and manufacturing. The composition is heavily weighted towards R & D at first but shifts towards operating people as full-scale operations are approached.

(ii) Corporate development units - these units have marketing staff and flexible pilot-scale facilities and can take new research technologies and exploit them. The unit can be a profit center deriving profits from the sale of new products. As products prove profitable operating groups want to take them on. Development is, thus, always forced to seek new technologies from research, and operating resistances are eliminated.

(iii) Outside companies - these may be used in special cases. The research laboratory may take a share of ownership in the new concern to exploit the technology or it may take license revenues. Large companies may give a small outside company exclusive rights under a royalty agreement for a specified period. At a later stage the larger company has the option of continuing the arrangement or introducing its own branded version of the product.

(iv) Staff groups at corporate level - these groups provide coordination of new technologies through existing divisional and functional organizations and are most effective when they have authority over key aspects of the line operations. Product managers perform this function successfully in some companies.

(v) Top executive with multifunctional line authority - integration is carried out by an individual at the upper levels of the hierarchy who can ensure that products move speedily from R & D into manufacturing. This approach only works well for small and medium size companies.

(vi) Research group with funds to buy time in manufacturing - this approach allows manufacturing to become familiar with the product or process and gives advance warning of problems before large scale production is undertaken.

(vii) Transfer of individuals - this approach is useful if the individual or individuals have the inclination to follow the idea to commercialization. There is also the opportunity for promotion and for a career change should the individual choose not to return to R & D.

(viii) Multilevel committee responsibility - some companies have used this approach. A research committee coordinates fundamental and early applied research. An R & D committee coordinates late applied and early developmental stages. In late development stages, a new product committee takes over. A full scale operation requires an executive committee approach.

(ix) An entrepreneurial group at corporate level - this approach needs individuals with the complex of skills and attitudes necessary to entrepreneur new products. Individuals must be replaced as they transfer to divisions along with products they have championed and in those cases where technical skills become outmoded. The group must have the resources to build small-scale facilities and to underwrite product introduction losses.

Regardless of the form of the transfer organization used, each company should develop a procedure to effectively link together the series of sequential steps to the market introduction. There must be adequate mechanisms for planning and then controlling the project (Quinn and Mueller, 1963).

The problems of transfer of technology have been discussed extensively with similar conclusions to those of Quinn and Mueller (Roberts, 1979). What is interesting is that the problem is still being discussed which is an indication that the solutions proposed either have not been successful or were not correctly implemented. Figure 48 illustrates in a general form the interfaces for technology driven and market driven innovation but it is recommended that simultaneous rather than sequential couplings be used (Galbraith, 1982). This allows the position of the project relative to the technology and the market to be continually updated. Continuous updating allows projects to be abandoned at an earlier stage before excessive funds have been expended (New and Schlacter, 1979). Case studies involving successes or failures in the transfer of technology have been reported (Cohen, Keller and Streeter, 1979).

Quinn and Mueller also have commented on the importance of the relationship between R & D and top management. "We are convinced that if a large company wants to effectively diversify through research or to make radical technical advances, the stimulus must come from the top corporate level" (Quinn and Mueller, 1963). Corporate managements often evince considerable

FIGURE 48

LINEAR SEQUENTIAL COUPLING COMPARED WITH SIMULTANEOUS COUPLING OF KNOWLEDGE

(a) Linear Sequential Coupling

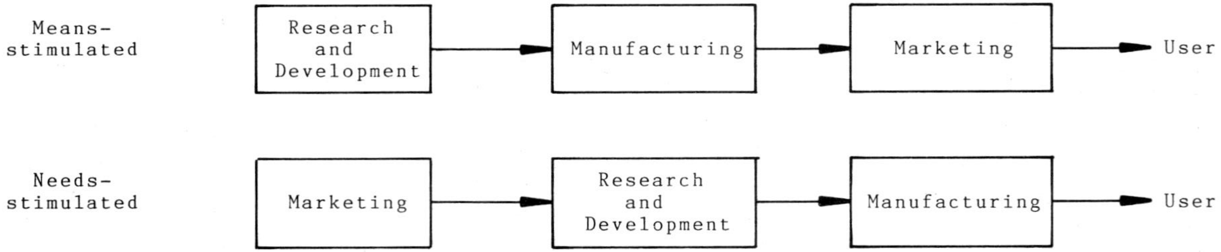

(b) Simultaneous Coupling

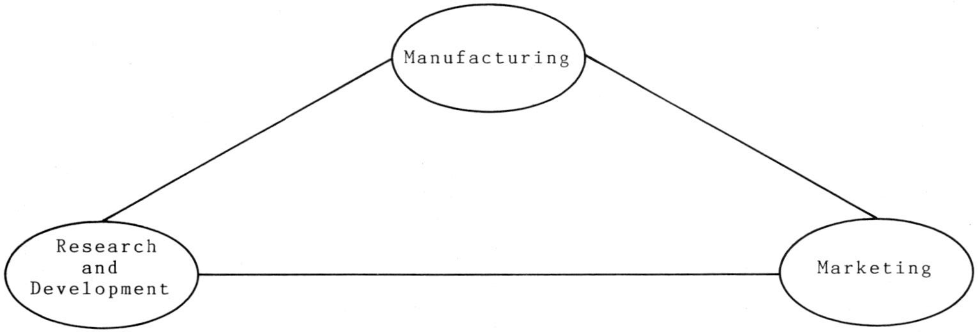

uneasiness about the activities of R & D (Biller and Stanley, 1975). There are four groups involved in the communication process, general managers, R & D managers, applied R & D and basic research (Roberts, 1972). These groups have different motivations, different ways of working and making decisions and different forms of personal recognition. The R & D manager is regarded as the key interface between the more or less personal goals of the research groups and the business and financial goals of the management groups. Corporate management has the responsibility to tell R & D management the goals of the corporation in clear terms. The R & D manager must pass this information to the research groups in a form understood by them and must pass R & D results to top management in a form that is comprehensible in business terms.

Examination of research success in companies using a questionnaire approach has led to the conclusion that "the critical determinant of R & D performance is the quality of management at the interface of R & D with senior corporate executives and with the executives in the marketing and production functions" (Gruber, Poensgen and Prakke, 1974). The thesis was that the lag in technology in the steel and automobile industries in the United States was not due to internal R & D activities but the low level of performance caused by failures at the R & D interface.

The conclusions of Gruber et al are based on the results of two surveys. A seminar, "Problems at the Interface between R & D and Other Management Functions" held at M.I.T. resulted in a questionnaire that was administered to the course participants. It turned out that about half the companies represented had studied the role of R & D and had analyzed the relationships between R & D and the rest of the organization. This had resulted in some form of plan; budget, reporting and control procedures had been established. Despite this 70 percent of all respondents believed that company management did not adequately understand R & D and 70 percent believed that R & D did not understand the company. The difficulties experienced at the interface prompted the expansion of the survey and the questionnaire was mailed to firms in the United States, Europe and Japan. The U.S. firms were chosen from the Standard and Poor

Index of Industrials. 660 mailings were made and 158 responded. Some of the larger firms made more than one submission resulting in 270 completed questionnaires. Only this U.S. experience was reported. The results of the survey are given in Tables 44-48.

The first finding (Table 44) was that the majority of R & D executives believed that non-R & D executives had either a medium or low opinion of the usefullness of the R & D function.

TABLE 44

The level of satisfaction with R & D maintained in several corporate groups as perceived by R & D executives

	Percentage of respondents rating the level of satisfaction			
Corporate group	High	Medium	Low	Total
Top management	44	46	10	100
Engineering	39	44	17	100
Marketing	37	42	21	100
Manufacturing	32	46	22	100
Finance	24	47	29	100

Thus R & D is seen to be misunderstood. R & D executives are themselves more satisfied with R & D performance as measured by the achievement of the company and R & D goals than they perceive non-R & D executives to be (Table 45 vs Table 44). The survey showed that non-R & D executives have a much shorter time horizon than R & D executives. The corporate need for major technical breakthroughs conflicts with the corporate need to meet this year's budget. Of all types of executives, the finance function was seen to be the area of most friction. Difficulties at the interface with finance were reported in the survey more widely than at the interfaces with other functions.

The evaluation of new technology transfer was made for high and low performing laboratories. R & D executives reporting high performance within R & D reported good technology transfer three times as frequently as did the executives in low performing R & D laboratories (Tables 46).

It is the opinion of Gruber et al that interface problems can be improved by management action. Evidence of the variation in management efforts to improve communications and planning is

TABLE 45

Achievement by R & D in meeting company and R & D objectives as perceived by R & D executives

Type of goals	High	Medium	Low	Total
Company goals	67	25	8	100
R & D goals	62	30	8	100

TABLE 46

Evaluation of transfer of new technology by recipients of the new technology for high and low levels of total R & D performance

Recipient of new technology from R & D	Mean scores of respondents by level of total R & D performance*	
	High	Low
Marketing	3.43	2.74*
Manufacturing	3.22	2.71
Engineering	3.49	2.95

*Mean scores calculated in a range of 5 for very good to 1 for very poor. All differences are statistically significant at least at the 99 percent confidence level.

given in Tables 47 and 48. Only the means are shown. The range in values is such that the very good companies had a much higher score and the very poor companies a much lower score than the mean values presented. A recognition of problems by executives on either side of the R & D interface can result in a rapid rate of improvement in interface problems.

Although the survey of Gruber et al showed that most friction occurred with the finance function most attention has been paid to the marketing interface. It has been pointed out that there is a need to improve the productivity of research. "To accomplish this, one of the major requirements is a close working relationship between the R & D and marketing components of the company" but "....the task of gaining mutual respect and understanding between R & D and marketing is much more difficult" (Monteleone, 1976). Monteleone states that it is important for both marketing and R & D to be given credit for successes. However, it is far more important to the relationship to share also the blame for the failures. O'Keefe and Chakrabarti studied

TABLE 47

Communications in the R & D relationship with top management: a comparison of high and low level R & D performance

Question	Mean scores of respondents by level of total R & D performance* High	Low
1. 'To what extent does top management understand the abilities of R & D?'	3.7	3.0
2. 'On major R & D projects how much co-ordination is there between R & D and top management?'	4.0	3.4
3. 'How much ability do you think top management has to evaluate the work of R & D?'	3.7	3.2
4. 'How satisfied are you with the quality of communications between R & D and top management?'	4.1	3.3
5. 'In your experience, how receptive is top management to ideas and proposals that you or your subordinates (R & D) might offer?'	4.1	3.6

*Mean scores calculated in a range of 5 for very good to 1 for very poor. All differences are statistically significant at least at the 99 percent level.

TABLE 48

The quality of planning and communication as reported by respondents with high and low levels of R & D performance

	Percentage of respondents reporting high scores on: High performance group	Low performance group
Index of quality of R & D communications with:		
Top management	71	39
Marketing	63	34
Manufacturing	50	21
Finance	20	6
Engineering	71	49
Index of groups present at reviews of R & D projects	72	59
Planning and goals index	38	16
Index of emphasis on long-run contribution of R & D to company goals	63	35

140 innovation projects within eight large U.S. corporations according to technical success and commercial success (O'Keefe and Chakrabarti, 1981). The project highlighted the difficulties in communication at the R & D interface especially with marketing. Difficulties at the interface adversely affected both technical and commercial success and this link could be quantified. At the heart of the R & D/marketing interface problem are difficulties that are common to those that characterize communication problems in other areas of human activity. Individuals are quick to characterize other individuals on the basis of stereotypes and then proceed to act as though these faulty and incomplete characterizations are valid. Table 49 gives the R & D view of marketing personnel and the marketing view of R & D.

TABLE 49

R & D and marketing stereotypes

R & D VIEW OF MARKETING
Impatient
Unable to understand the difficulties of doing good research
Unable to comprehend the technical difficulties
Unable to comprehend the real problems
Interested in shoddy temporary solutions to problems
Untrustworthy about products in embryonic states
MARKETING VIEW OF R & D
Too narrow
Too specialized
Unaware of real world problems
Too slow
Unable to provide a solution when needed
Not cost conscious
Too jargonistic

Further considerations regarding the marketing interface with R & D have been reported by Souder and others (Souder and Chakrabarti, 1978; Souder, 1980; Souder, 1981). In studies of innovations in companies it was found that a high degree of R & D/marketing interaction, participation in problem definition and integration all relate to technical and commercial project degree of success. In fact the degree of interaction seemed to be extremely important for determining project success (Souder and Chakrabarti, 1978). Variables that were not directly significant in determining commercial success included: R & D capability, competitor actions, technological opportunities and representation of R & D in the long-range planning process. Apart from interface difficulties the main reason for project failure was lack of fit between the project and the usual line of business of the firm. Tables 50 and 51 give the results of the examination of two propositions:

1. The degree of R & D/marketing integration is directly related to the degree of success of the project.

2. The effectiveness of information transfer and the clarity of understanding of the problem and the user's needs directly relates to the degree of success of the project.

Table 52 gives the results on the examination of the following proposition:

3. Effectiveness of integration is directly related to the degree of legitimization of the integrator role and the presence of joint reward systems. That is, effectiveness increases

TABLE 50

Results for proposition 1

	Dependent variables			
	Degree of Commercial success		Degree of Technical success	
Independent variables	Kendall Tau	Statistical significance level	Kendall Tau	Statistical significance level
Degree of R & D marketing interaction	.381	.001	.288	.001
Degree of participation in problem definition by R & D	.273	.001	.263	.001
Degree of participation in problem definition by marketing	.204	.006	.145	.034
Effectiveness of integration between R & D and marketing	.427	.001	.325	.001

TABLE 51

Results for proposition 2

Independent variables	Degree of Commercial success		Degree of Technical success	
	Kendall Tau	Statistical significance level	Kendall Tau	Statistical significance level
Clarity of problem definition	.546	.001	.514	.001
Clarity of understanding of understanding of user needs	.599	.001	.483	.001
Completeness of information exchanged	.604	.001	.443	.001

TABLE 52

Relationship between legitimization of integrator role and effectiveness of integration between R & D and marketing

	Legitimization of integrator role	Effectiveness of integration		
		Low	High	Total
Low ↑	No integrator	32	9	41
	Integrator role performed due to feeling of personal responsibility	4	2	6
	Informal authority as information gate keeper	26	19	45
	Formal authority of the integrator	4	9	13
↓ High	Formally appointed as an integrator	1	8	9
	TOTAL	67	47	114

FIGURE 49

SUMMARY MODEL OF SIGNIFICANT RELATIONSHIPS FOUND

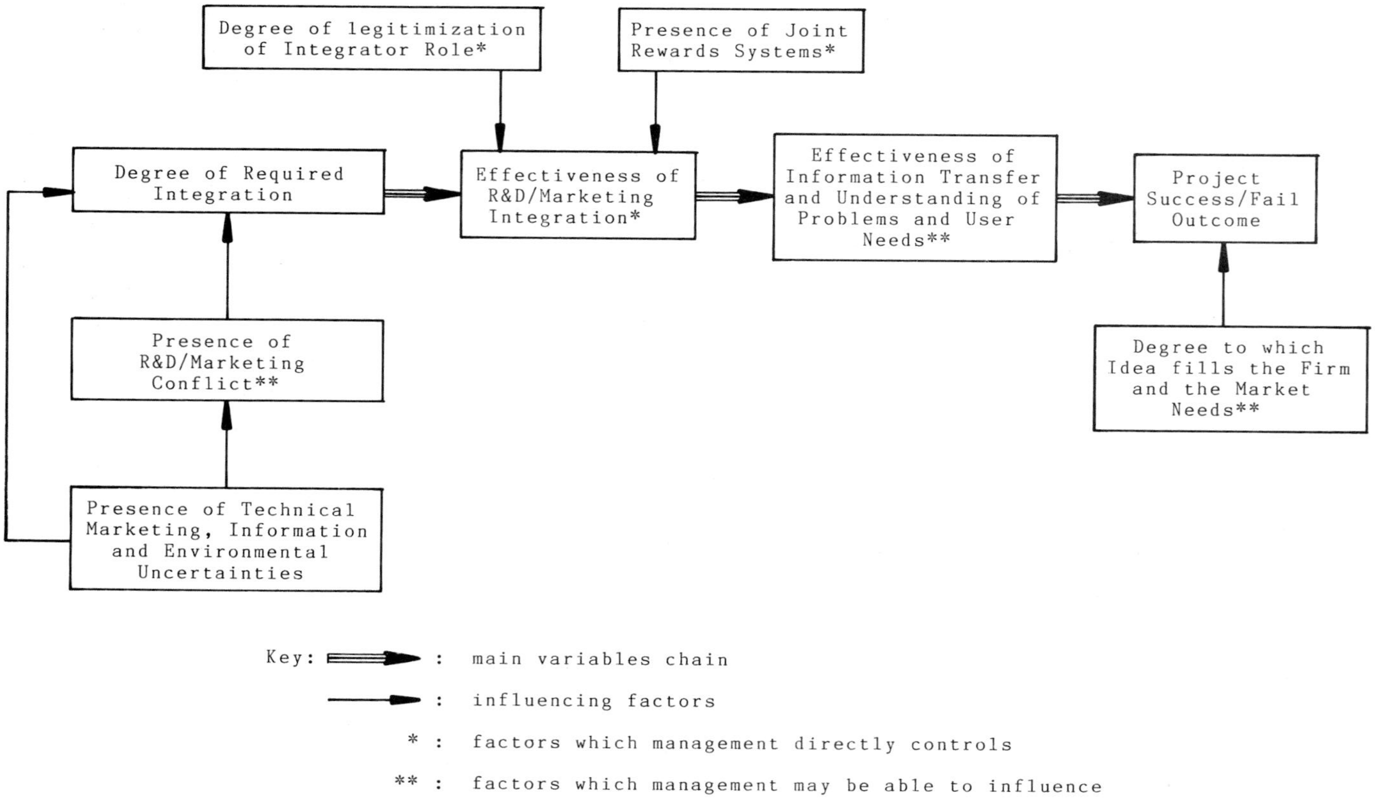

*© 1978 IEEE

(decreases) as the degree of legitimization increases (decreases).

The employment of some form of integration mechanism gives an improvement in the effectiveness of integration. Integration can be promoted by two actions: carefully legitimizing the integrator's role and by setting up joint reward systems. The findings are summarized in the model shown in Figure 49 which illustrates the main variables chain, the influencing factors and those factors that management can directly control or which can be influenced by management.

The effective management of the R & D/marketing interface is a complex problem (Souder, 1980). However, several effective approaches were identified and 10 guidelines devised for overcoming most R & D/marketing interface problems. A total of 312 in-depth interviews were conducted on 150 randomly selected projects at firms participating in an Industrial Research Institute study. The interviews involved top-level, middle-level and project-level personnel. The degree of R & D/marketing harmony experienced for each project was measured along three dimensions: the cooperativeness in the behaviors of the two parties; the feelings of warmth expressed by each party toward each other; the sense of mutual commitment felt by the two parties toward each other. Four distinct types of R & D/marketing interface problems were found. These were characterized as follows:

(i) Lack of Communications - in this type of problem the R & D and marketing parties maintained verbal, attitudinal and physical distances between each other. These feelings and behaviors were fostered by the normal time pressures and work deadlines. The situation is summarized in Table 53.

(ii) Lack of Appreciation - in this type of problem the R & D and marketing parties purposely avoid each other. Each has negative emotions about the other that stand in the way of collaboration. No single cause was found to be responsible for this situation. In some cases, there were histories of earlier ineffectiveness on the part of one of the parties that colored later attitudes. In other cases, organizational climate and structure contributed to the lack of appreciation. The

TABLE 53

Characteristics of the lack of communications problem

Behaviors

- There are few joint decision meetings
- Marketing is not fully informed of what R & D is working on until very late in the cycle
- Neither party attends the other's staff meetings
- Working documents, salesmen's call reports and progress reports are not circulated between the R & D and marketing personnel

Attitudes

- Marketing feels they cannot afford to get too involved with R & D for fear of getting too bogged down in technical details
- Marketing feels that it takes too much time and there is little reward for becoming intimately involved with R & D

---

characteristics of the lack of appreciation problem are given in Table 54.

(iii) Distrust - this problem reflects an extreme set of negative attitudes (Table 55). There were deep-seated negative attitudes, jealousies and hostile behaviors. There was no single cause but the distrust problem began either as a lack of appreciation or as a lack of communication problem; with time, the problem worsened. Many of the distrust cases were characterized by long-standing personality conflicts. The situation had often

TABLE 54

Characteristics of the lack of appreciation problem

Behaviors

- Marketing sometimes contracts out its R & D work, as needed
- R & D and marketing are jealous of each other's prerogatives
- R & D fails to consult marketing concerning new product ideas
- Marketing attempts to exercise close control over R & D whenever they work together
- R & D sometimes attempts to by-pass marketing and to directly market their new ideas

Attitudes

- Marketing feels that R & D is too sophisticated in their approaches
- R & D feels that marketing is too simplistic in their approaches
- Marketing feels that R & D should not visit customers because they will "talk over their heads"
- R & D feels that marketing does not really have a good grasp of the kinds of products that are needed

---

TABLE 55

Characteristics of the distrust problem

Behaviors
Marketing attempts to dictate exactly what, where, when and how to do the project, allowing no room for rebuttal and no tolerance for suggestions from R & D
R & D is seldom invited to participate in the planning and formulation of the projects
Marketing often goes outside the firm and sub-contracts out its R & D needs
Marketing brings R & D in only after they know exactly what they want done, so that R & D will not have any reason to argue for their own ideas
Marketing and R & D personnel purposely avoid each other, sometimes refusing to sit down together at the same table
Attitudes
Marketing fears that R & D will give away too many secrets and promise too many things if they talk to customers
Marketing feels they lose control when R & D gets involved; the project disappears and they never see it until it is completed, at which point it is seldom what they wanted
Marketing feels that R & D cannot be trusted to do what they are instructed to do
R & D feels that marketing muzzles them with regard to customers
R & D fears that marketing wants to liquidate them
R & D does not feel they can trust marketing to tell them what the customers really want, with any degree of accuracy
R & D feels they are blamed if products fail, but marketing is credited if products succeed

become institutionalized with the original reason for the conflict forgotten. In nearly all cases, there was an imbalance of power and prestige between R & D and marketing due to forcefulness of personality of one of the executives, budgetary powers, expanding authority or due to unequal attention and praise from top management.

(iv) Too-Good Friends - this type of problem is characterized by too much harmony. The parties had so much regard for each other that they were complacent and did not challenge each other's judgment. Information of importance could be overlooked (Table 56).

The too-good friends and lack of communication problems were found to be relatively mild types. These problems could sometimes be resolved by a moderate amount of joint effort from the involved personnel. In contrast, the lack of appreciation and

TABLE 56

Characteristics of the too-good friends problem

Behaviors
Neither party challenged the other or questioned the other's judgments and assumptions
There is open communication between the parties, but potential problem areas are glossed over and there is no real depth of communication between them
Both parties avoid entering into conflict with each other or arguing over details
The R & D and marketing individuals appear to be "good friends" - they see each other socially, they occasionally make joint visits to customers, they go to lunch together, they meet with each other frequently
Attitudes
Marketing feels that R & D "can do just about anything"
Neither party wants to hurt the other's feelings
Each feels that the other "has his own turf", and that they should "stay off each other's turf"
R & D relies exclusively on the marketing personnel for judgments and information about the marketing aspects

TABLE 57

Number of projects experiencing each problem type

Type of interface problem	Number and percent of projects experiencing this problem: Number	Percent
Mild problems		
Too-good friends	10	6.7
Lack of communications	21	14.0
Total	31	20.7
Severe problems		
Lack of appreciation	30	20.0
Distrust	17	11.3
Total	47	31.3
No significant problems	72	48.0
Totals	150	100%

distrust problems were much more severe types which rarely could be completely resolved. Attempts to solve these problems often led to the re-emergence at a later date. Table 57 shows that over half (52 percent) of the projects studied experienced some type of interface problem. Lack of appreciation was the most prevalent single problem (20 percent). Lack of communications was the second most prevalent problem (14 percent).Distrust occurred in 11.3 percent of the projects and the too-good friends problem in only 6.7 percent of the projects.

Table 58 shows the relationship between interface problems and project success. In 6 (60 percent) of the 10 projects which experienced the problem of too-good friends, important information was overlooked which diminished the effectivenss of the end product. In 14 (66.7 percent) of the 21 projects which experienced a lack of communications problem, the new product either failed to match the market needs or arrived too late to capture a fast-moving market. In 25 (83.3 percent) of the 30 projects where a lack of appreciation prevailed, either no new products were developed, the products did not perform adequately, the products were too costly or the products did not

TABLE 58

Percentage distribution of project successes

	Percentage distribution of projects within each problem category		
Problem category	Complete commercial success[1]	Partial commercial success[2]	commercial failure[3]
No significant problems (N = 72)	66.7%	19.4%	13.8%
Mild problems (N = 31)	32.2	51.6	16.1
Severe problems (N = 47)	14.8	21.3	63.8

[1]Met or exceeded all the commercial targets and expectations that were originally set for the product.

[2]Met most but not all of the commercial targets and expectations that were originally set for the product.

[3]Met few (or none) of the commercial targets and expectations that were originally set for the product.

match the market needs. Of the 17 projects where a distrust prevailed, 15 (88.3 percent) were terminated for lack of achievement. As Table 58 shows, about two-thirds (63.8 percent) of the projects experiencing severe problems (lack of appreciation or distrust) were commercial failures. For projects with mild problems (lack of communication or too-good friends) the commercial failure rate was 16.1 percent; only one out of three (32.2 percent) of the projects with mild problems was a complete success, however. Half of the projects with mild problems (51.6 percent) were only partially successful. When no significant problems existed the commercial failure rate dropped to 13.8 percent and two-thirds of the projects (66.7 percent) became complete successes. Thus, mild problems reduce the successfulness of new products while severe problems are likely to cause a complete failure of the project.

From the study, Souder developed ten guidelines involving actions that had been demonstrated to overcome interface problems:

(i) Break large projects into smaller ones - it was found that projects with less than 7 persons experienced far fewer and less severe problems. This is probably due to the increased ease of contacts.

(ii) Take a pro-active stance toward interface problems - in effect this boils down to resolving conflict by confrontation.

(iii) Eliminate mild problems before they grow into severe problems - early attention to problems at the interface can avoid much more serious problems later.

(iv) Encourage open relationships - this involves free communication and sharing of information.

(v) Promote dyadic relationships between R & D and marketing - strong interpersonal alliances have been shown to be especially fruitful in overcoming interface difficulties.

(vi) Use a New Products Committee - this is an integration device heavily weighted towards R & D in early stages of the project and progressively changing towards engineering as the project progresses.

(vii) Appoint only highly qualified individuals as project managers.

(viii) Involve both parties early on.

(ix) Try to obtain agreement on the decision authority in each meeting between R & D and marketing.

(x) Set up a new product development organization structure that is appropriate for the nature of the technology and the market. This is shown in Table 59.

TABLE 59

Appropriate organization structures

		Nature of the technology	
		Poorly-defined	Well-defined
Nature of the market	Poorly-defined	Project management	New products committee
	Well-defined	Project management	Line management

There has been less attention paid to quantifying the effect of interface problems between R & D and manufacturing than between R & D and marketing (Moore, 1970; Bergen, 1982). Presumably in the study of Souder described above there must have been an influence of the R & D/manufacturing interface on the product success. It is tempting to speculate that an R & D organization that adopts tactics to solve interface problems with marketing will also solve interface problems with manufacturing. The approach to solution of problems at the manufacturing interface is similar to those at the marketing interface.

5.4.3 The External Interface

The preceding discussion has focussed on the interface between R & D and other company units. The existence of the interface is often indicated by conflicts that arise or by project failures, and mechanisms were suggested to integrate the activities of differentiated units. For R & D the interface with the environment external to the firm is extremely important in view of the need for the latest information on technological developments. The approach here has been to focus on individual activities that bridge the interface between R & D and the technical environment (Allen, 1977). This work will be discussed with the understanding that the approach can also be employed in the discussion of interfaces internal to the company.

A good review of boundary (interface) roles in the innovation process is given by Tushman (Tushman, 1977). A critical aspect of the innovation process is the ability of the innovating unit to gather information but there is a substantial literature suggesting that communication across organizational boundaries is both inefficient and prone to bias and distortion. Table 60 gives a process approach to innovation. During the first phase, idea generation, information pertaining to new ideas or new approaches to technical problems can be found most effectively outside the innovating organization. It is during this first phase that relationships across the interface to the external environment should be emphasized. During later stages in the project interface relationships internal to the company assume greater importance. Information flow across the interface is a two-step process through a set of key individuals who channel this information to their colleagues within the laboratory.

TABLE 60

Innovation phases and corresponding key communication domains

	INNOVATION PHASES	
I. Idea generation	II. Problem solving	III. Implementation
Developing a design or proposal by integrating market need with technical means	Research, development, and engineering to develop a technical solution to the proposal	R & D, manufacturing, market tooling up, coordination, and administration to bring the new product to first use or to the market
	KEY COMMUNICATION DOMAINS BY PHASE	
I. Extra-laboratory, extra-organizational communication	II. Intra-laboratory communication	III. Laboratory and functional area communication
With universities, suppliers, vendors, literature	With functional areas as well as technical areas outside laboratory	Particularly between R & D and manufacturing and marketing

Internal communication stars are those people most frequently approached as technical discussion partners; internal stars having a substantial amount of extra-organizational communication are termed gatekeepers. Gatekeepers do not attend to all external communication areas but may specialize in those external areas most critical to the work of their sub-unit. For the research manager, the idea of gatekeepers emphasizes the importance of key individuals in the system communication network. By encouraging and rewarding boundary-spanning activity, managers increase the performance of the R & D organization by making better information available.

The concept of gatekeepers has been applied not only to external extra-organizational contacts but also to internal gatekeeping activities (Roberts, 1979). It may be noted that gatekeepers are not necessarily idea generators (Frohman, 1978). According to Frohman gatekeeping is "the collection and channeling of information about changes in the internal and external environments. Gatekeepers stay abreast of events in their respective environments through personal contacts primarily and through formal media secondarily." Besides gatekeeping a variety

TABLE 61

Percent performing each managerial role for innovation in two R & D organizations

Role	R & D 1	R & D 2
Idea generating	2	16
Technical* gatekeeping	12	8
Championing	7	4
Coaching	4	0
Project leading	5	10
Market* gatekeeping	15	4

Percentages do not add to 100 because some performed none of the managerial roles and others performed more than one.

*Two separate types of Gatekeeping were identified. One focused on technology, the other on the use of the research results or market information.

of roles were given as critical and Table 61 shows the percent performing each "managerial" role for innovation in two R & D organizations. The distribution of effort will depend upon the type of research and whether it is technology or market driven. Table 62 indicates those factors that encourage or inhibit performance. A questionnaire was employed to look at individual profiles as compared to the requirements of the job. Figure 50 shows an example in which the interest and skills of the person are not in line with the job requirements. Most respondents showed peaks in one or two roles. This suggests that an individual is predisposed to perform certain types of activity. For example, a person who is theoretically inclined and comfortable with abstractions is better suited to the idea generating role than someone who is very practical (this is connected to the idea of a learning style as discussed in Part 3).

TABLE 62

Factors supporting and inhibiting performance

Role	Factors supporting the performance of the role	Factors inhibiting the performance of the role
Idea generating	Well-recognized role Reasonable freedom Good linkages with the market	Tight resources Focus on the short-term
Technical gatekeeping	Well known by peers Freedom to travel	Emphasis on measurable performance Emphasis on creating new ideas Physical design - distances Poorly recognized function
Market gatekeeping	Well recognized role Ready market access	Poor customer credibility due to premature introduction of products
Championing	Freedom to act Recognition of role	High cost of failure Limited reward for risk taking Receives poor coaching
Project leading	Strong management Freedom to act Clear goals	Low acceptance by outside units Lack of management skills High cost of failure
Coaching	Clear business goals	Role not recognized Lack of a growth strategy "Firefighting" orientation

FIGURE 50

INDIVIDUAL PROFILE: RELATIVE JOB COMPONENT, PREFERENCE, AND SKILLS FOR EACH MANAGERIAL ROLE

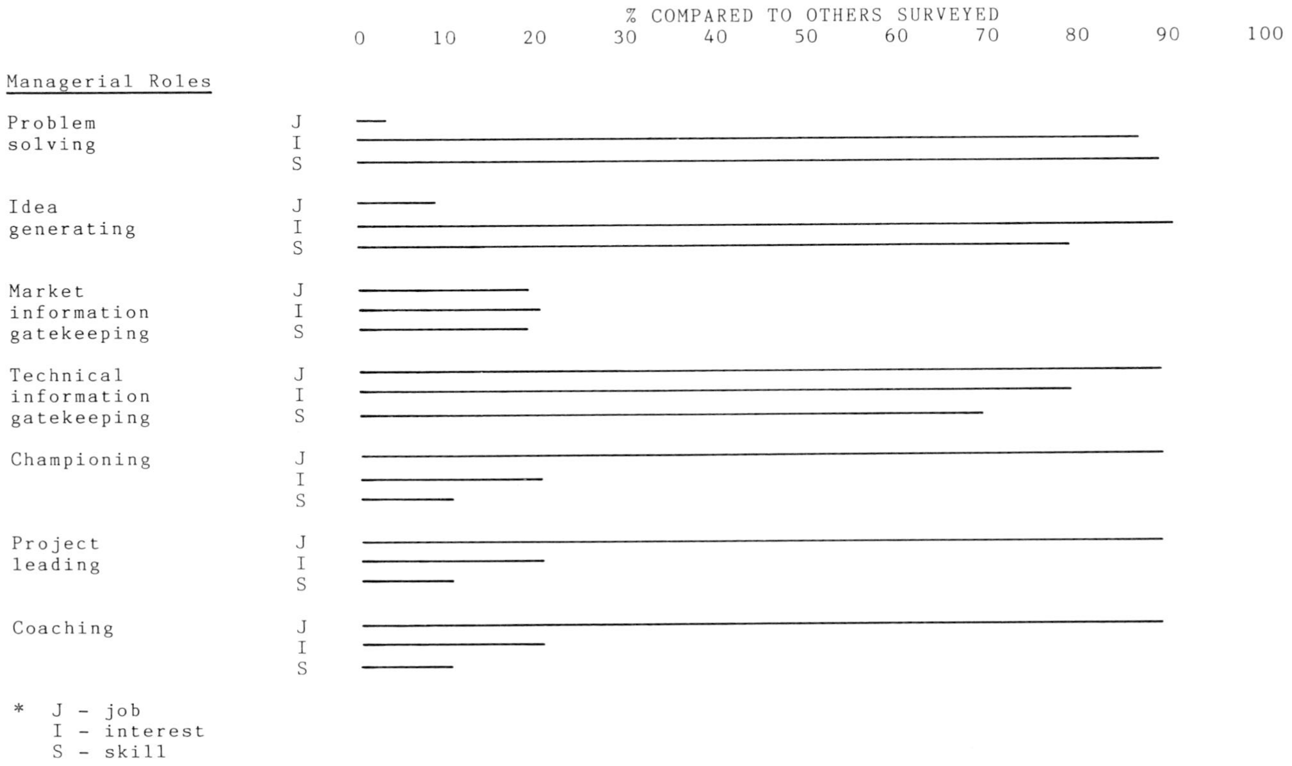

The role of gatekeepers has been studied as a function of project success (Miller et al, 1980). It was found that information transfer as regards technical and market aspects of the project were vital. The chief difficulty was in how to organize the environmental gatekeeper function. In many situations the information transfer function operates on an informal basis. Many of the projects studied were deficient in the information transfer area. Miller et al argued that a more structured environmental monitoring function was desirable whether or not this employed the gatekeeper function. The influence of gatekeepers on project success has been studied as a function of task nature (Katz and Tushman, 1980). The success of projects in the research, product development and technical service areas of a major R & D facility were studied. The results were surprising: research projects with gatekeepers were significantly lower performing than projects without gatekeepers. Technical service project success was independent of the presence or absence of gatekeepers. On the other hand the success of development projects was strongly increased by the presence of gatekeepers. It may be noted that the definitions employed for characterizing the nature of the task (research, development or technical service) are not universally acceptable. The nature of the tasks in development are more akin to applied research and those in technical service are more akin to development. Be that as it may, Katz and Tushman argue that the beneficial effects of technical gatekeepers are strongly contingent on both the nature of the task and the stability of the technology involved. Locally oriented projects with more dynamic technologies where the task expertise is located within the project, are more effective if technical gatekeepers link with external information areas. Direct external communication with information areas by all team members is more effective for more universally oriented research projects. Locally oriented projects but with stable technologies have no use for gatekeepers but can make use of the formal hierarchy for external contacts.

Boundary spanning individuals are not necessarily gatekeepers. An individual must be strongly connected both internally and externally. Individuals with substantial boundary spanning activity but who do not interact internally are often found to be low performers (Katz and Tushman, 1980). Latent gatekeepers have been found in many R & D organizations (Fischer and Rosen, 1980). Information stars, individuals who are repositories of information that is shared with colleagues, and gatekeepers, stars with boundary spanning roles, appear to make up only 5 to 10 percent of the laboratory population. Information stars have the following characteristics:

(i) they read far more than do their colleagues,

(ii) their readership of technical journals is far greater than that of the average technologist,

(iii) they have a much higher level of formal education,

(iv) they publish more,

(v) they have a higher degree of patent productivity,

(vi) they attend more scientific meetings,

(vii) they are chosen more often by their peers as technical discussion partners,

(viii) they enjoy higher degrees of credibility,

(ix) they appear to be better able to connect seemingly unconnected ideas.

The existence and functioning of information stars and gatekeepers is important for the flow of information throughout the R & D group. However, there is evidence that the roles of star and gatekeeper may have negative aspects. Individuals playing a boundary spanning role can utilize their unique external relationships to improve their personal power in the organization but there is a high degree of role ambiguity and anxiety associated with such positions. There is anecdotal evidence that the idea-generating component of information star behavior is not rewarded and might even be discouraged in some companies (Fischer and Rosen, 1980). It has been found that the correlation between idea generation and organization rewards is close to zero. Idea evaluation, however, was found to be positively associated with the organizational reward system. Fischer and Rosen confirmed these conclusions in a study of the staff in a large U.S. government laboratory. They recommended implimentation of a variety of approaches to increase involvement, participation and information sharing among research and development professionals.

The study of stars, gatekeepers and other roles emphasizes the informal nature of the information acquisition and information sharing processes. It would appear that individuals approach those who they see as technically competent regardless of formal status (Tushman and Scanlan, 1981). This study also indicated that there was substantial role overlap with 39 percent of the boundary spanning individuals linking to both laboratory and organizational areas. Similarly one-third of the gatekeepers linked to laboratory and organizational areas as well as to the external environment.

Informal organizations in R & D have been studied in detail (Farris, 1980). The term informal organization refers to the set of interpersonal relations in the organization that affect decisions but which are not included in the formal hierarchy. An informal organization arises to satisfy individual members' needs. In a formal organization the salient goals are those of the organization; in an informal organization the salient goals are those of the individual. Table 63 gives some contrasts between formal and informal organizations. According to Farris there are two important functions of interpersonal relations: information exchange and the exercise of influence. Three individual needs are task achievement, task integration and personal development. This 2 x 3 classification allows the identification of six functions of informal organizations in R & D as shown in Table 64. A description of each functions follows:

(i) Technical Communications - this involves the exchange of information relevant to task achievement and is concerned with information stars.

TABLE 63

Some contrasts between formal and informal organizations

	Formal	Informal
Salient goals	Organization's	Individual's
Structural Unit	Office/position	Individual role
Basis for communication	Offices formally related	Proximity (Physical, professional, task, social, formal)
Basis for power	Legitimate authority	Capacity to satisfy individual's needs
Type of hierarchy	Vertical	Lateral

TABLE 64

A classification of functions of the informal organization in R & D

		Interpersonal relations function	
		Information	Influence
Individual Need	Task achievement	Technical communication	Problem solving
	Task integration	Boundary spanning	Resource allocation
	Personal development	Grapevine	Socialization

(ii) Boundary Spanning - this is the provision of information from the outside environment. One thrust of work has focussed on the technical gatekeeper.

(iii) Grapevine - this is the exchange of information relevant to personal development.

(iv) Problem Solving - this function is concerned with the exercise of influence on task achievement and involves various forms of collaboration.

(v) Resource Allocation - the exercise of influence on task integration with the outside environment.

(vi) Socialization - the exercise of influence on personal development.

The basic structural unit of informal organizations is the individual role. The playing of roles involves interpersonal relationships. The interconnections among individuals by role transactions creates role networks that involve individuals throughout the organization. It has been found that participation in the informal organization is high skewed creating degrees of centrality, clusters and cliques and reciprocal roles. Some individuals are active in more than one role network while others are relatively isolated. Informal organizations are characterized by lateral hierarchies and some individuals attain higher degrees of status and influence in the informal organization regardless of their status in the formal organization. Five factors are important in determining patterns of interaction in informal organizations:

(i) Physical Proximity - interpersonal relations are more

likely between individuals who are physically close. There is an exponential decay in frequency of communiction between people. About two-thirds of the likelihood of communication disappears with 100 feet between people and another two-thirds suppression with another 100 feet (Allen, 1977).

(ii) Professional Proximity - other things being equal, R & D personnel are more likely to seek contact with colleagues of similar professional backgrounds. Such contact may provide security but as has been shown earlier interaction with persons of differing expertise and background is associated with higher technical performance.

(iii) Task-Created Proximity - other things being equal R & D personnel will interact with persons working on the same task in preference to those working on different tasks.

(iv) Social Proximity - R & D personnel who have social contact are more inclined to engage in work-related discussions on the job.

(v) Formal Organization-Created Proximity - R & D personnel will tend to interact with personnel in the same unit in preference to personnel in other organizational units (this is one reason for the existence of interfaces and the need for activities to span the interface).

Successful management involves the recognition and use of the informal organization as well as the formal organization. The manager may choose to organize around key individuals who have achieved high status in the informal organization or around key functions and roles. The lateral hierarchy may be utilized to integrate the isolates by encouraging them to interact more with their colleagues. The manager may wish to develop more key individuals. Finally, the manager may wish to take advantage of physical proximities to organize the informal organization. Too much overt manipulation of the informal organization is likely to be counterproductive since it will become identified with the formal organization.

5.5 POLITICAL RELATIONSHIPS

5.5.1 Introduction

In the earlier presentation of strategy and structure and the role of R & D in the company the subject of political relationships was mentioned but the subject was not pursued. The positioning of R & D in the company hierarchy and the implications of strategic decisions for R & D were assumed to be capable of rational evaluation. In fact, much of the strategy-structure literature assumes rationality but the reality of organizations is different and politics in the sense of power and influence must be considered even if political actions are not intended. The more recent treatments of organizations have emphasized political relationships.

There is a similar situation in the different approaches to interface relationships. The treatment of integration through progressively comprehensive mechanisms culminating in the establishment of a coordinating department represents a rational approach. The solution to the problem of differentiation is sought through changes in the organizational structure. A department (or individual) halfway between the orientations of two differentiated departments should logically be able to bridge the gap and thus resolve conflict but the fact remains that the conflict may be purposely sought and at least one of the protagonists may only want resolution on certain terms. No amount of logical and rational discussion will solve this type of problem. In organizational interventions the pitfall of assumed rationality has been emphasized (Zaltman and Duncan, 1977).

Rationality in organizational strategy means that perfect information is available on each of the strategies under consideration, that there is an unbiased means of choosing the strategy that is best for the organization and that, once a strategy has been chosen, all in the organization will work together to achieve that strategy. There are many reasons why rationality is a goal that cannot often be achieved in organizations. Perfect information is not available. There are usually too many factors to be considered to allow a quantitative assessment of alternative strategies. Disagreements as to the choice of strategy can lead to delay in decisions or compromise and to less than total dedication to the achievement of the strategic goals. There can also be disruptions due to political action. The following is a definition of politics that can serve as a working definition: political action is that which takes place with the primary aim of self-interest. It will be seen that there is ample opportunity for political action in organizations. There is, of course, always the chance that individual A will suspect individual B of political action simply because A does not understand or have knowledge of the basis on which B operates. There may, in fact, be a genuine difference of opinion between A and B as to what is the best way to achieve the goals dictated by a chosen company strategy. Thus the existence of conflict does not necessarily mean that politics is being practiced. Political action need not be confined to actors within the organization but also occurs at the level of the company itself (Thompson, 1967). Most companies operate with the primary aim of self-interest; objectives such as achieving power over competitors, attaining a dominant position in a chosen market and so on are common.

It is interesting to enumerate some of the ways in which political action manifests itself. Empire building within the organization is a well known phenomenon. Posturing to those with power for career advancement is also widely practiced. Conflict not rooted in genuine differences of opinion is often met. Much of this action provides the undercurrent for life in organizations and is not overt. An example of political action that is obvious is the chief executive officer (CEO) who proceeds on a path of selective acquisitions so that he can "make his mark" on the company. Or the CEO who is a model of inaction because he has only two years to retirement.

The subject of politics in organizations is complex and a full treatment cannot be presented here. However, the existence of politics has implications for problems at the R & D interface and some remarks will be made. Burns and Stalker have pointed out that at least three systems operate within companies. There is the system connected with the functioning of the firm - the formal organization. In addition, there is a career system and a system of social relationships (Burns and Stalker, 1966). A recent study asserts that strategy follows structure and not, as rationalists would have it, the other way around (Hall and Saias, 1980). Here the aspects other than the company structure are examined. These aspects such as ethnic, societal, experience and outside relationship factors lead to "private agendas" within the organization and consequently to political action.

Political actions may lead to an organization which only sees what it wants to see in the external environment. The processes by which information is transferred within the organization are not neutral; information is often censored and changed by the manner in which it is transmitted, Hall and Saias conclude that ".... one can infer that few organizations succeed in identifying the problems with which they are faced in a realistic manner". An aspect of this same problem is discussed by Adizes when he states that change for a corporation in decline must come from the outside - for political reasons the change agent cannot be part of the organization (Adizes, 1979).

It must be stated that the thrust of the argument of Hall and Saias is that the structure of an organization conditions the strategic options that are open (structure is taken to include process here). However, the informal structure and company climate are presented and it is in these areas that the political relationships become apparent.

Political relationships may also be seen in the formal structure. For example the relationship between units engaged in different activities within the company has been discussed (Thompson, 1967). It may be noted that the interface is very important in these relationships. The power and influence structures in the company are important in determining what activities are legitimate. Thus a company not committed to innovative activity may well regard the development of new products as an area to be avoided. Similarly, an organization that developed through acquisition in the past may pay lip service to internal growth but regard growth through external means as the legitimate area for expansion.

There has been a recent tendency to examine organizations on a micro or local basis and in this way to take into account the differing aspirations of the different parts of the system. This is said to be the political approach. Thus a spectrum is visualized in which one end is occupied by the rationalists with theories of organizations based on logic and the other end concerned with the politics of organizations. If one end of the spectrum is rational can the other end be said to be irrational? The answer to this is "No". As envisaged the political treatments are still rational in nature - the rationality is localized. It is assumed that each person or unit has his or its own rational reasons for acting. Political action describes acts carried out on the basis of self-interest and not with the well being of the company in mind. Hence it turns out that the recent political treatments do not represent a departure from rationality at all. The spectrum should run from the analytical rational organizational frameworks, through micropolitical theories to reality of organizations in which irrational acts as well as rational actions occur. The reason why the political theories are regarded as irrational by some is that in the strict definition of organizational rationality the baseline of rationality is taken as the goals of the organization. Hence, individuals or groups with their own goals at variance with the organizational goals may be regarded as acting irrationally in an organizational sense. To take an example, marketing may refuse to cooperate with R & D in the introduction of new products because the head of marketing wants to maximize the sales of the present product range; high sales lead to the promotion of the head of marketing so that he does not have to bother about the long term goals of the company. Obviously if the goals of local individuals or groups are totally at variance with the company goals, the actions may be regarded as irrational since the company may go out of business. But the distinction is hardly ever so clear cut.

In large corporations there is sufficient slack for groups to have goals that confound company goals especially if it is the set of long-range goals rather than the shortrange goals that are confounded. It is noted in the literature that small companies are more efficient than large companies in the introduction of innovative products. It may be hypothesized that interface problems in a small company are less because there is less slack and hence less opportunity for individuals and groups to have goals that are much at variance with the goals of the company. None of the theories of organizations effectively deal with the irrational act at the micro level and this is because irrationality is tied to an individual and a specific situation. Here irrationality at the local level is taken as a conflict not only with the goals of the company but also with the rational goal structure of the individual. Irrationality stems from emotion (greed, jealousy, fear, revenge) and has no place in general theories of the organization. However, such actions do occur in companies.

It may be argued that the possibilities for irrational acts at the local level are maximized at the interface. There is another point to be examined here, however. Suppose that there are two organizational units, A and B. Both of these units have goal structures which differ somewhat from the organizational goal structure and also differ from each other. Conflict at the interface will occur since each unit uses its own framework as a reference to judge the other unit. However, the conflict may be regarded as based on micro rationality as long as it is realized that each unit acts in accordance with its own goal structures and as long as each set of goal structures A and B does not threaten the survival of the organization. Irrationality may creep in if unit B acts against unit A not because of the goal structure but merely to "get even". Much as organization theorists would like to ignore this behavior, it does exist.

Returning now to the problem of interface communications, it will be assumed that units interact on the basis of individual goal sets. Whether or not these goal sets are consistent with the goal set of the company can be taken up later. The question of the interface will be taken up in the context of motivation. This places the interaction on an individual level but contacts at the interface are often made on this level (cf. the literature on boundary spanning and gatekeepers). According to Atkinson aroused motivation is the product of the basic motivation, M, with the expectancy of attaining the goal, E, and the perceived incentive value of the particular goal, I (Litwin and Stringer, 1968) or:

$$\text{Aroused Motivation} = M \times E \times I \qquad -(1)$$

In this equation the basic motivation, M, can be regarded as one of three types: need-achievement, need-power or need-affiliation. From this point of view the interaction across the interface may be examined.

(i) Individual - if there is only one individual involved then the behavior will be directly governed by Equation (1) and will depend on the basic motivation of the individual, the expectation of success and the reward. An example of this type of situation concerns the extra-organizational interface where an individual is motivated to carry out a literature search to develop new ideas that can be worked on in R & D. Here the basic motivation is probably needachievement, there is an expectancy of developing new ideas and there is a reward (perhaps in the longer term) for the new ideas (monetary, status or promotion).

(ii) Couple with no constraints - most interface interactions involve two or more individuals. The case of two individuals will be considered. Each individual is taken to reside in a different unit and to have a goal structure that reflects the goal set of that unit. Furthermore, it is assumed that there are no constraints on the actions of the two individuals. Each individual will have an aroused motivation given by Equation (1) but the basic motivations may be different e.g. individual A is need-achievement motivated whilst individual B is power motivated. The expectations and incentives will be different unless there is some coupling of the reward structures of the two units.

To take a concrete example it will be assumed that unit A is R & D and that unit B represents marketing. It will be further assumed that the individual in R & D is motivated in the sense given above and that marketing has been approached regarding cooperation in the development of new products. The response of B will depend upon the way in which the success or failure of A is viewed.

Suppose that the outcome is a success for A. There are three outcomes for B: success, no effect and adverse effect. If B has the expectation of success for A which is also a success for B then there will be cooperation; if the success of A does not influence the position of B then B will make little effort to cooperate. Where success for A will have an adverse effect on B the likelihood is that B will actively work against A. A similar set of circumstances can be envisaged if B regards the projects as a likely failure for A. There are again three outcomes for B: failure, no effect and beneficial effect. If a failure of A also will be viewed as a failure of B then B will work with A to avoid failure (it may be that B would not want A to have a success and so cooperation will continue until the failure is avoided but not thereafter). If the failure of A does not affect

B there will be little incentive to cooperation and if the failure of A will have a positive effect on B then B is encouraged to work against A.

The above interactions may be modified by at least two factors. The interaction will depend upon the competence of B. If B is professionally competent it is likely that, in those neutral cases in which the outcome for A does not affect B, there will be cooperation from B. Personality factors will enter into the evaluation. If B likes A then it is likely that cooperation will be better and an adversary situation may not develop even where success for A detracts from the standing of B.

It may be noted that in the above the goal structure of the organization has not been considered. This will be evaluated next in the context of its effect on the interaction of A and B.

(iii) Couple with organizational constraints - if there is an intrusion of the organizational goal set the interaction between A and B will be modified. There will be limitations on the actions. For example, even where B perceives that the success of A will have an adverse effect on B, the organization may demand that B work with A. The reward structure may be such that B will be rewarded for cooperation with A even where B sees his unit as being adversely affected. An example here would be a new product from R & D (A) to be introduced to the marketplace through marketing (B). This product originated in R & D which saw a market need. The product is needed to fulfill one of the goals in the organizational goal set. Thus if the product is successful B will get rewarded by the organization but would feel that marketing looks bad since the market need was not identified by marketing. It may be pointed out that the influence of the organization is not always beneficial. It is possible that two units would interact to produce an innovative product that would not be allowed if the organizational hierarchy were involved. Thus there could be pockets of innovation in a largely bureaucratic organization that would benefit from isolation. In general, however, it is expected that the involvement of a hierarchy in the sense of linking together the goal sets of different units (integration) is of benefit.

The above examples are highly simplified since only pairs of individuals are assumed to interact. In addition the interaction is only taken at one level. The goal sets of A and B may not be consistent with the goal sets of their individual units. The expectation of A will be affected by his estimate of the likelihood of obtaining the cooperation of B and so on. Interactions will depend on the previous history of the degree of cooperation between A and B.

What has been shown above is that the interactions at the interface are complex and depend on many factors. It is little wonder that the R & D interface is given as a major problem in the literature. Of its nature the interface gives rise to political actions. These actions are normally rational in the sense that, at least locally, there is coherence between the action and a goal set. Of course from an organizational viewpoint the action may be regarded as irrational. However, irrational actions do occur even at the local level and usually represent emotional factors.

One further point may be noted. Actions at the interface involve risk. The activities of a unit are exposed when these activities cross the interface. All too often political actions in companies result in winners and losers and the consequences for the losers may be severe. This state of affairs can act as a modifier to Equation (1). An individual may see that the incentive is not worth the effort since there is a penalty in the case of failure. The expectancy will be modified by the prevailing climate. The expectancy will be reduced if it is seen that most actions at the interface will be against the active unit. All of this means that the aroused motivation will be low and the only actions that will be taken are those that are non-political (and in this kind of climate every action is viewed as political). Thus it will be seen that the demise of an organization is likely to begin at the interface. The differentiation that gave the organization an edge over the competition now becomes a positive disadvantage since the concommitant integration is missing.

The subject of politics in organizations covers a large area and has many facets. An exhaustive treatment cannot be given here but a limited presentation will be given on power and influence and the effect of politics on strategic decision making. As with every area involving personal interactions there will be a mix of activities. Just as it is very rare for true organizational rationality to be achieved, so it is rare for actions to be purely political in nature. Survival of the organization is usually a constraint limiting action that is entirely for self-interest.

5.5.2 Power and Influence

There are differences in terminology regarding power and influence. The treatment here will follow that of Handy (Handy, 1976e). Influence is an active process in which an individual A modifies the attitudes or behavior of individual B. Power is the ability to influence and is a resource. In order to influence B, A must have some understanding of the motivational calculus, the psychological contract and the role of B. A must also be aware of his sources of power and which methods of influence are consequently available.

There have been various categorizations of power. Whatever the source of the power there are general considerations that apply:

(i) Relativity - if A has a power source that has no saliency for B then A will not be able to influence B on the basis of this particular power source. The saliency depends on many factors. The effect of time or the number of times a source of power is invoked has an influence on its saliency. The repeated use of threats, for example, eventually leads to this source becoming ineffective.

(ii) Balance - the exercise of power by A depends on the power possessed by B. Money may overrule expertise. If the power equation is very unbalanced, i.e. A has considerable power and B has little power, then A may not be able to exercise his power due to the outcry from elsewhere (fair play, concern for the underdog).

(iii) Domain - power is rarely a universal commodity. Power depends on the domain. Thus the expert who has power in the R & D setting is not regarded as legitimate in, say, the production setting. The executive who is important at the office often is powerless at home and cannot control his children. Power has to be regarded as legitimate in order to be effective. Power depends on the setting.

The categories of power are as follows:

(i) Physical Power - this is the threat of physical coercion. It is the power of the superior force. Such power is exercised in prisons, law enforcement and sometimes in armies and perhaps schools. It is not usual to exercise physical power in industrial organizations. The use of physical power is frowned upon in Western society but appears to be on the increase in confrontational situations such as strikes and is even advocated in some "law and order" circles.

(ii) Resource Power - possession of needed resources is an extremely potent form of power and is especially relevant to the industrial organization in which much of the conflict is due to the need for scarce resources. Apart from having control of the resources, the potential recipient must also desire the resources. Calculative contracts often involve resource as a power source, the ability to give money, promotion or even power. Resource power may not be very popular since most people do not like to acknowledge that they can be bought. The granting of resources is not a one-time activity but must be periodiically exercized e.g. a pay increase every year. To the extent that the increase in pay is fixed in size from year to year it becomes a part of the basic expectation and is seen more as a satisfier than as a positive motivator of the individual.

(iii) Position Power - this type of power is a result of the position of the individual in the organization. It is important to note that the legitimacy of the power comes from the position rather than from the individual. This is often an unpleasant surprise for an individual who loses a certain role only to discover that his power is lost at the same time. The power has to be guaranteed by the organization in order to be seen as effective and has to be ultimately <u>underwitten</u> with resource or physical power.

Position power gives the individual control over invisible assets. A flow of information is often directed to a certain position. Information is power and the access to more or better quality information often provides a considerable advantage. A position often gives the individual the right of access to certain networks and to those parts of the organization that make decisions. An individual in a specified position also has the opportunity to structure the organization under him, to arrange the organization of work and the flow of communication. These invisible assets are potentially very productive.

(iv) Expert Power - this type of power is the least offensive type of power because it belongs to an individual because of acknowledged expertise. In parts of the organization that depend for successful functioning on expertise there is no resentment in being influenced by the acknowledged expert. This type of power does not require sanctions but must, in effect, be given by those over whom it will be exercised. It may be noted that expert power is relative; an individual with a small increment of additional knowledge can become the new expert.

(v) Personal Power - as distinct from position power this type of power resides in the individual. Sometimes personal power is called charisma. Personal power is fleeting. It is enhanced by success but evaporates with failure. Personal power is only effective when the person is present and the effect may be short-lived. It is often enhanced by position or expert status.

(vi) Negative Power - negative power is the illegitimate use of power. It is the capacity to obstruct, to delay or to distort. This type of power can be out of all proportion to the position of the individual. For example, subordinates can deliberately withhold information from their superior so that the boss only sees what they want him to see. In corporate headquarters, staff can delay decision-making by inactivity or by requiring more and more information. This has two effects. It is a means of demonstrating frustration but also can be a means of reinforcing the position in the organization and increasing the importance of the incumbent. The use of negative power is found at times of low morale, stress or frustration and results in a lack of trust and the setting up of checking procedures to thwart the use of negative power. However, this action usually leads to further exercise of negative power which makes matters worse.

The bases of power described above allow the individual to exercise one or more methods of influence. Handy has divided the methods of influence into two classes: overt and unseen. The relationship between the bases of power and the methods of influence are shown in Table 65. The methods of influence are as follows:

(i) Force - this is the crudest method of influence and has as a base physical power or sometimes resource power. Such a method of influence is not accepted as a legitimate method and is not employed in industrial organizations except when emotion triumphs over reason. The immediate effects of force or the threat of force are usually very satisfying but the longer term implications are sufficiently damaging to preclude this method of influence except as a last resort.

TABLE 65

The relationship of bases of power to available methods of influence

Power source	Physical	Resource	Position	Expert	Personal
Influence (overt)	Force	Exchange	Rules and Procedures		Persuasion
Influence (Unseen)				Ecology	Magnetism
Adaptation	Compliance		Internalization		Identification
Socialization	Mortification	Schooling	Co-option		Apprenticeship

*Reprinted by permission of Penguin Books Ltd.

(ii) Rules and Procedures - this method may be general or specific. A may influence B directly by laying down a rule that has to be obeyed by all persons in B's position. To turn this into an indirect method A can maintain that the rule is not specifically directed at B. In order to influence B, A must have the perceived right to institute the rules and procedures and must have the appropriate power base for enforcement. Rules and regulations, therefore, are based on position power backed by resource power. It should be noted that rules and procedures provide an efficient way of structuring much activity especially those activities that are of a routine nature. Rules, in themselves, can be neutral and may not be perceived as specifically singling out an individual however unpopular the rule.

Rules and procedures may proliferate especially as an organization moves from a dynamic growth structure to a bureaucratic structure. In a bureaucracy, rules and procedures become an end rather than a means to an end. The institution of rules and procedures may be employed by individuals to strengthn their position in the organization. This is often found for staff positions and is one of the reasons for the excessive time for decision making in large corporations; paradoxically the smaller the expenditure the larger the decision-making process. It may be remarked that rules and procedures may be used in a negative sense by employees. "We were just following orders," is an example. This situation occurs where the actions of subordinates are hemmed in by rules and procedures. In those instances where individual action should be taken, the prescribed action is still employed even though the person knows what will happen.

(iii) Exchange - this method of influence includes bargaining and negotiating. A agrees to give something to B in exchange for a desired action on B's part. Resource and position power are the most usual bases for this type of influence. It is obvious that B must desire what A has offered (or has to offer because the offer does not have to be explicit) and A has to be in the position of being able to come through. An example of an explicit exchange is the offer of a pay increase by A if B can complete a specified task on schedule. However, B may carry out tasks in the belief that high performance will lead to a reward such as a promotion and A does not have to explicitly offer the promotion in order for it to be effective.

Exchange, in general, has a limited effect. The reward is tied to the transaction. For further desired performance a further reward is needed and this reward may become an expectation no longer tied to performance. Thus the exchange method of influence can be very effective in the short term under explicit conditions but cannot be expected as a long-time influence of behavior.

(iv) Persuasion - this method is the first method of choice because it relies on logic and the power of argument. The method is supposedly neutral but is often not seen as neutral either because the actors have different viewpoints, e.g. R & D versus manufacturing, or because the position of one of the actors is such that the relationship is seen as boss/subordinate i.e. the method is contaminated by position or resource power.

The four methods described, force, exchange, rules and

procedures and persuasion are overt methods of influence. Unseen methods are ecology and magnetism.

(i) Ecology - the environment plays an important part in determining performance. The environment may be adjusted to remove constraints or to facilitate some aspect of behavior and this is a method of indirect influence. The method is powerful because it is often unrecognized. Position power is a base from which to alter the environment by choice of organizational structure, design of information systems, structuring of tasks and delineation of reward and control systems. The conscious use of the environment to influence behavior is an extremely useful but little recognized tool. Unfortunately environmental considerations are only addressed after problems have arisen. Examples have already been cited of conflict arising at the R & D interface.

(ii) Magnetism - this method of influence depends on personal or sometimes on expert power. It is an individual thing since the attraction is not based on reason but on emotion. Trust, respect, charm and enthusiasm allow an individual to influence people without apparently trying. Admiration may be based on the individual's past successes even when the individual is personally not a very likable character. Magnetism as a method of influence may be easily abused. Magnetism is fragile. One failure or false step can lead to a loss of trust and respect. It is often impossible to undo the damage caused.

The method of influence chosen depends on the power base(s) available and also on the individual recipient of the influence attempt. Expert and personal power must be granted by the recipient. In a way resource power must also be granted since the recipient must desire the promised resources. There is more leeway if the influencer has built up credibility through prior success, recommendations of others and previous behavior. The R & D manager, and indeed any manager, must deal with a number of different constituencies. The approach taken in the R & D laboratory will not work with the production department or with the sales staff.

Influence attempts that are based on more than one power base are usually more effective than attempts dependent on only one power source. The advantage of multiple power sources is that the influencer can change the emphasis on each depending on the response of the person being influenced. The emphasis can also be changed during the course of the period of influencing. A relationship that has more than one strand to it is stronger than a relationship with only one strand. The expert who also builds a close personal relationship with the individual to be influenced has a much better chance of succeeding than the expert who relies on expert knowledge alone.

The recipient of an influence attempt can reject it, ignore it or accept it. As a whole persons do not like to admit that they have been influenced. The change of behavior may cause dissonance between what we are now doing and what we would do if not influenced. The response of the influenced individual can be classed into three areas: compliance, identification and internalization. These psychological responses will now be described:

(i) Compliance - the recipient agrees to carry out the agreed tasks because he has to. This may be because of an exchange method such as increased pay, because of rules and procedures or because of force.

There are long-term consequences of compliant behavior. The individual is doing what he is doing because he has to. There is no commitment to the task and the onus for ensuring that the task is carried out correctly lies with the influencer. Provided the power source is available, compliance is reliable in the short term and the response is rapid. However, compliant behavior is not desirable in those cases where the individual must make decisions in order to successfully carry out the task. Even when the individual performs well initially the level of performance will decline over time necessitating maintenance activities. Compliance implies checking; it is accompanied by the need for rules and procedures. Compliance is not accompanied by trust.

In industrial organizations the need for compliant behavior depends on the routine of the task. The greater the repetitiveness of the task the easier it is to accept compliance. However, the result may not be satisfactory as evidenced by the low quality of many of the goods produced on the production line. The response has been to place more emphasis on commitment through ownership of the task by the worker. In order to do this the individual must be given more freedom and the task more flexibility. This can be a very threatening situation for insecure management, who can find compliant behavior in subordinates a very satisfactory state of affairs. In the R & D laboratory the tasks are complex and demand a great degree of non-routine action. Often extraordinary efforts are needed to bring a project to fruition. This cannot be accomplished through compliant behavior.

(ii) Identification - here the recipient adopts the proposal or agrees to act in a certain manner because he admires or identifies with the initiator of the influence attempt. This is extremely satisfying for the initiator but it is necessary to maintain the identification. This can, in fact, circumscribe the actions of the initiator since he must conform to the idea or model seen by the recipient. On the other hand, the recipient is dependent on the initiator and loses flexibility. The recipient will obey rather than initiate. The success of the task lies with the initiator, all others are followers. This is no real problem as long as the leader knows what has to be done and is correct. In many situations no one person has all the answers and it is necessary to obtain information and opinions from others. This information may not be forthcoming in a leader and follower situation. It should also be pointed out that identification may be confined to a particular project. Once the project is over the degree of identification disappears.

(iii) Internalization - this is the most desirable response because it is self-maintaining. The individual adopts the task or project as his own and so has the necessary commitment. It is more difficult to obtain internalization than compliance or even identification and it takes longer to achieve this state. If internalization is desired then the individual must be free to accept the influence. There can be no pressure put on the individual, otherwise the response will be compliance or identification. The recipient is more likely to adopt the idea or task as his own if the source of the influence is respected.

Internalization is very valuable in those situations where great commitment is needed, for example, in the R & D laboratory. Once the recipient has internalized an idea he will be convinced that it is his own and will even deny that an influence attempt took place. This can be very hard for the R & D manager since he has to let all of the credit go to the recipient.

Organizations often seek to make an individual more amenable to the mode of influence. This process is called socialization. There are four examples of the process. Schooling is formal instruction in the history and traditions, in the ethics and practices and in the language and technology of the organization. This is practiced in law, medicine and the other professions. New members of an organization are often subjected to an "indoctrination" to describe the structure, rules and procedures and values of the company. Apprenticeship is the process whereby the details of the art are passed on by having an individual assigned to another individual or group well versed in the practices. Along with the knowledge, the values are also passed on. Law, medicine, engineering and crafts use this form of socialization. In co-option the individual is made a member of progressively inner groups with all the attendant privileges, e.g. a partner in a law firm, access to the executive washroom. The individual progressively has a larger stake in the organization and is more and more inclined to continue the customs and practices to maintain the status quo. Even before promotion, the individual will have adapted his behavior to that of the desired group so that his selection becomes more likely. In mortification, the individual is harassed, deprived of his identity and forced into conformity. Military academies provide an example of this form of socialization. Individuals who survive the initiation have developed an identification with the norms of the institution. Small allowances of privilege are then highly prized and the individual will actively seek to enforce the norms of the institution for newcomers.

The individual will reject the socialization process or will employ one of the three response mechanisms if the influence is accepted. A compliant response goes with schooling and mortification. Identification goes with apprenticeship and co-option. Socialization does not of itself lead to internalization; it is necessary to have the recipient develop a share of ownership.

Behavior change is relatively easy to accomplish. It is much more difficult to change attitudes. Of the responses discussed, it is only internalization that leads to a long-lasting change. Identification can lead to attitude change but the change may not last as it is tied to the initiator.

The psychological contract between actors A and B depends on the source of power and the method of influence employed. Alienation results from the use of physical or resource power although the judicious use of resource power leads to a calculative contract as does the employment of position power and sometimes expert power. Personal power and sometimes expert power lead to a cooperative psychological contract. The full model is shown in Table 65. The source of power determines everything else. For example, position power allows rules and procedures to be employed and results in a calculative contract. The method of adapting will be compliance or internalization. The socialization process employed is likely schooling.

It has to be remarked that in any interaction there will be a mix of power sources. The perception of the initiator A may be quite different from that of the recipient B. Whereas A may try to speak from a base of expert power, the perception of B is of the use of a resource or position power base. The method of influence chosen will depend on the power base and on the circumstances. Other things being equal, it is preferable to employ methods that lead to self-maintenance, especially where the task calls for a large measure of commitment and individual decision-making. This is the case in R & D.

5.5.3 Politics of Strategy Formulation and Implementation

Some aspects of politics in strategic decision-making have been recently discussed (Fahey and Narayanan, 1982). Politics in organizations is regarded as inevitable in view of resource limitations, uncertain and dynamic environments, organizational differentiation and hierarchies; organizations may be regarded as loose structures of interests and demands. Organizational politics is defined as "those activities and behaviors engaged in by actors to acquire, sustain and exercise power and influence over others in order to have their preferred goals, options or choices enacted." As will be discussed later the strategic goal setting process becomes incremental in nature with political considerations impacting at each step.

TABLE 66

Assumptions of the rational and political perspectives

Assumptions of rational model	Organizational characteristics		Assumptions of political model
1. Goals are consistent and consensually shared	1. Differentiation—> a. vertical b. horizontal		1. Goals are inconsistent across social entities and therefore not consensually shared
2. The decision process is orderly and substantially rational	2. Interdependence of subunits and levels——>	Leading to——>	2. The decision process is disorderly; characterized by push and pull of interests
3. The norm of optimization	3. Environment——> a. places constraints on resources b. creates ambiguities in information		3. The norm of frequent play of "market forces" conflict is legitimate and expected
4. Extensive and systematic information search			4. Information search and usage triggered by political reasons leads to distortions
5. Cause-effect relationships are known at least probabilistically			5. Disagreement in cause-effect relationships
6. Decisions flow from value maximizing choice			6. Decisions are result of bargaining and free flow of political forces
7. Underlying ideological theme of efficiency and effectiveness			7. Ideologically, notions of struggle, conflict, "winners" and "losers" prevail

Table 66 emphasizes the contrasts between the rational and the political models of organizational behavior. In the political model decisions are a result of bargaining with the notions of winners and losers (zero sum encounters). Table 67 gives a listing of the political activities across phases of strategic decision-making. The different phases are shown in Figure 51.

TABLE 67

Political activities across phases of strategic decision making

Phases of strategic decision making	Focus of political action	Examples of political activity
Issue identification and diagnosis	Control of a) issues b) cause-effect relationships	Control of agenda Interpretation of past events and future trends
Alternative development	Control of alternatives	Mobilization Coalition formation Resource commitment for information search
Choice and evaluation	Control of choice	Selective advocacy of criteria Search and representation of information to justify choice
Implementation	Interaction between winners and losers	Winners attempt to "sell" or co-opt losers Losers attempt to thwart decisions and trigger fresh strategic issues
Evaluation of results	Representing oneself as successful	Selective advocacy of criteria

The activities outlined are not limited to any one decision-making phase.

Part 2 emphasized the need for strategic planning. Management theory has concentrated on the analytic aspects of the process. However, the reality is far different. Political considerations require that statements of goals are achieved only through

FIGURE 51

PHASES OF STRATEGIC DECISION MAKING

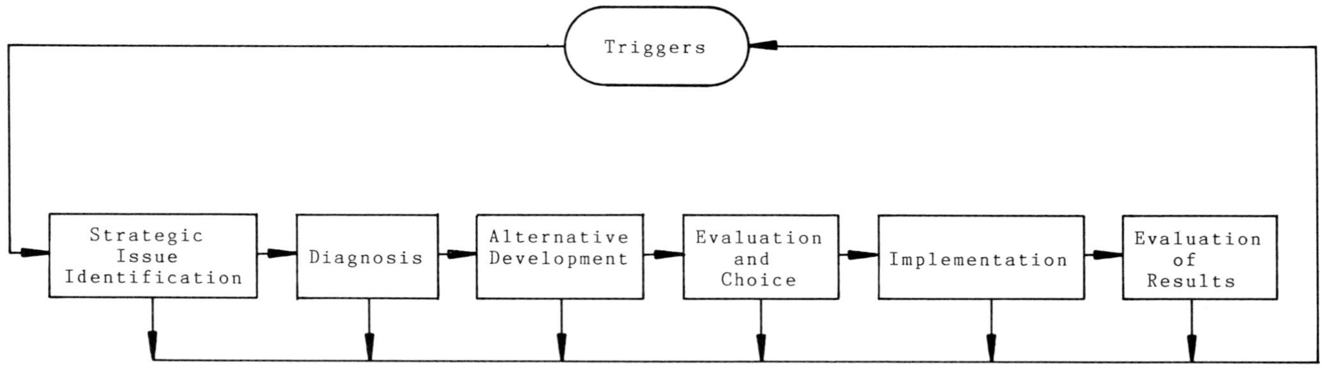

processes that appear to be muddled to the outsider. The process and politics of defining and achieving strategic goals has been well described by Quinn (Quinn, 1977). This process will now be covered.

Executives are always under pressure to define specific goals and objectives for their organizations. These goals should be given clearly, explicitly and quantitatively. Responsibility for the goals should be assigned to individuals or to units of the organization. Progress towards goals is monitored and corrective action taken if progress is not to plan. In practice, however, it is found that successful executives announce few goals and those goals that are announced are usually broad and general. It is also found that managers arrive at the specification of goals through an incremental "muddling" process rather than the analytical structured process specified in management literature.

There are several reasons why precise goal announcements are rarely made by successful managers:

(i) Centralization - quantitative goal announcements give signals to subordinates that a decision has been made. This can cut off the flow of valuable information since subordinates have been told that certain issues are closed. Exclusion of subordinates eliminates their genuine commitment and participation. The organization becomes centralized about the goal position, positions become rigid too soon and creative options are precluded.

(ii) Focus for Opposition - explicitly stated goals, especially on complex issues, provides a focus for opposition. Fragmented forces may be powerless against a diffuse goal but can unify in their opposition to an explicit goal. Experienced executives do not put forward complete goal proposals but progress by building agreement around one or a few parts of the proposal. In this way, opposition may remain fragmented, each constituency can be dealt with separately until the complete proposal is passed.

(iii) Rigidity - once a top executive announces a goal it can be very difficult to change. The executive and the supporters of the goal become identified with the goal. Changing the goal leads to a loss of credibility for all concerned. Goals may be adhered to long after it is clear that the goal is unattainable or even wrong for just this reason. Experienced managers keep their options open as long as possible so that changes can be made depending on the latest information available. This may be accomplished by specifying only broad directions until it is clear that a specific statement can be made.

(iv) Security - a further reason why highly specific goals are not announced is the mobility of top executives. Specific information on future directions is not provided to executives who might join competitors. Unforeseen changes in political stability or the economy can make nonsense of specific quantitative goals. Management do not want to announce specific acts such as acquisitions, divestitures or plant closings for fear of triggering an undesirable response. Goals that are announced tend to reflect or build a consensus, are broad enough to allow opportunism and are sufficiently long-range that achievement is possible from a number of alternatives.

There are many situations in which goals should not be specific, measurable or quantitative. Broad goals are often more effective than narrow quantitative goals. A certain generality in goals promotes cohesion. Many people can support goals such as superior products, high quality and environmental awareness. It is often impossible to obtain consensus if the goal is highly specific and quantitative; there is always someone to object to a detail. Once agreement has been achieved on a broad proposal, the details can be worked out. Broad goals create identity and enthusiasm. Quantitative goals are often difficult to identify with. Just as a specific goal allows the unification of the opposition, so a general goal allows the unification of fragmented support while keeping the opposition fragmented. The unity that is achieved can be employed to later make the general goal specific enough for the attainment of a worthwhile tangible end-result.

There are, of course, those situations in which specific goals are indicated. The announcement of specific goals at the right moment can create desired action. Specific new goals can be employed to indicate a major transition from the past. After a major trauma an organization often requires specific goals; general goals indicate the direction while specific goals attainable in the shorter term allow confidence to be gained by the organization. The annoucement of specific goals allows different emphases to be placed with the constraints of the broad goals.

Effective goals are important. Goals must be set at the right time, with maximum input from those with relevant information and with commitment from those responsible for achieving the results. If people have a common purpose, actions can be self-directed and commitment will be self-sustaining. Quinn points out that this is especially important for creative groups such as R & D. The existence of effective goals improves morale. People can share common performance goals. A sense of common purpose can resolve much inter-personal conflict in the organization. Organizations without a clear mission often have low morale. This is true of some government organizations. Finally, goals allow problems to be identified and defined. Goals that are not effective dissipate the energies and resources of the company. Effective goals that are internalized are required to move the organization forward.

In setting goals genuine participation of many persons in the organization is required but the manager must retain control of the goal-setting process. There is indeed a danger in the "bottom-up" process advocated for goal-setting. Time and energy can be expended with overmuch discussion of issues that could easily be resolved. People will regard the whole exercise as empty and manipulative if their goals are rejected by top management. On the other hand there are considerable shortcomings to goals that are announced from above. It seems best to provide a broad framework of goals from above with middle and lower management filling in the details. Unfortunately this latter process does not allow significant changes in organizational direction and is not effect for companies in a more dynamic environment.

New strategic goals rarely emerge from the "bottom up" process or from corporate planners. The process of identifying is complex and does not have a well identified beginning or end. A new consensus is created through a continuous, evolving, incremental and highly political process. In the beginning there is considerable uncertainty to specify a new direction for the enterprise. There is, therefore, a process of information gathering. There is active experimentation to identify and quantitate environmental changes specifically in the technical and marketing areas. Resources must be acquired. Different groups must be committed to the new direction. To minimize risk resource commitments are delayed as long as possible and are coupled to the increase in information. This is the prime reason for incrementalism.

Although the actual process of arriving at effective strategic goals is not analytical in nature but involves a muddling, incremental approach, this is not to say that the process cannot be managed. The process, when well managed, can be purposeful, politically astute and effective. It appears that the process consists of the following stages:

(i) Sensing Needs - effective top executives often sense needs for strategic change in non-specific terms. This can be due to the wide network of contacts that the top executive fosters along with the careful reading of the underlying signals. These contacts are not exclusively extra-organizational and probing of the organization itself may be carried out to avoid screens that block valuable information.

(ii) Building Awareness - once a vague need has been identified, it is necessary to build awareness of this need in the organization. This is done by setting up a study group, using selected staff or hiring an outside consultant. Usually this results in an intensified indication of the need without identifying specific actions. Although the lack of direct action can be frustrating, the organization is not yet ready for change.

Key actors are not yet ready to commit to the new direction because information is not yet persuasive.

(iii) Broadening Support - discussion of the information spreads more widely through the organization. Options are explored but decisions are not made. The guiding executive encourages others to give ideas and opinions while maintaining identification with the original idea. The purpose is to begin constructive movement without threatening major centers of power. The executive does not directly oppose strongly supported ideas. Goals remain broad and unrefined at this stage.

(iv) Creating Pockets of Commitment - options are still kept open but options are tested. Projects may be small and the executive may run programs through subordinates to avoid a loss of credibility should the project fail. The right moment to launch into the desired direction is sought. Opportunities to do this include reorganization, a crisis or loss of credibility of an opponent. On occasion it is possible for the executive to manufacture the crisis for which he alone has the answer.

(v) Crystallizing a Developing Focus - the executive can again employ a committee. The membership, charter and timing can be elected to influence its direction. The objective is to have the committee broaden support and increase support for the new goals. However, the committee can act as a focus for the opposition. There is also the danger that the recommendations can be overruled. If the timing is right the goal begins to appear as legitimate with statements in public, directions to divisions and so on.

(vi) Obtaining a Real Commitment - it is now necessary to have an individual responsible for the goal. This individual must have real commitment to the goal. This is the project champion. On identification of this champion the goal is formally stated and the necessary budgets, personnel resources and systems are put in place. Supporters are put in positions of responsibility and, if possible, persistent opponents are reassigned.

(vii) Continuing Dynamics - effort is needed on a continuing basis to ensure that the goals are achieved. Over a period of time the new strategic thrust becomes the established direction. Once the organization achieves consensus the executive must ensure that it does not become inflexible. The effort and commitment to achieving the goals may result in an over-commitment once the goals have been achieved. It is extremely difficult for the executive to introduce change in a new direction.

The implications of the political process have been considered elsewhere (Narayanan and Fahey, 1982). The role of coalitions has been emphasized. Coalitions evolve in organizations due to the existence of limited resources, interdependence of tasks, limited information, differences of opinion and self-interest. Usually a single individual cannot impose his will upon the organization and must seek the support of others by forming a coalition. There are conclusions that can be drawn from the coalition model of strategy:

(i) The nature of alternative(s) sponsored and the extent to which they are accepted, modified or rejected depends on the power/influence distribution within and across the relevant coalition(s).

(ii) Failures in strategic issue recognition, development and resolution may be ascribed to political infeasibility or the insufficient power and influence of organizational actors.

(iii) Coalitional influences in strategic decision-making may be both substantive and symbolic.

(iv) Commitment to a strategic decision begins to evolve during the early phases of decision-making rather than after the decision has been made.

(v) Formulating the content of strategy inevitably entails managing its context and process.

5.6 SUMMARY

Organizations are composed of people. People create policy, define strategic goals, assign tasks and carry them out. Ultimately, the success of an enterprise depends on its personnel. Personal factors at the individual, group and organizational level are extremely important. For the scientist and engineer the art of dealing with people, inter-personal skills, do not often come naturally but must be learnt.

At the personal level motivation is of prime importance. The method used to motivate depends on the saliency of the individual need, the expectancy that the effort will achieve a given result and the instrumentality of that result in satisfying the need. Leadership qualities can be employed for motivation and for other tasks. There are trait, style and contingency theories of leadership. Of these, contingency theories seem to have the most predictive power. The leadership style is chosen depending on the leader, the subordinate and the task. Flexibility in leadership fits best with unstructured tasks found in R & D. Of course many tasks in the organization are undertaken by groups. There is a whole theory covering group interactions, the clash of group social needs versus performance requirements. Group processes have been well studied and contingency theories again stress the fit of the group to the task. Thus in the management and social science literature there is a bank of experience that can be reviewed regarding other personal factors in addition to motivation, leadership and group dynamics.

There have been some studies regarding the performance of scientists in organizations. Major studies in this area are, however, 10 to 20 years old. Contingency theories have come to the fore. Of special interest is the examination of scientific performance as a function of age. It was found that performance slumped at mid-career and this was ascribed to motivational factors, either complacency due to success or discouragement with lack of achieved goals. Those who relied overmuch on supervisors had performance below par while those individuals who were self-motivated were highly effective. With groups the performance improved up to a group life of two years, was constant for 2-4 years and fell after four years. The explanation is that the group is shaking down for the initial time period and then works well but will develop too comfortable a relationship between the members and a "not invented here" mentality stifling innovation. All of the above factors can be addressed through mid-career reviews, restructuring groups at appropriate intervals and so on.

The performance of individuals in R & D has been compared to that of individuals in manufacturing both for successful and for less successful organizations. Individuals in R & D and manufacturing were found to differ in time orientation, goal orientation, patterns of influence and control and in coordination. The factors needed for success in R & D were not those required for success in manufacturing. There was also a difference between successful and not so successful R & D departments and manufacturing units regarding the factors listed. For successfully performing R & D units there were less formal reviews, more scientific goal orientations, lower formality of structure and less formal coordination mechanisms that did not inhibit significant interactions. Conflict was settled via confrontation.

The discussion of differing orientation of personnel in different units naturally leads to the consideration of interfaces. R & D must interact with company units to define programs and pass on results or new products. Continual information flow is needed to ensure that the programs are relevant to the organization's needs. It is now realized that specialized roles evolve in R & D that allow the bridging of the interface either to company units such as marketing and production or to the external environment of the company. Gatekeepers and boundary spanning activity are terms that are now recognized regarding the persons and processes that must take place for R & D to be effective. Although it has been realized for some time that interface relationships are important it was only recently that the relationship was quantified. Problems such as lack of communication, distrust, lack of appreciation and "too good friends" cause failures in the new product development chain. With severe problems it has been found that only 15 percent of new products were an unqualified success as compared to 67 percent where there were no problems. Again steps can be taken to reduce and hopefully eliminate severe problems at the interface thereby greatly increasing the rate of new product successes.

Organizations are composed of career and social systems besides the purposeful set of activities ensuring organization success. Political processes have various degrees of impact depending on the circumstances. For example, organizational climate and the position of the company in its life cycle.

Political action is that which takes place for reasons of self-interest. Individuals seek power bases so as to be able to influence others. There are various types of power source including physical, resource, expert and position power amongst others. The path of influence depends on the source of power available. Political processes cannot be ignored and have an impact on many activities. Strategic decision-making rather than following the analytical, logical path laid down by rationality often appears to have a muddled pattern to the outsider for just this reason.

The technical manager is often superbly trained in a narrow technical area. Often one task of managing is to oversee a much broader technical domain, which involves new technical education of the manager. At the same time, the manager must learn about the business including elements of finance, marketing and so on. It is little wonder that the manager pays little attention to the personal factors, to the informal networks of information, to political processes and interface factors. Yet it is precisely these factors that can make or break the manager. A technical achievement is found to have no market, R & D personnel are demotivated, there are budget and time overruns and top management is disillusioned. The technical manager ignors personal factors at his peril.

REFERENCES

Adizes, I. 1979 "Organizational Passages - Diagnosing and Treating Life-cycle Problems of Organizations", Organizational Dynamics, Summer, 3-25.

Allen, T.J. 1977 "Managing the Flow of Technology", M.I.T. Press, Cambridge, Mass.

Bergen, S.A. 1982 "The R and D/Production Interface - United Kingdom and West German Practices and Achievements in the Scientific Instrument Industry", R and D Management, 12, #1, 21-25.

Biller, A.D. and Stanley, E.S. 1975 "Understanding the Conflicts Between R and D and Other Groups", Research Management, XVIII, 16-21.

Burns, T. and Stalker, G.M. 1966 "The Management of Innovation", Tavistock Publications, London. See Preface to 2nd Edition.

Cohen, H., Keller, S. and Streeter, D. 1979 "The Transfer of Technology from Research to Developmnent", Research Management, XXII, May 11-17.

Dessler, G. 1976 "Organization and Management", Prentice-Hall, Englewood Cliffs, New Jersey.

Duncan, R. 1979 "What is the Right Organization Structure", Organizational Dynamics, Winter, 59-80.

Fahey, L. and Narayanan, V.K.	1982	"The Politics of Strategic Decision Making", in "Handbook of Corporate Strategy", McGraw-Hill, New York.
Farris, G.F.	1973	"Motivating R and D Performance in a Stable Organization", Research Management, XVI, #5, 22-27.
Farris, G.F.	1980	"Informal Organizations in Research and Development", International Conference on Industrial R and D Strategy, Manchester Business School, England.
Fischer, W. and Rosen, B.	1980	"The Latent Gatekeeper: Search for Motivation and Behavior in Gatekeeper Activities", International Conference on Industrial R and D Strategy, Manchester Business School, England.
Frohman, A.L.	1978	"The Performance of Innovation: Managerial Roles", California Management Review, XX, #3, 5-12.
Galbraith, J.R.	1977	"Organization Design", Addison-Wesley, Reading, Mass.
Galbraith, J.R.	1982	"Designing the Innovating Organization", Organizational Dynamics, Winter, 5-25.
Gruber, W.H., Poensgen, O.H. and Prakke, F.	1974	"Research on the Interface Factor in the Development and Utilization of New Technology", R and D Management, 4, #3, 157-163.
Hall, D.J. and Saias, M.A.	1980	"Strategy Follows Structure", Strategic Management Journal, 1, 149-163.
Handy, C.B.	1976a	"Understanding Organizations", Penguin Books, Ltd., Harmonsworth, Middlesex, England.
Handy, C.B.	1976b	Ibid, Chapter 2.
Handy, C.B.	1976c	Ibid, Chapter 4.
Handy, C.B.	1976d	Ibid, Chapter 6.
Handy, C.B.	1976e	Ibid, Chapter 5.

Harvey, E. and Mills, R.	1970	"Patterns of Organizational Adaptation: A Political Perspective", in Zald, M.N. (ed.), "Power in Organizations", Vanderbilt University Press, Nashville, Tenn.
Katz, R. and Tushman, M.L.	1980	"The Influence of Gatekeepers on Project Performance in a Major R and D Facility", International Conference on Industrial R and D Strategy, Manchester Business School, England.
Katz, R. and Allen, T.J.	1980	"An Empirical Test of the Not Invented Here (NIH) Syndrome: A Look at the Performance, Tenure, and Communication Patterns of 50 R and D Project Groups", International Conference and R and D Strategy, Manchester Business School, England.
Kottcamp, E.H., Jr. and Rushton, B.M.	1979	"Improving the Corporate Environment", Research Management XXII, November 19-22.
Lawrence, P.R. and Lorsch, J.W.	1967	"New Management Job: The Integrator", Harvard Business Review, November-December, 86-95.
Litwin, G.H. and Stringer, R.A.	1968	"Motivation and Organization Climate", Harvard Business School, Boston, Mass.
Lorsch, J.W. and Morse, J.J.	1974a	"Organizations and Their Members: A Contingency Approach", Harper and Row, New York p. 14.
Lorsch, J.W. and Morse, J.J.	1974b	Ibid, Chapter 2.
Lorsch, J.W. and Morse, J.J.	1974c	Ibid, Chapter 3.
Lorsch, J.W. and Morse, J.J.	1974d	Ibid, Chapter 5.
Margerison, C.	1978	"Managing the R and D Group, Management Decision, 16, #1, 52-63.
Marquis, D.G. and Myers, S.	1969	"Successful Industrial Innovations", National Science Foundation, U.S. Government Printing Office. p. 1.

Miller, D., Pearson, A.W. and Ball, D.F.	1980	"Environmental Gatekeeper - Product of Hindsight or Tool of the Future", International Conference on Industrial R and D Strategy, Manchester Business School, England.
Monteleone, J.P.	1976	"How R and D and Marketing Can Work Together", Research Management, XIX, #2, 19-21.
Moore, R.F.	1970	"Five Ways to Bridge the Gap Between R and D and Production", Research Management XIII, #5, 367-373.
Narayanan, V.K. and Fahey, L.	1982	The Micro-Politics of Strategy Formulation", Academy of Management Review, January.
New, D.E. and Schlachter, J.L.	1979	"Abandon Bad R and D Projects with Earlier Marketing Appraisals", Industrial Marketing Management, 8, 274-280.
O'Keefe, R.D. and Chakrabarti, A.K.	1981	"Coordination and Communication in Industrial Innovation: The R and D/Marketing Interface Problem", Baylor Business Studies, 12, Pt. 1, 35-43.
Quinn, J.B.	1977	"Strategic Goals: Process and Politics", Sloan Management Review, 19, #1, 21-37.
Quinn, J.B. and Mueller, J.A.	1963	"Transferring Research Results to Operations", Harvard Business Review, January-February, 21-38.
Pelz, D.C. and Andrews, F.M.	1966	Scientists in Organizations; Productive Climates for Research and Development, John Wiley and Sons, New York.
Roberts, E.B.	1979	"Organizational Approaches", Research Management, XXII, November 26-30.
Roberts, G.A.	1972	"The Commnication Imperative Between Management and R and D Research Management, XV, 67-72.
Sarett, L.H.	1979	"The Innovative Spirit in an Industrial Setting", Research Management, November 15-18.
Sayles, L.R.	1974	"The Innovation Process: An Organizational Analysis", Journal of Management Studies, October, 190-204.

Smith, C.G.	1971	"Scientific Performance and the Composition of Research Teams", Administrative Science Quarterly, 16, #4, 486-495.
Souder, W.E. and Chakrabarti, A.K.	1978	"The R and D/Marketing Interface: Results from an Empirical Study of Innovation Projects", IEEE Transactions on Engineering Management, EM-25, #4, 88-93.
Souder, W.E.	1980	"Promoting an Effective R and D/Marketing Interface", Research Management, XXIII, #4, 10-15.
Souder, W.E.	1981	"Disharmony Between R and D and Marketing", Industrial Marketing Management, 10, 67-73.
Stein, M.I.	1982	"Creativity, Groups and Management", in "Improving Group Decision Making in Organizations: Approaches from Theory and Research", Academic Press, New York, Chapter 6.
Strategic Planning Institute	1982	"Is R and D Profitable?" Strategic Planning Institute, Cambridge, Massachusetts, p. 7.
Thompson, J.D.	1967	"Organizations in Action", McGraw-Hill, New York.
Tushman, M.L.	1977	"Special Boundary Roles in the Innovation Process", Administrative Science Quarterly, 22, #4, 587-605.
Tushman, M.L. and Scanlan, T.J.	1981	"Characteristics and External Orientations of Boundary Spanning Individuals", Academy of Management Journal, 24, #1, 83-98.
Zaltman, G. and Duncan, R.	1977	"Strategies for Planned Change", John Wiley and Sons, New York, p. 26.

PART 6

TRENDS IN RESEARCH AND DEVELOPMENT

6.1 INTRODUCTION

At the time of writing there is no recession for R & D spending in the United States. For 1985 spending is estimated at $110 billion which is an increase of 13.4 percent over 1984 (Mosbacher, 1985). Inflation for 1984 was between 4 and 5 percent giving a real growth of about 8 percent. Industrial spending on R & D is estimated at $55 billion, with the Federal Government spending $52 billion (GDP, 1985). The increase in government spending is 13 percent in real terms over 1984. However, the largest increase in government spending was for defense at 26 percent over 1984; for 1985 defense spending is estimated at $37 billion.

Industrial support has been spread more widely with the tendency to support programs outside the company. Late in 1983 the law was changed to remove major restrictions on cooperative R and D. This allowed non-profit organizations such as Battelle to seek wider support for cooperative projects. Industry has formed new groups such as the Semiconductor Research Foundation. Funding from industry for the academic area is scheduled to increase for 1985; this trend began in 1983.

Although funding of R and D is increasing in the United States the picture in Europe is not so clear. There has been little recovery from the recession and it has been reported that budget cutbacks threaten the EEC research program. Budgeting constraints are expected to last until the end of 1985. The Commission proposal was for spending of $490 million but the Council of Ministers wants to reduce this to $426 million (Research and Development, 1985a). A sharply reduced spending program of $853 million over 4 to 5 years was agreed for the following project areas: biotechnology, fusion, radioactive waste disposal and reactor safety. The original request was for $2.61 billion (Research and Development, 1985b).

At the same time that R & D expenditures increase in the United States there are major changes in economic structure that are expected to continue and accelerate through the 1990's (Minshall and Moody, 1984). The results from technological activities in general and R and D in particular have already wrought considerable changes. It is stated that 8 out of 10 high growth activities fall into the high technology area (as defined by R & D expenditures). In 1960 over 24 percent of the labor force was employed in the industrial sector but this had fallen to 23 percent in 1970, 20.6 percent in 1980 and is predicted to decline to 19.5 percent in 1990 and to less than 18 percent in the year 2000.

The national growth industries in the 1970's ranged from drugs to women's outerwear, from computers to structural products and from surgical implants to miscellaneous fossil fuels. Not all of these are high technology areas and there is a tendency to equate growth with high technology. The descriptive definitions that can be used to define a high technology area fall into three categories: evolutionary factors, descriptive factors and qualitative factors. The use of evolutionary factors relates technological advances to a key discovery e.g. the semiconductor industry developed from the invention of the transistor. Descriptive factors indicating that an industry is high technology include concepts that are radically new, procedures that include a high level of R & D, processes that incorporate a high level of automation and systems that integrate scientific knowledge with production. Qualitative definitions relate to the type of training and skills needed and the relationship of the technology to the "state of the art".

The importance of the high technology area has been emphasized as a way of meeting foreign competition (Merrifield, 1983). Over the last decade high technology businesses grew at a rate of an average of 7 percent per year as against 3 percent for all U.S. industry. For this 1970-1980 period price inflation was 2.5 percent per year for the high technology sector versus 7 percent annually for all U.S. industry. High technology businesses produced a $30.5 billion positive trade balance in 1980 against a negative trade balance of $54.7 billion for all other businesses. Productivity increased six times faster for high technology businesses than the average productivity of all U.S. businesses. Each job created in the high technology sector creates eight additional jobs in the other sectors supplying it.

The definition of an area as high technology can be made on a quantitative basis. It is found that by far the most important characteristic of high technology activities is the level of investment on research, development and engineering. Non-high technology companies seldom spend more than 1 to 1.5 percent of sales in R & D activities whilst high technology companies typically spend 4 to 6 percent of sales on R & D.

The tremendous investment in R & D by companies in the high technology area results in a certain structuring of that activity. There are more likely to be free-standing corporate research centers. In addition there are sophisticated research and problem-solving centers that are often located next to production facilities. There are of necessity high numbers of technically trained personnel which typically runs at a level of 18 to 30 percent including production. Currently professional and technical workers represent about 15 percent of the total work-force in the United States (14.5 million). It is expected that by 1990 the number will be 17 million. This is mainly due to the anticipated increases in R & D activity.

The continued importance of R & D has been demonstrated. Expenditures, at least in the United States, will continue to increase. Areas that are of especial importance include robotics, artificial intelligence, biotechnology, electronics, materials and bioengineering. In this concluding section the intention is to re-examine some of the topics discussed in earlier sections with the aim of highlighting R & D problems of current interest and with the hope that future trends in R & D can be forecast.

6.2 INNOVATION

The subject of technological innovation has recently been reviewed (NSF, 1983a). This review is timely in that it covers many of the factors thought to influence innovation such as structure of the organization and interface problems. There is continuing interest in innovation ranging from pleas for increased levels of innovation (Kanter, 1984) to examination of the level of innovation at the national level (Wilson, 1984; Chakrabati et al 1982). Thus Chakrabati et al examined the level of innovation in the United States and other countries using 500 technologically significant innovations over the period 1953-73. It was shown that competitiveness in the international marketplace increased during the time period but that the United States maintained a uniform position in terms of the number of innovations. However the relative share decreased. The pattern of innovations varied between countries - Japan, for example, had 25 percent of its innovations for internal utilization (about double the level of other countries). Only about one--fourth of the innovations were radical while about one-third involved major technical shifts. The United Kingdom had the highest proportion of radical innovation while Japan had the least. The structure of the market was shown to influence the level of innovation with the level increasing as the concentration of the market increased. Wilson has challenged the notion that the United Kingdom has a high level of radical innovations that are poorly implemented and states that the original ideas might have been less outstanding than originally believed; the point is that an innovation that does not satisfy a market need or create a market need that it can satisfy will not be successful in commercial terms no matter how radical the innovation. It appears, in fact, that Britain pays less attention to customers and does not employ customer suggestions for improved products; in this way R & D becomes isolated from the marketplace.

There has recently been interest in looking at innovation in terms of a long time scale and as a component in the cycles of boom and recession that have been followed in industrial nations for the last 120 years. An excellent summary has been made by Wilkinson (Wilkinson, 1983), who pointed out that the developments of the 1970's emphasized profitability and cash management (BCG matrix, PIMS approach) and resulted in an emphasis on finance and marketing. R & D no longer, if it ever really did, makes a serious contribution to the company planning process.

The rate of change of technology has accelerated but at the same time financial constraints have led many companies to maintain their status quo in technology thus resulting in an increasing technology gap. The reaction of mature companies to technological threats can be summed up as "too little too late". There are, therefore, two conflicting tendencies, the concentration of companies into narrower product/market segments in response to economic pressures and the technological changes that will continue into the future. There is great interest in seeing whether future trends can be predicted so that the impact on current strategies can be assessed.

There is evidence, admittedly incomplete, that economic activity proceeds in cycles having a period of about 50 years, the long-wave business cycles first commented on by Kondratieff. This cyclic behavior is regarded as being superimposed on the underlying natural growth of the particular economy. Wilkinson cites many examples, for the United Kingdom, of activites that proceed in cyclic fashion e.g. consumption of polyethylene, production of steel ingots, fixed capital investment in selected industries and bank rate. These cycles are of much shorter period but lend credence to the notion that cyclic variations can occur.

The Kondratieff cycles number four up to the present with the fourth cycle to complete about the year 2010. The cycles may be described as follows: I. Industrial Revolution; II. Bourgeois U.K.; III. Neo-mercantilist; IV. Western World. The present cycle peaked about 1955 and shows the bottom of the depression about 1980 with high prosperity at the end of the cycle. From the point of view of innovation, the cycle has a significance as it is observed that the rate of innovation peaks in the depression phase. It may be argued that the growth out of the depression is due in large part to the development of these innovations. The rationale for this behavior is that in the depression phase there is spare cash for investment, surplus capacity and surplus manpower leading to increased risk-taking. It should be noted, however, that there is evidence to suggest that high company profits stimulate the amount of R & D carried out (Bosworth and Westaway, 1984). At times of depression R & D is expected to be low in activity which implies that the innovations do not appear via formal R & D activities. This would be an interesting avenue for research.

The concept of Kondratieff cycles has been applied to the change in industrial activity (Wilkinson, 1983). It appears that industries fall into two types: those that are really hurt by the depression and those that are unaffected. The former businesses are either eliminated or are restricted to smaller specialty areas. Elimination may be total if the product no longer has relevance or may occur with reference to the West by transfer overseas to locations with lower labor rates and/or which lie closer to raw material sources. Industries that are not really affected by the depression may move into the former category during the next depression; Wilkinson argues that many industrial sectors have a natural lifespan of two Kondratieff cycles.

The options that are open to a company are as follows:

(i) If the industry is declining then the inevitability of this should be recognized. Cash yield should be maximized. It may be possible to extend the period of decline by finding a market niche.

(ii) Identify the technology that will supplant the existing technology and buy or merge into it.

(iii) Acquire the know-how and go it alone.

(iv) Start a new program via R & D to get into the new technology.

(v) Put a limited effort into several new areas so that a better selection of know-how to acquire can be made at a later date.

It is interesting to note that approach (v) involves spending research funds outside the company whereas approach (iv) involves internal R & D. The tendency for companies is to move to the external research approach as noted in the Introduction.

The subject of invention, innovation and economic cycles has been examined by Sahal, who developed a model to predict the probability of a given number of innovations per year (Sahal,

1983). The model was of the negative binomial type and predicted that, as observed, technical discoveries cluster in time rather than being uniformly distributed. However, there were two not one periodicity to the cyclic generation of discoveries; one cycle has a 24 year period and the other a 48 year period (the same period as the Kondratieff cycle). Sahal states that generic innovations are responsible for long wave response in the economy while short waves are due to the development of a characteristic innovation within an industry. Economic waves and innovation are tied together but contrary to popular opinion what is needed is not unprecedented expenditure on basic research but technological development of a few key fundamental innovations and the diffusion of developments across broad areas of industrial application. As distinct from the simplified picture given by Wilkinson, Sahal points out that fundamental innovations occur toward the end of prosperity and anywhere during a recession and toward the end of depression and anywhere during recovery.

There are implications for "R and D management during the 1980's and beyond" (Sahal, 1983). Success depends on following a variety of innovative activity rather than one lead to the exclusion of all others. R and D activity, by its nature, is not orderly and there is little to be gained by rationalizing the process. Any attempt to routinize R & D activity is doomed to failure. This attitude goes against the trend in the management of R & D towards greater specification of tasks and activities and the adherence to time schedules (see Part 4). Of course, this trend may be a reflection of the trend towards greater financial emphasis in corporations.

A variation on the concept of economic waves of activity has recently been presented (Marchetti, 1985). A mathematical model derived from ecology is presented. Specifically the model is one developed in the 1920's to describe the interactions of populations of predators with their prey. The model is applied to diverse areas such as length of transport infrastructure in the U.S., car populations for major countries and technology of energy generation. The common factor is saturation; every technology is born, grows and ultimately dies. Innovation is the mechanism that should provide "new jobs for old".

The themes in current innovation studies are concerned with the pattern of innovative activity on a geographic and on a temporal basis. There seems to be a reciprocal relationship between economic waves of activity and the emergence of innovations. These waves of activity are responsible for clusters of innovations that in turn cause the economic cycle to repeat.

6.3 INNOVATION AND FIRM SIZE

The increased interest in the promotion of innovation has led to government attempts to influence the level of innovation by formal programs. Much recent attention has been focussed at the small firm as current opinion has it that the locus of innovative activity is at the level of the small firm. This view is contentious, however, as there is no clear evidence to support this position. A discussion of the influence of firm size on innovation has recently been covered (National Science Foundation, 1983b). The conclusion was that size turned out to be a much more ambiguous variable than might be expected at first sight. The point made is that size is correlated with other organizational variables (vertical hierarchy, internal complexity, decision-making processes) but the correlation is often not direct. However, it would appear that firms below a certain size (say 20 employees) could be treated as a homogeneous group since it would be difficult to run such a company with different types of hierarchy and different degrees of internal complexity.

The literature does not usually place the cut-off for smallness at 20 employees and studies may consider as small firms with several hundred employees. There are indications, despite the above reservations, that small R & D based firms are highly involved in innovation, have higher R & D productivity than large firms and produce more in terms of innovations, patents and products for a given unit of resource input (National Science Foundation, 1983c). It has to be pointed out that only a small proportion of small firms engage in formal R & D, probably about 10 percent, with the remainder engaging in service activities, standard types of production and so on.

Other studies of the influence of firm size on innovation have examined many facets of the relationship. Economic theoreticians have argued that economies of scale in R & D cannot be theoretically justified but recent formulations of theory state that it is possible to relate firm size to R & D employment, output and value added (it is noted from the discussion in Part 1 that the measurement of output is fraught with difficulty). The theory to be tested is a variation on the Schumpeterian economy of scale (Kohn and Scott, 1982).

The role of external sources of funding has been emphasized (Oakey, 1984). It has been pointed out that small firms perform less well in development areas. A study of this fact led to the realization that the most innovative of small firms sought external funding and thus the presence or absence of such funding played a critical role in the survival and growth of the small firm.

Some small firms engage in international sales. A study conducted on small and medium sized firms in the United States and Canada suggested that substantial R & D had little relevance to success in the international marketplace (Kirplani and MacIntosh, 1980).

Statistics from 1953-1973 were employed to look at the level of innovation as a function of firm size (Chakrabati, et al, 1982). In this study firm size was based on the annual level of sales. On an aggregate basis over six countries it was found that large firms, medium firms and small firms produced 55, 14 and 31 percent of innovations respectively. Thus small companies have a significant innovative output. The proportion depends on the country chosen; in the United States small firms produce 35 percent of innovations but in Japan only 4 percent. Small firms also tend to be important in determining the type of innovation especially in the United States, where 40 percent of the radical innovations and 38 percent of the innovations involving major technological shifts were introduced by small firms. It will be interesting to see if a different pattern has appeared in the post-1973 period.

A balanced view of the need for small and large companies has been put forward (Rothwell, 1983). Because of certain behavioral and organizational factors small firms are better adapted to innovation. On the other hand large size and monopoly power have advantages in resources allocation and market control. It is argued that there are essentially two types of innovation, entrepreneurial innovation and managed innovation. In the first type, new basic technologies appear that are coupled in an unspecified way to new scientific developments; entrepreneurial innovations largely fall outside existing companies and market structures. Entrepreneurs found small, dynamic companies that grow rapidly as the product or technology takes off. As the technology and markets mature firms increase in size and the inventive activity becomes progressively formalized as in-house R & D leading to managed innovations that are largely incremental in nature. Later in the life cycle price becomes of prime importance. Internal R & D has shifted from a product to a process focus and at the last stages of the life cycle there is no need for R & D at all.

TABLE 68

Advantages of small and large firms in innovation

	Small firms	Large firms
Marketing	Ability to act quickly	Comprehensive market power
Organization	Lack of bureaucracy	Can be bureaucratic Risk-averse
Process	Efficient communication	Often have political overtones
Personnel	Often shortage of people	Amply qualified people available
External communication	Often do not have time or resources	Resources for library, literature searches, external contacts
Finance	May have problem in attracting capital	Ability to borrow Often publicly held
Growth	Cash flow problems due to growth	Problem is to identify and act on opportunities
Economy of scale	Can cause problem unless product benefits outweigh cost	Capacity to reduce price
Regulatory	Must cope with patent law, government and other regulations	Resources to handle regulatory area

Small firms and large firms have complementary activities. Often small innovative companies are formed by individuals who have left large firms. In the short term small companies provide only limited employment but in the larger term successful small companies grow and can provide substantial levels of employment. The advantages and disadvantages of small and large firms are given in Table 68. One factor not commented on is the low rate of innovation for medium-sized firms (Chakrabati et al, 1982). Presumably such firms are engaged in the development of innovations and do not have resources to develop new products until sufficient cash has been generated.

6.4 BCG AND PIMS

In Part 2 the use of the BCG matrix was described. The use of the matrix with the dimensions of market growth rate and market share allows a business to be categorized as a Question Mark (Problem Child), Star, Cash Cow or Dog. Positioning of the business allows strategic options to be formulated and these options depend on the business position. Also described in Part 2 was the PIMS data base whereby businesses (strategic business units) annually submit data regarding business performance. Recent developments have been attempts to use the PIMS data base to empirically explore the performance tendencies and strategic attributes of businesses in the four cells of the BCG matrix.

The main theme of the BCG approach, that the four cells of the matrix have different tendencies to generate or consume cash, was corroborated (Hambrick et al, 1982). Significant differences among the four cells on performance measures such as return on investment, return per risk and market share change were also observed. There were, however, some variations on the advice of BCG. For example, it was found that the average Dog has a positive cash flow that can be employed to fund a Question Mark. Contrary to the advice of BCG, Dogs should not be unthinkingly divested or liquidated. Another conclusion was that businesses may not face sharp tradeoffs between share building and cash flow or profitability. However, not much new information emerged concerning R & D. Question Marks and Stars have high R & D expenses while Cash Cows and Dogs have low R & D expenses.

A further analysis of the BCG matrix via the PIMS data base focussed on profitability (MacMillan et al, 1982). It was found that in the short-term product R & D has an adverse effect on ROI; process R & D, on the other hand, neither helps nor harms businesses in three of the BCG boxes but adversely affects Cash Cow profitability. From Part 2 it will be remembered that the payback period for product R & D is longer than for process R & D. The business that is focussed on the short-term will tend to neglect R & D. Profitable businesses probably will spend more on R & D especially if in areas of technology change. It is interesting to note that high share businesses need not have low prices but can charge premium prices; this is in contrast to the BCG strategy of driving down costs and prices in order to discourage low share competitors.

Other studies have explored the strategic options for Dogs (Christensen et al, 1982; Gelb, 1982). The point made is that in recessionery times many businesses become Dogs. However, it does not necessarily follow that the business be unprofitable. It has been found that effective businesses (ROI greater than 20 percent) with low market shares were more likely to be found in low growth markets than in high growth markets. Businesses with low market shares do not necessarily have a poorer cost position than higher share companies. PIMS research has identified high purchase frequency and an average price as factors augmenting cash flow in most Dog businesses. Table 69 indicates opportunity prospects for Dogs. Some businesses may be in markets with negative growth. Although not considered by the BCG method even these businesses may not have totally bleak prospects (Gelb, 1982). These businesses are classed as Under-Dogs and Buckets respectively depending upon having low or high market share. In a declining market there may be sufficient fall-out of competitors to allow the Under-Dog to move to a high market share position. There is more likelihood of this happening for the Under-Dog than for the Dog. However, it has to be emphasized that none of the strategies proposed involve R & D. Even in those cases in which the decision is to exit an industry it may not be possible to do so due to substantial exit barriers (Harrigan, 1982).

TABLE 69

Opportunity prospects for Dogs

Capability of implementing change	Applicability of BCG assumptions	
	High	Low
High	Follow BCG strategy	Look for opportunities
Low	Exit	Position may be acceptable

On a smaller scale the life-cycle concept has been applied to factories. There is, of course, often a close correspondence between the demise of a factory and the demise of a Dog business. However, this is not necessarily so. During the 1970's the average plant age was 19.3 years and the median age only 15 years (Schmenner, 1983). One-third of plants closed were no more than 6 years old. The reasons most commonly given for closing a plant are: inefficient or outdated production technology, fall off in sales volume, price competition, high labor rates and competition from superior products. Plant closings may sometimes be necessary but the best strategy is not simply to set up a plant and forget about it until it is too late but rather to plan the activities and upgrading of the facility so as to maximize the life-cycle.

One other area that has been examined via the PIMS data base is that of vertical integration (Buzzell, 1982). Vertical integration provides a way of controlling costs, assuring supplies of materials (see Thompson, 1967 in Part 2), improving coordination and increasing technological capabilities. There are pitfalls, however, and vertical integration often requires high investment, can result in reduced flexibility and loss of specialization and may, in fact, lead to unbalanced throughput if the different operations have different scales. Analysis of the data base indicates that profitability is highest at both ends of the spectrum i.e. at very low or at very high levels of integration. The results also indicate that highly integrated businesses generate more new products - this may be due to synergy between the differing components of the vertically integrated chain.

One study has investigated the effect of strategy on product R & D spending (Hambrick et al, 1983); the PIMS data base was used and businesses were divided into two groups, those in growth markets and those in mature markets. The responses found were different in the two groups. Mature businesses demonstrate an inertia effect. Dollar expenditures for R & D are relatively stable irrespective of sales; there is a significant tendency to adjust R & D intensity (size of budget to sales ratio) in the opposite direction to past adjustments. Prospectors (see Part 2) in mature markets tend to cut back on R & D intensity but, in general, mature businesses do not appear to be responsive to competitive pressure. Although the data were not presented for a long enough time period, it is likely that large changes in R & D intensity occur over a relatively long time scale. On the other hand growth businesses are more complex and dynamic in adjusting their R & D budgets. Changes in R & D intensity are partly in response to changes in sales but also depend on the level of technological output relative to competitors. Past changes in R & D intensity are likely to carry into the future as there is commitment to programs. Stance toward innovation (prospecting), emphasis on product quality and market share also play a role in determining the R & D intensity.

The trend in strategic areas is to test the largely unproven recommendations of strategists, such as BCG, empirically using the PIMS data base. Many of the recommendations turn out to be supportable based on results. There are few implications for R & D except for the one study cited due to the short-term orientation of the focus. It is to be expected that empirical tests will continue as the data base grows. The PIMS data base itself is not without criticism (Hambrick et al, 1982).

6.5 CORPORATE INNOVATION

There are many difficulties with innovation in the larger firm. There has correspondingly been an increasing interest on organizations that promote innovation. The strategic posture of the firm plays an important role (Iyer and Ramaprasad, 1984). It has been demonstrated that the strategy-structure causal sequence is differentiated by radical versus incremental innovation (Ettlie et al, 1984). In other words, unique strategy and structure will be needed for the radical innovation process as demonstrated by an aggressive technology policy and the concentration of technical specialists. Traditional strategy-structure arrangements tend to support new product introduction and incremental process adoption; these arrangements tend to be found in large, complex, decentralized organizations that have market dominated growth strategies.

Specific prescriptions have been given for the firm that must manage the daily operations while simultaneously developing new products (Sands, 1983). The suggestions appear to apply to incremental rather than radical innovation and cover the organizational options for location of a new product group, staffing requirements, coordination mechanisms and so on. Most of these options have been given earlier (Part 2 and Part 5).

There has been renewed interest in the external acquisition of new products as against the internal generation of new products via R & D (Link et al, 1983). For specific industries at a certain stage in the life-cycle, it is found that the best way to stimulate innovation is by stimulating the R & D efforts of the suppliers to that industry. This finding is especially true for process innovation for industries characterized by low technological intensity and those exhibiting high levels of product standardization.

An alternative mechanism for growth involves merger and acquisition. The characteristics of mergers and acquisitions in the United States in the 1970's have been reported (Chakrabarti and Burton, 1983). There were 1,680 cases in the sample 1974-75 and 1,192 in the 1979 sample. The emphasis here will be on the influence of R & D technological intensity on merger/acquisition activity. The results of the survey showed that acquisition is predominantly an activity of firms in industries of low and medium R & D intensity. Firms in low R & D intensive industries became more active as acquiring firms in the 1970's while firms in medium R & D intensive industries became less active. Firms in very high R & D intensive industries did not change their level of activity in acquiring firms but firms in high R & D intensive industries did increase their efforts. The majority of the acquired companies were in low and medium R & D intensive industries. While there was a decrease in the number of companies in the very high R & D intensive industries being acquired there was a 5.34 percent higher rate of acquisition of companies in high R & D intensive industries.

Companies in R & D intensive industries acquired companies in R & D intensive industries to a larger extent in the early 1970's than in the late 1970's. However, firms in low R & D intensive industries tended to acquire companies in low and medium R & D intensity areas. Medium and high R & D intensive companies had a preference for acquiring the same type of company as regards R & D intensity. This behavior indicates that companies in low R & D intensity industries, generally taken to be more mature with older technologies, do not move into higher technology areas via acquisition.

Reasons for acquisition varied depending on R & D intensity and there were changes in acquisition behavior during the 1970's. Companies from high R & D intensive industries showed a decline of vertical acquisitions but companies from very high R & D intensive industries increased their acquisitions of the vertical type. The latter companies increased efforts at diversification.

It was also noted that during the 1970's acquisitions/mergers became the province of large companies. The size of the acquired company increased and the ratio of the size of the acquiring to the acquired firms decreased. Less R & D intensive industries see more acquisitions because acquisition competes with internal development. Companies with large corporate R & D efforts depend more on internal growth while growth by acquisition strategy is practiced by companies with low R & D expenditures. It is interesting to speculate that this situation may have changed. Firms that are R & D intensive but cash rich may want to hedge their bets by spending externally on R & D, licensing or acquisition.

The other side of the coin to acquiring technology is the sale of technology (licensing-out). Large technology producers get very little payoff from licensing (Roberts, 1982). Data were obtained on the activities of 33 large American corporations. The results indicate that many ideas are generated in R & D groups, many of these ideas are filed, most files are patented, few patents are licensed and few licenses generate much income (results differ a great deal from industry to industry with the greatest payoff in areas such as pharmaceuticals). It was found that over the 33 companies an average of 28.7 ideas per 100 technical employees were generated annually and submitted for possible patent application. Some 21 of the 33 organizations had from 90 to 100 percent of filings issued as patents with an average of 86.1 percent over the whole sample; filings for patents represented an average 22 percent of ideas but the spread was very wide. Overall almost 20 percent of ideas originally filed for possible patent action do eventually issue as patents.

Obtaining a patent is only the first step in the licensing process. The time to find the first licensee differed widely from minus one month (one month before the patent issued) to 41 months. The bulk of the experience was with annual incomes in the range $1,000 - $10,000. Some 20 percent result in less than $1,000 per year; 40 percent less than $5,000 per year; 60 percent less than $10,000 per year; 90 percent less than $50,000 per year; and 95 percent generate less than $100,000 per year. Clearly licensing is not a major generator of revenue except for the atypical case. Greater opportunities occur in the licensing of process technology than for new products. Attractive areas are where a process produces a unique product and processes that achieve considerable advantages for an existing product via decrease in cost or increase in quality.

A range of options for achieving commercial benefits has been given by Roberts:

(i) internal product/process development
(ii) joint ventures
(iii) creation of spin-off companies
(iv) sale of technology
(v) licensing of patents.

These options range from active corporate involvement for internal development to passivity as far as licensing is concerned. It is suggested that there be more involvement upstream so that activity in R & D can be directed towards the development of products and processes for sale rather than waiting for development to be a fait accompli before looking at licensing possibilities.

As has been shown the maintenance of viability of large companies is a matter of some interest. There is no universal panacea for overcoming the inertia of the large company so that innovation is facilitated. Prescriptions have been given for organizational hierarchies to promote innovation. The subject of external versus internal development is still being debated as is the best way to get a return on technology that was developed but is not needed by the firm.

6.6 PERSONAL FACTORS

Topics addressed in Part 5 and those mentioned in Parts 1 and 3 continue to be of interest. The subject of scientist/engineer job performance evaluation continues to be of great importance due to the high cost of R & D personnel. Of its nature R & D is labor-intensive and will continue to be so into the future despite the trends to automation in manufacturing. A recent review of methods of evaluation has been given and a decision modeling approach is suggested (Stahl et al, 1984). There is little evidence, however, that the suggested approach will be any more effective than other techniques tried. One interesting facet of evaluation was the finding that managers evaluated performance differently than did non-supervisory personnel and placed greater emphasis on the long-term effectiveness of the work performed but were less concerned with characteristics such as "works hard" or "works well with peers". Not unexpectedly managers are results-oriented in performance evaluation of professionals.

Top management roles in R & D, to some extent, have been shown to support the above statement regarding interest in long-term performance (Rafael and Rubenstein, 1984). In fact, top management is concerned with shorter-term R & D matters as well as with the effects of R & D on long-term strategy. The influence is more marked at early stages of projects as well as later. Top management intervenes at irreversible decision points and in areas that affect other parts of the company.

Leading on from the view of top management is the question as to what motivates individuals to become managers (Miner and Smith, 1981). Based on perceived management organizational roles, the motivational base for each role was proposed and a questionnaire formulated so that individuals could be tested for their suitability for managerial positions. This questionnaire was used as far back as 1960 and more than 25 different samples have been obtained. The results are given in Table 70. The figures shown are the percentage of students showing a high tendency. There has been a considerable drop in the overall motivation to manage which may reflect changing attitudes and also may be due to a change in the mix of students. For example, by 1980 the percentage of women had tripled to 43 percent. Naturally, there is a need to examine the basis of the questionnaire and to determine how accurately the questions reflect the requirements of the "ideal" manager. It is quite possible that managerial roles should have changed over a 20 year period along with the change in values of society, impact of technology and task uncertainty. A manager clinging to a value set of 1960 may well be regarded as reactionary in today's world. Miner and Smith indicate that organizational structures and processes may

TABLE 70

Motivation to manage. Data for business administration students, 1960-1980.

Motivation	1960-61	1967-68	1972-73	1980
Overall motivation to manage	64	39	23	26
Favorable attitude to authority	70	53	41	40
Competitiveness	50	36	22	25
Assertiveness	69	62	44	35
Power need	43	46	42	44
Desire for a distinctive position	47	43	43	44
Sense of responsibility	53	37	27	27

well have to change to accommodate the changing motivations of potential managers.

In Part 5 the effect of research team age on performance was examined. Many of the findings were from the 1960's but recent work has tended to confirm the findings (Katz, 1984). Data were gathered in a large American corporation employing 345 engineering and 61 scientific staff. As has been noted earlier group members pass through stages of socialization, innovation and stabilization. Innovation is the productive stage. As research teams grow older there is a trend to increasing isolation, selective exposure of group members via limited communication networks and selective perception of outside information. Management actions to upset the comfortable equilibrium appear to be necessary using rotations, promotions or other mechanisms. The data presented indicated that the best performing groups were those with longevities between 1-1/2 and 5 years. However, a more recent data accumulation from 12 different technology-based organizations with over 1200 R & D groups indicated that a large number of high performance groups had a tenure in excess of 5 years (Katz, 1984). Although the analysis of the data has not been published the indication is that job supervision plays an important role in determining group performance and that project managers should place less emphasis on participative management and more emphasis on direction and control. A full report is awaited with interest. What the results may be saying is that the tasks become more certain for groups with longevity over 5 years so that tasks can be programmed by the manager. It then matters less what the characteristics of the individual members are as adherence to the schedule provides good performance of the group.

The interface between R & D and marketing and manufacturing has received further attention (Kiel, 1984; McDonough III, 1984). The approach is not to extend the work already carried out on the "interface problem" but rather to present the nature of the problem in the management literature and to indicate potential solutions.

The area of politics in organizations has now been reviewed and applied to the R & D situation (Dill and Pearson, 1984). The rational actor model is contrasted with the organizational politics model (see Part 5) and it is pointed out that several strands of R & D research suggest that political activity should be common. For example, R & D personnel have strong professional identification and will form common interest groups, have informal information networks, have diffused authority and are subject to conflict over resource allocation and with other groups. For the R & D project manager the implications of the political model are that there must be an emphasis on personal power, the manager must be effectively placed in the hierarchy and the manager must develop tactical skills so that the power of the person and the position can be effectively utilized.

6.7 SUMMARY

The latest information indicates that R & D is more important than ever for the economic well-being of the nation. Innovation is linked to cycles of recession and recovery in a way that is not well understood. Many of the issues that have been addressed are still unresolved. It is not really known how to stimulate innovation. What is the relationship of firm size to rate of innovation? How does the large corporation maintain or, if lost, develop an atmosphere enhancing creativity. Strategy-structure process relationships and the impact of technology has been incompletely investigated. There is a continuing interest in the choice between internal product development and the external acquisition of products. The impact of personnel factors on the performance of R & D is still being studied. In the late 1970's there has been a flurry of interest in key R & D personnel and the development of informal networks of communication. In an organizational context the role of political action has only just been accepted; the wide application of these findings to R & D is a task for the future. It will be noted that the subject of creativity was not covered in the discussion of R & D trends. It is interesting to note that widespread interest in creativity peaked in the late 1960's. Creativity programs exist at a small number of institutions and the thrust at the present time is towards the use of creativity techniques for new product development. Apart from anecdotal evidence the effectiveness of these efforts is not known.

REFERENCES

Bosworth, D. and Westaway, T. | 1984 | "The Influence of Demand and Supply Side Pressures on the Quantity and Quality of Inventive Activity", Applied Economics 16, 131-146.

Buzzell, R.D. | 1983 | "Is Vertical Integration Profitable?" Harvard Business Review, #1, Jan.-Feb., 92-102.

Chakrabarti, A.K. and Burton, J. | 1983 | "Technological Characteristics of Mergers and Acquisitions in the 1970's in Manufacturing Industries in the US", Quarterly Review of Economics and Business 23, #3, 81-90.

Chakrabarti, A.K. Feinman, S. and Fuentivilla, W. | 1982 | "The Cross-National Comparison of Patterns of Industrial Innovations", Columbia Journal of World Business, 17, #3, 33-38.

Christensen, H.K., Cooper, A.C. and DeKluyver, C.A. | 1982 | "The Dog Business: A Re-examination", Business Horizons, 25, #6, 12-18.

Dill, D.D. and Pearson, A.W. | 1984 | "The Effectiveness of Project Managers: Implications of a Political Model of Influence", IEEE Transactions on Engineering Management, EM-31, #3, 138-146.

Ettlie, J.E., Bridges, W.P. and O'Keefe, R.D. | 1984 | "Organizational Strategy and Structural Differences for Radical Versus Incremental Innovation", Management Science, 30, #6, 682-695.

GDP | 1985 | See pp. 8296-8298, March, Government Data Publications, Brooklyn, New York.

Gelb, B.D. | 1982 | "Strategic Planning for the Under-Dog". Business Horizons, 25, #6, 8-11.

Hambrick, D.C., MacMillan, I.C. and Day, D.L. | 1982 | "Strategic Attributes and Performance in the BCG Matrix-A PIMS-Based Analysis of Industrial Product Businesses", Academy of Management Journal, 25, #3, 510-531.

Hambrick, D.C., MacMillan, I.C. and Barbosa, R.R. | 1983 | "Business Unit Strategy and Changes in the Product R and D Budget", Management Science, 29, #7, 757-769.

Harrigan, K.R.	1982	"Exit Decisions in Mature Industries", Academy of Management Journal, 25, #4, 707-732.
Iyer, E.S. and Ramaprasad, A.	1984	"Strategic Postures Toward Innovation", IEEE Transactions on Engineering Management, EM-31, #2, 87-90.
Kanter, R.M.	1984	"SMR Forum: Innovation- The Only Hope for Times Ahead", Sloan Management Review, Summer, 51-55.
Katz, R.	1984	"As Research Teams Grow Older", Research Management 27, #1, 23-28.
Kiel, G.	1984	"Technology and Marketing: The Magic Mix?", Business Horizons, May-June, 7-14.
Kirpalani, V.H. and Macintosh, N.B.	1980	"International Marketing Effectiveness of Technology-Oriented Small Firms", Journal of International Business Studies, Winter, 81-90.
Kohn, M. and Scott, J.T.	1982	"Scale Economies in Research and Development: The Schumpeterian Hypothesis", The Economics, XXX, #3, 239-249.
Link,A., Tassey, G. and Zmud, R.W.	1983	"The Induce Versus Purchase Decision: An Empirical Analysis of Industrial R and D", Decision Sciences, 14, 46-61.
MacMillan, I.C. Hambrick, D.C. and Day, D.L.	1982	"The Product Portfolio and Profitability - A PIMS-Based Analysis of Industrial Product Businesses", Academy of Management Journal, 25, #4, 733-755.
Marchetti, C.	1985	"Swings, Cycles and the Global Economy", New Scientist, May 2, 12-15.
McDonough, III, E.F.	1984	"Needed: An Expanded HRM Role to Bridge the Gap Between R and D and Manufacturing", Personnel, 61, #3, 47-52.
Merrifield, D.B.	1983	"Technology and the Management of Rapid Change", Research Management, May-June, 10-13.
Miner, J.B. and	1981	"Can Organizational Design

Smith, N.R. — Make Up for Motivational Decline", Wharton Magazine, 5, #4, 29-35.

Minshall, C.W. and Moody, C.W. — 1984 — "An Assessment of Trends in High Technology Activities, Growth Industries and Occupational Structure", Economics and Policy Analysis, Occasional Paper #43, Battelle, Columbus, Ohio.

Mosbacher, C.J. — 1985 — "R and D Funding Will Rise to $110 Billion in 1985", Research and Development 68-70, January.

National Science Foundation — 1983a — "The Process of Technological Innovation: Reviewing the Literature", National Science Foundation, Washington, D.C., May.

National Science Foundation — 1983b — Ibid, 67-69.

National Science Foundation — 1983c — Ibid, 176-180.

Oakey, R.P. — 1984 — "Finance and Innovation in British Small Independent Firms", International Journal of Management Science, 12, #2, 113-124.

Rafael, I.D. and Rubenstein, A.H. — 1984 — "Top Management Roles in R and D Projects", R and D Management, 14, #1, 37-46.

Research and Development — 1985a — See p. 47, January, Research and Development.

Research and Development — 1985b — See p. 42, March, Research and Development.

Roberts, E.B. — 1982 — "Is Licensing An Effective Alternative?" Research Management, 25, 20-24.

Rothwell, R. — 1983 — "Innovation and Firm Size: A Case for Dynamic Complementarity; Or, Is Small Really So Beautiful?", Journal of General Management, 8, #3, 5-25.

Sahal, D. — 1983 — "Invention, Innovation and Economic Evolution", Technological Forecasting and Social Change, 23, 213-235.

Sands, S.	1983	"Problems of Organizing for Effective New-Product Development", European Journal of Marketing, 17, #4, 18-33.
Schmenner, R.W.	1983	"Every Factory Was A Life Cycle", Harvard Business Review 61, 2, 121-129.
Stahl, M.J., Zimmerer, T.W. and Gulati, A.	1984	"Measuring Innovation, Productivity and Job Performance of Professionals: A Decision Modeling Approach", IEEE Transactions of Engineering Management, EM-31, #1, 25-29.
Wilkinson, A.	1983	"Technology - An Increasingly Dominant Factor in Corporate Strategy", R and D Management 13, #4, 245-259.
Wilson, A.	1984	"Innovation in the Marketplace", Management Today, June, 78-82.

AUTHOR INDEX

The number directly after the initial(s) of the author is the page number on which the author (or his work) is mentioned in the text. Numbers in brackets indicate that an author's work is referred to, although his name may not be cited in the text. Numbers underlined give the page on which the complete reference is listed.

SUBJECT INDEX